The Russian Oil Economy

This book is dedicated to

Melvin A. Conant

Alexander Kemp

James D. Gaisford

and Richard P. Kendon

The Russian Oil Economy

Jennifer I. Considine
Associate Editor, Geopolitics of Energy

William A. Kerr
Van Vliet Professor, University of Saskatchewan, Canada

Edward Elgar
Cheltenham, UK • Northampton, MA, USA

Published by
Edward Elgar Publishing Limited
Glensanda House
Montpellier Parade
Cheltenham
Glos GL50 1UA
UK

Edward Elgar Publishing, Inc.
136 West Street
Suite 202
Northampton
Massachusetts 01060
USA

A catalogue record for this book
is available from the British Library

Library of Congress Cataloguing in Publication Data
Considine, Jennifer I., 1962–
    The Russian oil economy / Jennifer I. Considine, William A. Kerr.
        p. cm.
    Includes bibliographical references and index.
    1. Petroleum industry and trade—Russia (Federation) I. Kerr, William A., 1947–
II. Title.

    HD9575.R82 C57 2002
    338.2'728'0947—21   2002019529

ISBN 1 84064 758 2

Printed and bound in Great Britain by MPG Books Ltd, Bodmin, Cornwall

Contents

Tables

Preface

A former colleague at the University of Calgary, Richard Kendon, always started his course on the International Petroleum Economy by informing his students that 'oil was a commodity like any other and subject to the laws of supply and demand'. In an 'oil city' like Calgary, which benefits (and suffers from) the booms (and busts) of international energy markets, petroleum is often treated as a 'special' good whose analysis requires a 'new' economics. This is largely mystique, and Richard Kendon is right. The study of petroleum markets is complex, however, because the ability of market participants to react to changing market conditions is constrained for both demanders and suppliers. On the demand side, consumers of petroleum products are often constrained in their ability to substitute among energy sources by the investments they have made in the past – vehicles, furnaces, storage, transport infrastructure, production processes and so on. This makes it difficult to alter their choice of energy purchased in the short run when prices rise or fall. On the supply side, long lead times are required for the expansion of output when prices are favourable – exploration, oil field development, infrastructure investments (in pipelines, refineries and so on). Typically, these sunk investment costs dwarf the actual production costs. This means that production does not decline significantly when prices fall. The productive life of oil fields, pipelines, refineries and so forth tends to be measured in multiple decades. Thus, decisions made in the distant past have a significant bearing on today's markets. The technologies chosen at the time when expansionary investments are made also determine the speed with which oil can be extracted, the proportion of reserves that can be exploited and the life of the fields. In few other industries is understanding the past so important for analysing the present. Oil fields developed a century ago are often still productive assets.

As in other sectors of their economy, the architects of the Soviet oil economy set out to defy 'capitalist' laws of supply and demand. In theory, they did this through meticulous planning and allocation by command. Of course, the truth is that planning was a failure because the information requirements of a complex economy exceed the resources available to planners to collect it and their ability to process it. As a result, plans were unrealizable and, rather than following an orderly course of development, the Soviet oil economy's history appears chaotic with investments dependent on priorities determined by events and political personalities. While planning was a failure, the Soviets developed allocation by command to a fine art. Command, at times, produced great achievements. Of course, those achievements were seldom accomplished in ways that would be considered resource efficient in market economies. In some cases the costs in terms of human suffering, much less the waste of more conventional resources, was enormous.

Development undertaken in chaotic fits and starts combined with the inefficiencies associated with allocation by command seldom produces results that take the long view. The most obvious example of this in the Soviet oil economy is the total disregard for the environments that accompanied the exploitation of petroleum resources. The pressure to achieve results in the short run, however, is manifest in a host of ways – in oil field exploration, in oil field extraction, in oil field management, in transportation infrastructure and in refining. Hence, the means by which the Soviets developed their petroleum industry still heavily influences the current potential of the industry that the Russian Federation inherited. Thus, understanding the industry's past is essential for understanding its future.

As with the entire communist experiment in defying the laws of supply and demand, eventually the Soviet oil industry could no longer garner sufficient resources to continue to offset its built-in inefficiencies. In a similar fashion to the wider economy of the Soviet Union, it suffered from a deep-seated malaise in the last decades of the communist era. The running down of the industry over the long term left the Russian Federation with a huge deficit in investments required to offset depreciation, much less for revitalization and expansion.

The post-command era has also been chaotic. While the attempts to create a planned economy can be seen as the 'great economic experiment' of the 20th century, the efforts to create a market economy out of the ashes of the failed communist experiment may well be the 'great economic experiment' of the 21st century. Despite consistent optimism by those with a deep faith in the capitalist system, the transition to a modern market economy in Russia or other parts of the old Soviet empire, or even China, is far from a sure thing. There are no road maps. Western economists do not know how to create a modern market economy any more than they know how to create the conditions whereby developing countries actually become developed. The real question is not 'How long will Russia's transition to a modern market economy take?' but rather 'Where is the transition leading?'. If one views the Russian oil economy from this perspective, more insights may be gained into the likely role of the Russian oil industry in the global petroleum market in the 21st century than if one views it through lenses that make it appear as if it conforms to a familiar model found in a modern market economy.

We think that the development and prospects for the Russian oil economy make a fascinating story. Much of the story had to await the opening of Russian archives. Old 'truths' based on Soviet era statistics have had to be re-evaluated in the face of new evidence. Despite the burden of its history and its current problems, the Russian oil economy is a major player in the global energy market as demonstrated in the fall of 2001 when Russia was able to act as a spoiler for OPEC's carefully orchestrated constraints on output leading to a dramatic price decline. Events that unfold in the Russian oil patch have significant impacts on international petroleum markets. We hope that this book provides new insights into the Russian industry and its operation.

The authors would like to thank Shari Boyd of the Estey Centre for Law and Economics in International Trade in Saskatoon, Canada, for her technical assistance in the preparation of the manuscript and the Estey Centre for its support of this project.

Jennifer I. Considine, Calgary, Canada
William A. Kerr, Saskatoon, Canada
February 2002.

1. Introduction

1.1 OIL WITHOUT MARKETS

> There is something about petroleum that is controversial and intriguing. There is something about Russia that is mystifying and absorbing. When the two merge in a study of Russian petroleum, the result is likely to be tantalising and engrossing . . . . There are so many precedents, similarities, and coincidences in a study of the history of Russian petroleum that discussion of the present generates a sense of deja vu.[1]

It can be argued that Russia has never been a market economy. While it is true that prior to World War I Czarist Russia's economy had a veneer of capitalism, the economy was just one step out of feudalism. The economy was overwhelmingly agrarian with peasant farmers, at best, one generation removed from serfdom. A small cadre of industrialists and merchants operated in the relatively free market for manufactured goods and imported consumer products but this embryo market system never extended to the development of resources. This does not mean that profit-oriented entrepreneurs were excluded from resource exploitation – on the contrary they were heavily involved. The rights to resources were, however, controlled directly by the Czarist regime. Resource rights were either delegated by royal patronage to the aristocracy or held by the government itself, although it is often difficult to disentangle the two given the large role played by the former in the government. In this system, the ability of entrepreneurs to gain access to resources, including petroleum resources, in which they saw potential depended upon political connections and the ability to curry favour rather than the ability to seize upon a market opportunity. The development of the Russian oil industry prior to the Bolshevik revolution in 1917 reflects this non-market allocation system.

When Lenin and his relatively small clique of Bolsheviks, to their surprise, actually gained control of Russia and had the opportunity to reorient the economy in the name of Karl Marx, they were groping in the dark. For all his predictions of the revolutionary overthrow of the capitalist system, Marx provided no insights into what would happen afterward. Marx was interested in the evolution of the capitalist system, not in providing road maps for the transition to communism.

While the Marxist intellectuals who formed the core of the new Soviet administration of Russia did not have a positive guide to creating their new 'workers' paradise', they certainly knew what they did not like about capitalism. With the benefit of hindsight and stripped of its optimistic and emotive rhetoric, the basis of the economic model that was to dominate Russia for over 60 years (and a large portion of the rest of the world's population for shorter periods) was simple. If the capitalist system has it; we will do away with it. If capitalism has profit (meaning workers don't receive their fair share of the fruits of production); we will abolish it. Capitalism has private property (that gives some rights to more resources than just their labour and, hence, leads to inequality in incomes); we will nationalize it. It has capitalists (who are forced by competition to exploit their workers in desperate attempts to survive); we will replace them with technocratic managers. Prices determine resource allocation (even when it is not in the best interest of society); we will allocate by commands made by altruistic bureaucrats. Markets create business cycles, instability, booms and busts (which impose hardship on the proletariat through unemployment); we will replace them with scientific planning. All of the institutions associated with capitalism were done away with – banks, interest, stock markets, individual initiative, value in human capital accumulation. The model had some appeal in theory, in practice it did not work.

In part, the failure of this negative model can be explained by human nature – the socially altruistic 'new socialist man' did not evolve and could not be created. More important, however, in their outrage with the negative aspects of capitalism, they were blinded to the true role and function of capitalist institutions. Over time, those in charge of the Soviet economy were forced to 'invent' many institutions that paralleled in function the capitalist ones they had abolished, although always dressed up in socialist rhetoric. To the very end, however, they were true to one central theme – markets would not be used to allocate resources.

Without markets to guide the allocation of resources, however, the Soviets developed an oil industry on a massive scale. From a low of 25 million barrels per year at the end of the Russian Civil War in 1920, production expanded to peak in 1987–88 at 4.5 billion barrels (Reinsch et al., 1992). This is an impressive achievement. How it was achieved, however, has a direct bearing on any assessment of the potential of the Russian oil industry today.

The workings of a command economy are not widely understood in modern market economies. It is not often taught in modern economics curricula in universities. In part, this is because since the fall of the Berlin wall and the collapse of the Soviet Empire at the end of the 1980s it has not been needed. It is an economic system that only survives as an anachronism in a few unreformed communist countries such as Cuba and North Korea. As a result, there is half a generation of university graduates who have no knowledge of the workings of a command economy. Even at the apex of the Cold War when communism and capitalism were (apparently) vying for the position as the globe's pre-eminent economic system, few students in the West bothered to study communist economies. This was because non-market economies were relatively closed and did not interact with market economies to any great degree. There was, for the most part, no need to know. With the fall of communism, these economies began their slow and painful integration (or re-integration) into the international economy. Understanding the workings of the previous system had little relevance for transition.

This is not the case, however, in a few sectors including the petroleum industry where the impact of previous investments is long-lived. In market economies, well managed oil fields have very long productive lives – West Texas and Pruhoe Bay come immediately to mind. In contrast, the vast 1948 Romashkino oil field, the flagship of the post World War II Soviet industry, was all but exhausted in 30 years. The Samotlor field discovered in the 1960s in West Siberia was depleted in 20 years (Reinsch et al., 1992). Thus, any current assessment of Russian oil reserves depends upon an understanding of the nature of the investments made in the Soviet era and how oilfields were managed. Further, given the relative chaos and high risks associated with the post-Soviet era in Russia, much of the infrastructure that currently exists is a holdover from investments made by the previous regime.

In theory, the non-market economy established by Lenin and subsequently refined over the years was based on two institutions. The first was 'scientific planning' whereby supply and demand were matched under broad priorities established by the Communist Party. The second was execution of the plans through orders (commands) given by bureaucrats. The problem with the theory was that the planners never had sufficient resources to collect and process the information they required in order to develop sufficiently comprehensive plans. It is not even clear if it is possible to collect sufficient information to develop an operational plan. As a result, the much-vaunted 'Five-Year Plans' were crude instruments at best.[2] They resulted in chronic mismatches between supply and demand. In particular, the shortages wreaked havoc along supply chains as too little output at one level meant too few inputs at the next level leading, in turn, to lower than planned output (Hobbs et al., 1997).

While 'scientific planning' was a monumental failure, there is little doubt that 'command' could accomplish much if a particular objective was considered a priority. Bureaucrats given the authority to garner and allocate resources to particular tasks could achieve results. The Soviet space programme is probably the best example but there is a wide range of examples: electrical power generation capacity, steel production and, at times, petroleum exploration. On the other hand, non-priority areas tended to languish starved of resources. Of course, the ability to command resources for priority projects disrupted the resource balances detailed in the plans, further contributing to the creation of shortages in other areas. Thus, while command could achieve specific results, the cost of achieving the results could be horrendous both in human terms and in inefficient resource use.

Beyond the problems created by the chronic mismatching of supply and demand, there was a more fundamental problem with economic allocation by command. The inability of central planning to choose rationally among allocation alternatives was central to the command economies' failure to provide for sustained improvements in living standards. This was the 'calculation problem' identified by Ludwig von Mises in 1922,[3] early in the Soviet era. Understanding the calculation problem is particularly germane to gaining insights into the development of the Soviet petroleum industry.

The essence of von Mises' argument is that once government intervenes in the economy to set prices by fiat, it is no longer possible to use prices as a means of allocation. Further, it is not possible to make allocation decisions based on an objective efficiency criterion. This is because the set of relative prices promulgated by the state does not reflect the relative value (based on opportunity cost) placed on goods and services by their users whether they be producers of goods or final consumers. According to von Mises (1981, p. 103):

> Each commodity produced will pass through a whole series of such establishments before it is ready for consumption. Yet in the incessant press of all these processes the economic administration will have no real sense of direction. It will have no means of ascertaining whether a given piece of work is

really necessary, whether labour or materials are not being wasted in completing it. How would it discover which of two processes was the most satisfactory?

Planners in command economies realised that their official prices could not be used for the purposes of allocating resources. Instead, they attempted to use a 'material balances' approach whereby technical input coefficients were used to determine the ratio of resources in production. While this could theoretically be used as a static allocation rule, it is not useful for dynamic decision-making processes that characterize a modern economy, particularly one with a preference for development such as that of Soviet Russia. Questions relating to when an industry should decline or expand as resource scarcity changes or consumer preferences evolve cannot be answered. The value of new products and the efficacy of new processes cannot be assessed objectively.

The problem of determining relative value, as identified by von Mises in 1922, was never solved by communist intellectuals or planners. The absence of a means by which to make informed allocation decisions meant that the targets set by policy makers in the Party were striven for with no thought to efficiency or management over the long run. With no means to value energy in the Soviet Union, there was no constraint on demand and the economy became energy intensive. The rapid and forced growth required ever increasing quantities of energy. The result was an industry fixated on the short run. According to Reinsch et al. (1992, p. 17)

> Since . . . the only constraints placed on Soviet industrial production were lack of investment funds and physical supplies, the Soviet Union developed a very particular type of development strategy, sometimes referred to as a shortage management strategy; that is, to get as much as possible, as quickly as possible, before competing demands drained off the available resources.

This emphasis on the short run can have considerable impact on both the management of existing reserves and exploration strategies. The effect is outlined by Reinsch et al. (1992, p. 13):

> Unfortunately, the focus on rapid oil recovery carried a heavy price. Forcing production beyond the Maximum Efficient Rate (MER) resulted in early and rapid decline in production volumes from existing fields. Technological deficiencies and general organisational decay forced the industry to move prematurely to new, less favourable producing regions.

The cumulative effect of a petroleum sector plagued with the 'calculation problem' for over 50 years means that it is inappropriate to apply Western interpretative norms to any Russian data; that is, wells drilled, reported reserves, depletion rates and so on. The Soviet record must be examined in detail. Thus, while the command economy built a petroleum industry without the use of markets, the configuration of that industry is a far cry from those that evolved in modern market economies.

The third period when the oil industry operated largely without markets is in the post-Soviet era. While 'scientific planning' has been abandoned and, at least officially, so has allocation by command, the Russian economy has not yet made the transition to being a modern market economy. As suggested above, the problem of determining relative value, as defined by von Mises in 1922, was never solved by communist intellectuals or planners. As a result, the former command economies were all faced with a dilemma. The existing set of prices conferred benefits to some members of society. For example, low energy prices provided a degree of energy security for those on fixed incomes such as pensioners. If they did not free prices, however, correct signals would not be conveyed upon which to base resource allocation decisions.

Freeing prices without the institutions necessary to support markets was, unfortunately, not likely to produce the set of prices which reflect relative values as was envisioned by those who put their faith in the process of allocation though market forces. The rapid rates of inflation that followed the freeing of prices further reduced the ability of individuals to discern relative value from available prices. Where markets are not developed, as in Russia, the ability to use prices as a decision criterion is severely limited.

The problem of resource allocation in the Russian energy sector is further complicated by the conversion of state monopolies into private monopolies. In market economies, the role of prices as a guide to resource allocation is based on the premise that they represent, for the goods in question, a convergence of the value of the opportunities foregone in the goods' production and the value placed on the goods by consumers. In other words, price is set where the marginal (resource) cost of producing a good equals the marginal valuation which consumers put on that good. This is the familiar intersection of supply and demand curves. Disequilibrium in a market sets economic forces in motion to reallocate resources. For example, when excess demand exists, rising prices provide an incentive for increased resource allocations to the goods supplied to that market. Falling prices suggest a reduction in the resources committed to the production of goods supplied to markets exhibiting excess supply. This basic tenet of market economies bears repeating in the case of liberalizing command economies.

In market economies it is recognized that monopolies lead to inefficient levels of output because price (and hence, valuation of consumers) exceeds the marginal cost of producing the good or service. The observed monopoly price does provide a signal for more resources to be transferred to the production of the good but barriers to entry prevent that transfer from taking place. Prices are not allowed to play their role as a guide to resource allocation.

In modern market economies, monopoly inefficiencies are often tolerated because the effect of their perceived distortions is small (or the gains from regulating such monopolies do not justify the resource costs associated with the regulator process). Where the inefficiencies or price distortions created by a monopoly (or potential monopoly) are considered to be unacceptable, government policy has generally been: (1) to prevent monopolies arising – anti-merger provisions in anti-trust or competitions legislation; (2) to break up existing monopolies; or (3) to regulate the output and price of monopolies. The intent, in each case, is to keep or move the observed price nearer to the price that would arise in a competitive market. It is recognized, however, that a price established through a regulatory process will only be an approximation to the theoretically based price. This is because acquiring the information to determine the regulated price is not costless. The problem of setting a regulated price is especially difficult when the firm whose price is being regulated has an incentive not to provide information.

In liberalizing economies such as Russia, monopolies are far more prevalent than in market economies. There are three reasons for this: (1) the former command system stressed large-scale production/distribution facilities; (2) the ability of firms to identify and conclude transactions with alternative suppliers/customers is limited by the lack of institutions to support the process of broadening markets; that is, the costs of broadening markets is very high; and (3) the government will not, or cannot, act to limit monopolies. In the energy industry, facilities were of a particularly large scale; for example, there were only 17 refineries serving an economy of 200 million people in the old Soviet energy system (Considine and Kerr, 1993). Further, the large-scale energy systems were kept largely intact in their Soviet form whether or not they have been privatized. These large monopolies have provided those who control them with the ability to become extremely wealthy – and to use that wealth to influence governments. They have been able to use that influence to strengthen their monopolies and limit competition. Further, rights to energy resources has, in part, been delegated to lower levels of government. The regional governments in the Russian Federation are notoriously corrupt, meaning that their resource allocations are unlikely to reflect market conditions.

The creation of monopolies, whether through privatization or reorganization of state firms, and their ability to maintain their monopoly position, means that prices are not able to carry out their resource allocating role to the same degree as is the case in market economies.

What does it mean when the 'planner's conscious hand' has been lifted but the 'invisible hand' does not yet exist? Fundamentally, it means that the prices which do exist will often give false signals. In the short run, they will give false signals about what to produce, in the intermediate run they will give false signals about where to invest. If producers and investors are unaware that existing prices are giving false signals, and do not learn, the market clearing price is not likely to represent a stable equilibrium and there will be considerable wasted investment. Little is understood about the actions of oil producers and investors when they are aware that prices give false signals. If they are risk averse, they are likely to under-produce and under-invest when they cannot believe the price signals they receive. Further, given the absence of secure property rights and the propensity of bureaucrats in Russia to use the tax system opportunistically to confiscate profits if they arise, doing business in Russia's energy sector has been extremely risky in the post-Soviet period. Not surprisingly, the industry was characterized by chronic under-investment during the 1990s.

Transition has not been smooth and, as yet, is a long way from completion. Some even doubt that the economy is transforming into a market economy but rather into a 'licensing' economy (Hobbs et al., 1997). Kerr and MacKay (1997) define a licensing economy as one characterized by an absence of secure property rights, endemic corruption and bureaucratic licenses required to engage in economic activity. If the income of officials with the ability to grant licenses comes largely from the ability to create and sell licenses, this leads to a vested interest in retaining the system of licenses. As a result, they will work to ensure that property rights remain poorly defined and enforced. The outcome is an economy trapped in a low investment/high cost of doing business equilibrium. Of all the Russian industries, the oil industry fits this characterization, particularly given the access to hard currency that sales to world markets have provided.

It seems clear that the Russian petroleum industry has never been guided by market forces as they are understood in the West. This means that the Russian industry cannot be approached from a market economy perspective and that, above all, history matters.

1.2 RUSSIAN OIL IN THE GLOBAL ECONOMY

While the Russian petroleum industry has never operated in a market system, it has still become a major player in the international energy industry. At the end of the Soviet era it was the largest oil producer in the world. While a combination of reserve depletion and the economic disruptions of the 1990s led to a decline in production, Russia remains the third largest crude oil producer, surpassed only by Saudi Arabia and the US. The industry's interaction with the international oil market has often destabilized prices. The latest example was in the fall of 2001 when Russia played the spoiler in OPEC's carefully constructed output reductions, leading to a dramatic decline in prices.

From a high of almost 590 million metric tons of crude oil production in 1988, Russian output declined to just over 300 million metric tons in 1996 and then remained stable for the rest of the decade. Strong oil prices in the first years of the new century have led to some modest increases in production. The break-up of the Soviet Union has meant that some oil producing areas are now independent countries. Kazakhstan, in particular, is a major producer of oil but Azerbaijan, Turkmenistan and Uzbekistan are also important producers that are no longer directly part of the Russian energy system. As a result, Russia's main oil fields are now located in Siberia. Approximately two thirds of Russia's output originates in the Tyumen region. Russian production is derived from approximately 135 000 wells of which over 35 000 are idle. Drilling activity has, however, declined considerably since the Soviet era, from 38 million metres in 1988 to just under 6 million metres in 1999. The number of wells drilled has declined to an even greater extent from 15 643 completed wells to assist in exploitation of reserves in 1988 to 2179 in 1999. Exploration wells drilled fell from 816 in 1988 to less than 300 at the end of the 1990s. Two thirds of Russian reserves require that secondary and tertiary recovery methods be used. Thus, even with the recovery of international oil prices in the early years of the 21st century, the sustainability of Russia's production is questionable. While some of the reduction in exploration and recovery drilling reflects the beginnings of market discipline being applied to the industry, far more important is the general malaise of the post-liberalization Russian economy and, in particular, the absence of clear property rights, poor predictability in the taxation system and government restrictions on foreign participation that kept the industry starved for capital, particularly foreign capital. The lack of investment funds has left the industry saddled with a large stock of ageing and technologically obsolete equipment and infrastructure. Thus, keeping production levels stable, even at the lower levels of the mid-1990s, has proved a considerable challenge. In 2000 and 2001, rising international oil prices removed this constraint to some extent but increased government taxes and the need to maintain existing infrastructure have siphoned off most of the windfall gains so that there has only been a limited expansion in activities aimed at locating new reserves.

Russian reserves, however, remain large. Although estimates differ, it has been estimated that Russia's original oil endowment was 262 billion barrels, second only to Saudi Arabia (377 bbl). Russia is followed by the US (260 bbl) and Iran (152 bbl). While the US and Russia are estimated to have similar endowments, the US has used up a far greater proportion of its endowment. The remaining US endowment is estimated at 92 billion barrels while Russia's is 196 billion barrels. Saudi Arabia has the largest remaining endowment at 302 billion barrels. In terms of proven reserves, however, Russia has approximately 49 billion barrels compared to 23 billion barrels for the US. In proven reserves, Russia lags behind Saudi Arabia (160 bbl), Iraq (91 bbl), Kuwait (86 bbl), Iran (69 bbl) Venezuela (64 bbl) and the United Arab Emirates (61 bbl). While the US is expected to be able to maintain current levels of oil production for less than ten years, Russia will be able to sustain production in excess of 50 years assuming no major technological changes and that sufficient capital is available to finance exploration, recovery and infrastructure expansion and refurbishment.

Russian oil consumption is considerably less than production, leading to considerable oil being available for export. In 2000, daily production in Russia was approximately 6.62 million barrels per day while domestic consumption was only 2.34 million barrels per day. Russian consumption has declined from over 2 million barrels per day since 1992. As a result, even though production has declined, from 7.86 million barrels per day to 6.62 million barrels per day, more oil is available for export. Consumption has declined for two reasons. First, the decline in industrial production and economic activity in general that followed the end of the communist era led to a decrease in demand for petroleum products. Further, while domestic prices have not risen to world levels, they have increased considerably. As a result, individuals, firms and government institutions have had to find ways to curb their energy consumption.

Since 1991, Russian oil exports have been increasingly shifted from the New Independent States of the former Soviet Union (NIS) and the Central and Eastern European Countries (CEEC), which used to fall within the Soviet Union's sphere of influence, to western markets. This largely reflects the economic reality that the NIS and CEEC countries have had difficulty paying for their energy imports. As a result, Russia's exports have been targeted toward western Europe where demand is strong, domestic production is limited and payment made at world prices in cash. The majority of Russian oil exports are being sold to the United Kingdom, France, Italy, Germany and Spain. The share of net exports to countries outside the former Soviet Union increased from approximately 50 per cent in 1992 to 90 per cent by the end of the century.

Oil exports are extremely important for the Russian economy. Prior to the oil price rise in 2000 and 2001, oil accounted for approximately 30 per cent of Russian hard currency export revenues. Tax revenues from oil and natural gas account for more than half of the Federation government's tax revenues. Given that oil production and exports are relatively easy to monitor, it is a relatively easy sector to tax. Hence, the government is unlikely to reduce the tax burden borne by the industry. The administration of President Vladimir Putin has kept a tight rein on the sector by insisting that it pay all its taxes in cash. The oil export tax was increased as international oil prices rose in 2000. The government's reliance on the oil industry for revenue means that the oil industry is financing those areas of the economy where it is hard to tax. As a result, the industry is short of retained earning with which to finance the replacement of its ageing capital stock and to expand production.

Further, given that foreign hard currency markets are more lucrative than domestic sales, the government has had to implement domestic delivery quotas to prevent oil companies from starving the domestic market of crude oil. While the domestic delivery quotas are a contentious issue given the hard currency foregone, no Russian administration has been able to seriously consider raising the domestic price to international prices. As a result, domestic consumers will have to continue to fight for crude, and industrial expansion will be constrained by the absence of secure sources of additional supplies. The majority of Russian oil is exported by tanker. The major export ports are on the Baltic sea and at Novorossiisk on the Black Sea. Black Sea exports must pass through the Bosporous Straits where there is considerable worry about the potential environmental damage of an oil spill. Black Sea facilities are being used near capacity while those on the Baltic are not, particularly since a new terminal was opened in Kaliningrad Oblast, a Russian enclave on the Baltic Sea, in late 2000.

The major overland route to Western Europe is the 1.2 million bbl/day capacity Druzhba pipeline. This pipeline has yet to be utilized to capacity. Russian export routes are far from secure, running, in some cases, through

NIS countries that suffer from political instability, lack of funds for maintenance or a predilection for practising hold up on transit fees. As a result, Russia is attempting to diversify its export routes though new pipelines.

In 1993, Russia initiated a privatization process for the oil industry. The first stage involved the organization of state owned joint stock companies and led to the establishment of a small number of vertically integrated oil companies. Since then, there has been some further consolidation through mergers. The second stage of privatization has been the selling off of government shares in companies.

The principal vertically integrated companies include LUKOIL, Surgutneftegas, Slavneft, ONAKO, Eastern Oil Company, Tyumen Oil Company and Rosneft, which is the only remaining firm owned solely by the state. In January 1998 two of the large vertically integrated companies, YUKOS and Sibneft merged to form YUKSI. It is one of the largest oil firms in the world. It has the most reserves under its control and follows only Exxon and Shell in extraction. These large companies are important in the Russian political process and are enmeshed in it. This makes reform difficult, particularly given the government's dependence on the sector for revenue.

The hold of the domestic industry over the regulatory process has made it difficult for foreign oil companies to prosper in Russia. The need for technology and capital, however, has led to some reforms that have improved the appeal of Russia for foreign firms. They also feel it is important to maintain a presence so as not to be shut out in the long run. Major foreign oil companies include BP Amoco, Chevron, Conoco, Exxon, Shell, Texaco, Mitsubishi and Mitsui among others.

While the Russian oil companies cannot yet compete effectively on a technological or managerial level with the other major firms, they operate relatively effectively in Russia. The future of the Russian industry will be determined partly by the Russian industry's ability to modernize both technologically and managerially. It will also depend upon what it has to work with – what it has inherited from what has gone before. For a considerable time to come, the latter will be as important as the former. It is to the determinants of that inheritance that we now turn our attention.

NOTES

1. See Goldman, 1980, p. 13.
2. For an accessible and readable account of how this is largely avoided in modern market economies see Friedman and Friedman's (1980) discussion of the parable of the pencil, pp. 11–13.
3. See von Mises (1981). He originally published his work in 1922 under the title *Die Gemeinwirtschaft: Untersuchungen über den Sozialismus*, Gustav Fisher, Jena. The first English translation (with a few additions) was published in 1936 as *Socialism: An Economic and Sociological Analysis*, Jonathan Cape, London (translator J. Kahane).

2. Risky Business – Oil in the Russian Empire

2.1 ROTHSCHILD AND THE RUSSIAN NOBELITY

Among the most promising markets for the 'new light' (kerosene) was the vast Russian empire, which was beginning to industrialise, and for which artificial light had a special importance. The capital city, St. Petersburg, was so far north that, in the winter, it had barely six hours of daylight. As early as 1862, American kerosene reached Russia, and in St. Petersburg, it quickly won acceptance, with kerosene lamps swiftly replacing the tallow on which the populace had almost entirely depended. The United States consul at St. Petersburg reported happily in December 1863 that it was 'safe to calculate upon a large annual increase of the demand from the United States for several years to come.' But his calculations could not take into account future developments in a distant and inaccessible part of the empire, which would not only foreclose the Russian market to American oil but would also spell the undoing of Rockefeller's global plans.[1]

The Russian crude oil industry had its modest beginning near the remote, and nearly inaccessible city of Baku, the territory of an independent duchy that had been annexed to the Russian Empire in the first years of the nineteenth century. Legends of the abundance of oil to be found on the Aspheron Peninsula – the landmass between the Caucasus Mountains and the Caspian Sea – date back to the end of the fourth Islamic century (the tenth century AD). Perpetual flames sprang from the earth, igniting fear and wonder among the earliest explorers of the Baku region. The Zoroastrians, adherents of an ancient Persian religion, worshipped these 'eternal pillars of fire' – flammable gas associated with petroleum deposits that had escaped from tiny fissures in the surrounding limestone (Hourani, 1991).[2] By the first years of the thirteenth century, the pillars had been recognized as having practical value: 'good to burn' and extremely useful for 'cleaning the mange of camels' (Yergin, 1991, p. 57)

By the nineteenth century, crude oil technology had developed sufficiently to attain international status as a viable, and lucrative, economic endeavour. At the same time, the acquisition of Baku provided the Russian Empire with a highly promising, albeit primitive, source of hydrocarbons. Nevertheless, despite the importance of artificial lighting to St. Petersburg, the initial development of the Russian crude oil industry was hesitant and sporadic. Entrepreneurial activity was restrained by the inefficiency and corruption of the Czarist regime whose aim was to maintain the tiny industry as a state monopoly. Technological progress was delayed by the remoteness and backwardness of the Baku economy. In 1829 the entire Empire's oil industry consisted of 82 hand-dug oil pits, all located in close proximity to Baku. The first hint of initiative, a group of Russian industrialists attempting to drill a number of commercial oil wells in the 1860s, was quickly repressed. Production trickled from hand-dug pits for well over a decade.

With 'American' kerosene rapidly becoming the fuel of choice in St. Petersburg, the Russian government was forced to examine alternative policies toward energy development. A decision to abolish the state monopoly system, thereby permitting competitive private enterprise, was reached in 1870. The resulting 'explosion of entrepreneurship' led to the rapid completion of the first of a series of very productive oil wells in 1871–72 (Yergin, 1991, p. 58). It is interesting to note that the first commercially successful drilling campaign was sponsored by foreign investors – the German energy company of Siemens and Halske (Reinsch et al., 1992).

The fate of the Baku region was sealed with the arrival of Robert Nobel, the eldest son of the renowned Swedish inventor Immanuel Nobel. On a mission to secure wood for a manufacturing contract, Robert enthusiastically joined the 'oil rush' of the late 1870s. When the first shipment of 'Nobel Kerosene' arrived in St. Petersburg in October 1876, Robert's brother Ludwig, a leading Russian industrialist, joined him in the petroleum endeavour. Immediate success, and the benign treatment of minority (non-Russian) labourers, attracted a huge multinational work-force. Four years later, the Russian crude oil industry was completely dominated by the Nobel Brothers' Petroleum Producing Company. The company – affectionately nicknamed the 'Nobelity' at a most inopportune moment in Russian history – proceeded to set standards for enterprise and efficiency that would prevail throughout the nineteenth century. Aided by an integrated network of wells, pipelines, refineries, barge tankers, and a railway, Russian crude oil production reached 10.8 million barrels in 1884, a volume equal to an incredible 30 per cent of the crude oil production of the, by then, firmly established US oil industry.

To the dismay of a growing number of ambitious 'new' competitors, the Russian domestic oil market was quickly saturated by supplies from the 'Nobelity' and, their major market rival, low-priced imports of American kerosene. The vast European market, clearly within striking distance, beckoned invitingly to Russian producers. In the late 1870s, the Russian producers Bunge and Palashkovsky obtained government approval to build a railroad from Baku to Batum, a newly acquired Russian Empire port on the Black Sea. Construction of the railroad, while delayed by a financial disaster accompanying a significant decline in Russian crude oil prices, was completed in 1883. The new line, a relatively modest accomplishment when compared to the spectacular expansion of the vast Nobel oil empire, had profound and permanent implications for the future of the Russian oil industry.

The French branch of the Rothschild family, enticed by the spectacular development of the Baku region and exciting new prospects for exports to Europe, had been more than willing to rescue the Bunge–Palashkovsky consortium in its darkest hour after the financial collapse. A substantial loan was exchanged for mortgages on a number of Russian oil facilities. As part of the deal, Bunge–Palashkovsky was guaranteed access to the European market at attractive prices via the Rothschilds' Fiume refinery on the Adriatic. In this way, the Rothschild family established a formidable presence in Baku. By 1886, the Rothschilds had created the Caspian and Black Sea Petroleum Company and invested in new storage and marketing facilities in Batum. This initiated a fierce, thirty-year international struggle for the oil markets of the world (Yergin, 1991).

The application of the Rothschilds' ingenuity to bountiful Russian oil fields, combined with the opening of the new gateway to European markets, triggered a unique sequence of events that would come to be known as the 'Oil Wars of the 1890s'. By 1888, just two years after the completion of the Baku–Batum railroad, Russian crude oil flows had reached 23 million barrels per annum, ten times higher that the meagre production recorded in 1879. The oil 'fountains' of Baku became legendary. To cite only one example: 'Droozba', the Baku 'Friendship' well, recorded flows as high as 43 000 barrels per day for nearly five months before gradually entering a period of decline (Yergin, 1991). An aggressive marketing campaign by the Rothchilds resulted in Russian petroleum products encroaching on what Standard Oil considered its exclusive domain – the well-established European market for 'American illuminating oils'. For the next ten years, the global oil market was dominated by a fierce battle for survival and expansion between four formidable rivals: the Standard Oil Company of America, the Rothschilds, the Nobel Brothers and a loose consortium of Russian oil producers.

At the same time, the Russian Empire was undergoing a period of extraordinary economic and political transformation. First and foremost was a rapid expansion of large-scale industrial facilities. Under the careful guidance of Count Sergei Witte, who was a trained mathematician and Minister of Finance from 1892 to 1903, the Russian economy began the delicate transition from a traditional rural society to a modern industrial state. The strategy of rapid development, permitting and even encouraging direct foreign participation, transformed the Russian oil industry and was considered the pinnacle of Witte's success. By 1901, crude oil flows had reached levels as high as 86 million barrels per annum, an incredible 63 million barrel increase over the 23 million barrels reported in 1888 (Reinsch et al., 1992).

The rapid increase in production was, in part, financed by significant injections of foreign capital and assisted by western expertise. According to Hassmann (1953, p. 58):

> The total amount invested in the Russian oil industry before 1914 was approximately $214 million (US). The Russian share of this amount was only around $85 million, while foreign investment in the industry amounted to only about $130 million (gold). . . . Particularly conspicuous is the high British share in this foreign investment, which may have been due to Britain's intention to become independent of the United States in obtaining oil.

The boom, however, would be short-lived. Concealed amidst the miracle of competition and efficiency lay the foundations for ruin, revolution and despair. Czar Nicholas II, well known for his contempt of all non-Russian minorities, had 'sanctioned' a regime of repression that would inspire generations of Bolshevik enthusiasts. As Yergin (1991, p. 61) suggests:

> The Caucasus – home of the Russian oil industry – was one of the worst-run parts of the ill-run empire. Living and working conditions in the area were deplorable. Most workers were in Baku without their families, and in Batum, the working day was often fourteen hours, with two hours of compulsory overtime.

Baku, the centre of the Russian oil industry, became the training ground for a host of revolutionaries.

2.2 *SMUTNOE VREMIA*: A 'TIME OF TROUBLES'[3]

The ensuing legacy of chaos and civil war is one of the major events of the twentieth century. The periodic outbreak of Bolshevik strikes and demonstrations from 1901–1903, the political revolution of 1905, German–Russian hostilities – including a struggle for Baku – World War I, the collapse of the Czarist regime in early 1917, the civil war of 1918–20 and the Polish invasion of Russia in May–October 1920, destroyed all remnants of economic and social order and left the great Russian oil empire of the late nineteenth century little more than a memory. By 1920, at the end of the Civil War, crude oil production had fallen to a mere 25 million barrels a year, less than 30 per cent of pre-revolutionary levels (Reinsch et al., 1992).

The response of the major oil interests to the winds of political transformation in Russia was varied and sometimes contradictory. The Standard Oil Company of America moved quickly to reclaim markets in Europe and the Far East that had been lost during the 'Oil Wars' of the 1890s. Russia's share of global petroleum exports fell dramatically, from 31 per cent in 1904 to only 9 per cent in 1913. The Rothschild family, concerned over both strong anti-foreign and anti-Semitic sentiments in Russia, and discouraged by the collapse of profits, sold their entire Russian oil holdings – a vast production, refining, and distribution system – to the Royal Dutch/Shell Company. Their acquisition of the Rothschilds' petroleum empire in 1912 gave Royal Dutch/Shell control over 20 per cent of the Russian oil industry. With a balanced portfolio of global oil supplies – 53 per cent from the East Indies, 29 per cent from Russia, and 17 per cent from Romania (Yergin, 1991) – the Royal Dutch/Shell group felt confident that it could survive even a prolonged episode of political insurrection in Russia.

The Nobel family, denounced as traitors and spies by the Bolshevik revolutionaries, fled to France. After 44 years of painstaking and highly successful corporate initiative, the fate of the immense Nobel oil empire was decided in a series of hasty 'family' negotiations conducted from the security of the Hotel Meurice in Paris. Assuming a Bolshevik victory in the civil war, there was only one solution open for consideration – a fire sale divesting the Nobelity of all equity interests in Russia. At the same time, there was a possibility, however remote, that the communists might still be defeated. To a growing number of interested participants – including Royal Dutch/Shell and Standard Oil – the potential rewards far outweighed the risks. A compromise arrangement, transferring 50 per cent of the Nobel oil empire to the Standard Oil Company of New Jersey at a bargain US$14 million was completed in July 1920, three months after the Bolshevik nationalization of the Baku oil fields.

By November, 1920, few doubts remained as to the resolve – and ultimate political victory – of the Bolshevik revolutionaries. The White (capitalist) resistance had, for all practical purposes, collapsed and War-Communism was being widely, and fanatically, implemented by the Bolshevik government. 'War Communism' is the term commonly applied to a period of 'extreme' communism that lasted from the middle of 1918 – eight months after the Revolution – to February 1921 and the beginning of the 'New Economic Policy'. The main elements of war communism included: (1) the nationalization of all private industry and enterprise; (2) a ban on private trade; (3) Prodrazverstka or the government expropriation of all surplus food supplies; and (4) the elimination of money. Nationalization was central not only to economic re-organization but also to the elimination of the influence of capitalists. It was often carried to extremes. To cite only one example:

> An industrial census [conducted in August 1920] counted over 37,000 nationalised enterprises. Of these, however, over 5,000 employed one worker only. Many of these 'enterprises' were, apparently, windmills. (Nove, 1972, p. 70)

The petroleum industry was no exception and it was nationalized during War Communism. War Communism, as an economic system, however, was a disaster although there is little doubt that its social consequences allowed the Bolsheviks to eliminate the traditional political influence of capitalists, both foreign and domestic, as well as landowners, thus allowing them to consolidate their hold on political power.

The Russian Soviet Federal Socialist Republic (RSFSR) ended up with a powerful government that defeated its domestic enemies. Based on its success, a number of new 'Soviet' republics had been established in the Ukraine, Belorussia and Transcaucasia. The official annexation of these Republics, and the creation of the Union of Soviet Socialist Republics (USSR) in December 1922, remained a mere administrative formality (White, 1991). Once political power was assured through victory in the civil war and the elimination of the economic power base of the former land owning and capitalist elites, the Bolsheviks under Lenin had to figure out what to do with the economy. While Marx had predicted the revolution, he provided no real guidelines for how the post-revolution economy was to be run. The petroleum industry was no exception and was to be transformed as the Bolsheviks began to feel their way. While often couched in grand and emotive rhetoric, the Communist Party was blindly feeling its way in its attempts to re-organize the Russian oil industry. It is to this 'great experiment' that we turn in Chapter 3.

NOTES

1. See Yergin (1991), p. 57.
2. The Cloister of the Eternal Fire stood at Surkhany on the Apsheron Peninsula. The temple served as a holy sight where the followers of Zoroaster could gather to worship until its abandonment in 1880.
3. 'One of the most traumatic periods in Russian history began four centuries ago with the death of the last member of the dynasty that founded Moscow. There followed a 15-year power struggle so bloody it has come to be known as the *smutnoe vremia*, or 'time of troubles' (Freeland, 1995, p. 17).

3. The Command Oil Economy

3.1 THE NEW ECONOMIC POLICY

> We were mistaken . . . We acted as if one could build Socialism in a country where capitalism scarcely existed. Before we can achieve a Socialist society, we must rebuild capitalism. (V. Lenin, 1921)[1]

Despite the theoretical aspirations of the Communist Party, the economy of the new Russian republic could not continue the policy of War Communism without risking a total economic collapse and the real political dangers that posed for them. After two decades of war and revolution:

> the Soviet Union was a country headed toward economic disaster, beset as it was by woeful industrial underproduction, inflation, severe lack of capital, and a widespread food shortage that was turning into famine. It desperately needed foreign capital to develop, produce and sell its natural resources (Yergin, 1991, p. 239)

In recognition of this dire economic reality, Lenin announced a dramatic reversal of policy. There was a recognition that revitalization and modernization of the economy would require foreign technology. To gain the required assistance, Lenin was willing to give extensive concessions 'to the most powerful imperialist syndicates'. His first two examples referred to oil – 'a quarter of Baku, a quarter of Groznyy'.[2]

The objective of the New Economic Policy was clear, and driven by the political reality. According to Alex Nove (1972, p. 78):

> The unbearable living conditions of the early 1920s inspired intense episodes of rioting among the peasant population of Russia. The issue reached crisis proportions on February 28, 1921 when the Kronstadt sailors 'rebelled against the miserable conditions of life . . . their slogans (reflecting) the peasants' hostility to the party's politics.'

In addition, there were severe fuel shortages, widespread famine and the impending foreclosure of vital state enterprises due to the absence of raw materials. As a result, Lenin was prepared to go to almost any lengths to restore economic prosperity, 'feeling, with justice, that this was essential for survival' (Nove, 1972, p. 89). In November 1920, the Party offered concessions to all foreign investors interested in the development of Russian resources, primarily oil fields and timber. A decree nationalizing small-scale industry was revoked on 17 May 1921. The leasing of enterprises in the possession of the Supreme Council of National Economy Vesenka (VSNKh)[3] was permitted on 5 July 1921. These three initiatives, and a host of similar decrees and resolutions, formed the basis for Lenin's New Economic Policy – an acknowledged retreat into the realm of a mixed market economy that would tolerate, and in some instances encourage, a limited volume of small private enterprise. The reprieve, honoured until 1926, would be sufficient to restore industrial production to pre-World War I levels (see Table 3.1).

Table 3.1: Industrial production in Russia, 1913–26

	1913	1921	1922	1924	1926
Industrial Production-GDP (Millions 1926 Roubles)	10,251.0	2,004.0	2,619.0	4,660.0	11,083.0
Electricity (Million Kwhs)	1,945.0	520.0	775.0	1,562.0	3,508.0
Grain Harvest (Million tons)	80.1	37.6	50.3	51.4	76.8

Source: Nove (1972)

The line was drawn at the 'commanding heights' of the new Soviet economy: banking, railroads, foreign trade and large industrial plants.

These would remain socialized, under the exclusive jurisdiction of the Communist Party, throughout the New Economic Policy. Needless to say, the outstanding property claims of Royal Dutch/Shell, the Nobels, and the Standard Oil Company of New Jersey were never formally acknowledged by the Bolshevik administration. Concessions, including the restoration of 'stolen' (nationalized) oil properties, were offered to anyone willing to negotiate.

The first 'deal' was made with the Barnsdall Corporation, a US company interested in sophisticated state-of-the-art drilling technology, in particular advanced rotary drills and deep well pumps. Hasty negotiations were concluded in October 1921. The agreement, signed one month in advance of the 'public' announcement offering Soviet concessions to foreign investors, was, by most accounts, highly successful. Barnsdall remained in the Soviet Union until 1924, restoring both crude oil flows and confidence in the Bolshevik administration. The Americans were soon joined by a host of eager foreign participants – British Petroleum, the Societa Minerere Italo Belge di Georgia and a Japanese consortium in Sakhalin. Ambitious joint ventures included the joint British–Soviet construction of a second pipeline from Baku to Batum, the French supply of a Schlumberger well-logging process, and American (Standard Oil of New York or Socony), German, and British support for refinery construction (Goldman, 1980).

Royal Dutch/Shell, Standard Oil, and the Nobels, outraged at the prospects of leasing their own production facilities, refused to participate. In the words of Walter Teagle,[4] head of Standard Oil: 'Affording the Soviet a market for petroleum not only is actually becoming a receiver of stolen goods, but operates to encourage the thief to persist in his evil courses by making theft readily profitable.' This group, newly united in the cause of common grievances, swore to fight the Soviet menace together. They agreed never to negotiate with the Russians individually and above all, never to purchase cheap Soviet oil supplies. *Front Uni*, which attempted an international boycott of Russian crude oil supplies, was formed in 1922 and was ultimately comprised of a consortium of twelve oil producing companies.

The boycott, organized one year after the path-breaking Barnsdall agreement, was too weak to negate the enticements to foreign firms of Lenin's New Economic Policy. By 1923, Soviet crude oil flows had reached 5.4 million tonnes (mt) per annum, a 54 per cent improvement over the 3.5 (mt) recorded in 1920. The world was, once again, awash with cheap Soviet oil exports. To add insult to injury, Standard Oil was forced to compete with European refiners supplied from crude obtained from Standard Oil's nationalized properties in the Caucasus. In a final effort to contain the damage, and hopefully out-manoeuvre the Bolsheviks, Standard Oil and Royal Dutch Shell began to investigate the opportunities for a 'joint' business arrangement with the Soviets. A Standard–Shell organization for the purchase of Russian crude was formed in 1924.

After twelve months of tedious negotiations a tentative arrangement appeared imminent. Five per cent of the original purchase price would be set aside to compensate 'former' owners, thereby satisfying the primary bargaining requirements of the *Front Uni*. Both Walter Teagle and Henri Deterding, the head of Royal Dutch Shell, remained sceptical. When the proposed agreement 'fell apart' in 1927 according to Yergin (1991, p. 242)

> Deterding was almost gleeful. 'I am so glad that nothing came of these Soviet deals' he wrote to Teagle 'I feel that everybody will regret at some time that he had anything to do with these robbers, whose only aim is the destruction of all civilisation and the re-establishment of brute force'.

The *Front Uni* was dissolved, and dreams of a united Western embargo forgotten later that year.[5] Despite heated protest by Henri Deterding, two prominent American oil companies, Vacuum and Socony, agreed to purchase a substantial volume of Russian kerosene for resale in India and Asia. To further aggravate matters, Socony was in the process of constructing a kerosene plant at Batum. Both Vacuum and Socony were Standard Oil successor companies and, in the eyes of the Europeans, irrevocably 'linked' with, and subordinate to, the Standard Oil Company of New Jersey. Incensed at the clear violation of 'western solidarity', Deterding organized a massive press campaign denouncing Socony, and by association Standard Oil. The campaign was accompanied by a fierce 'price war' in India, Asia, and ultimately Europe. As expanding volumes of Soviet crude oil supplies began entering the international oil market, crude oil prices crumbled. By the end of 1927, the posted price for West Texas intermediate crude had reached levels as low as US$1.28 per barrel, a significant US$0.62 reduction from the US$1.90 recorded in 1926.

Even the best arguments of the determined press campaign of Deterding were destined to fall on deaf ears as western firms moved to take advantage of the opportunities provided by the New Economic Policy. Future prospects for even the most successful foreign concessions, however, were grim. The hostility of the Communist Party to profit, private property, and above all foreign capitalists was made abundantly clear throughout the period of the New Economic Policy. The extent of this antagonism – often reflected in the discussions, writings, and public speeches of the Lenin/Stalin administration – can be measured by the limited acceptance of the multitude of foreign concession proposals. According to Nove (1972, p. 98) by 1924–25, the apogee of the New Economic Policy, a mere '4,260 workers were engaged in thirteen significant 'concession enterprises'. All sixty-eight concessions that existed in 1928 accounted for 0.6 per cent of industrial output.'

In short, however important foreign investment might have been as a catalyst to the revitalization of the Soviet oil industry, it could not, by any stretch of the imagination, have provided the mainstay of its development. Both crude oil flows, and the Soviet control over the essential oil industry, increased steadily throughout the decade of the 1920s. In 1920, the Chief Oil Committee authorized the creation of three local trusts to oversee the development of the primary producing regions. Azneft was given control over the Baku region, Grozneft was assigned to Grozny, and Embaneft was assigned to the oil fields of Emba. In 1922, the three trusts formed Neftesyndikat, a commercial syndicate with full monopoly powers over oil exports and foreign activity. Neftesyndikat was succeeded by Soiuzneft, and eventually in 1926 by Soiuznefteeksport.

By 1927 Soviet oil production had reached 10.67 million tonnes per annum, a 7.17 million tonne improvement over the 3.50 million tonnes recorded in 1920, and only 1.05 million tonnes below the previous industry peak, 11.72 million tonnes in 1901. The impact on global oil markets was staggering. Years of war and revolution had taken a severe toll on the Soviet economy, reducing the domestic demand for oil and refined petroleum products significantly, meaning there was more available for export. According to Ebel (1970, p. 15):

> In addition, as pointed out (by the party) in defence of expanding exports, the population of the Soviet Union in the 1920s was nearly 20 per cent smaller than the population of the Russian Empire in 1913. This loss of potential demand had its own particular influence on the export surplus.

Soviet exports exceeded 2.0 million tonnes in 1927, effectively surpassing peak annual exports by Czarist Russia (see Table 3.2). The flood of surplus oil soon exceeded existing export capacity, inspiring a new wave of ambitious Soviet construction projects. The second pipeline between Baku and Batum was completed in 1928 and later the port of Tuapse, also on the

Table 3.2: Soviet oil exports

(000 tonnes per annum)	
Year	Volume
1918	Not available
1919	Negligible
1920	31.3
1921	169.7
1922	382.9
1923	815.1
1924	1,505.4
1925	1,685.3
1926	2,097.1
1927	2,787.0
1928	3,625.0
1930	4,712.0
1931	5,224.0
1932	6,011.0
1933	4,894.0
1934	4,315.0
1935	3,368.0
1936	2,665.0
1937	1,930.0
1938	1,400.0
1939	500.0
1940	900.0

Source: Ebel, (1970).

Black Sea, was linked by pipeline with the producing fields of Grozny (Ebel, 1970).

The Soviets were simply biding their time, willing to tolerate a degree of private enterprise until the economy was strong enough to return to the process of creating a 'true socialist economy'. Similarly, foreign firms were accepted as a necessary evil because of the technology and expertise they provided. The death of Lenin eventually ushered in the era of Stalin and a will to continue with economic experimentation.

3.2 STALIN'S GREAT LEAP FORWARD

The events of 1929–1934 constitute one of the great dramas of history. They need much more space than they can possibly receive here, and a more eloquent pen than the author's to describe them. They also need a sounder base in reliable data than is available at present to any historian, in the East or West. For we are now entering a period in which the lines dividing propaganda from reported fact tend to disappear, and statistics too often become an adjunct of the party's publicity office. Official statements and pronouncements by leaders can no longer be checked against counter-arguments made by contemporary critics, since criticism is silenced (Nove, 1972, p. 160)

The idea that 'great changes' were about to occur in the Soviet Union was introduced gradually, and without fanfare throughout the NEP years. Stalin's succession to party secretary in 1922, the gradual phasing out of NEP policies, the 'emergency' confiscation of 'surplus' grain in the Urals and West Siberia in 1928,[6] and an accelerating intolerance of any form of dissent, disagreement and private enterprise (for example, Kulacs and Nepmen) all suggested a radical departure from established Party tradition. No one was prepared for the 'great turn' or 'revolution from above' which was to shake Russia to its foundations (Nove, 1972). The rapid collectivization of agriculture, liquidation of the Kulacs as a class, and great party purge of the 1930s, left a trail of fear and devastation that would torment Russia, and, indeed, the rest of the world for generations.

At the same time, the Soviet economy was undergoing a period of profound structural change, and, by the mid-1930s, unparalleled industrial achievement. Ironically, the foundations for this 'Great Leap Forward' were laid at the height of the New Economic Policy. As early as 1925 one finds in planning documents such declarations as: 'The state is becoming the real master of its industry' or 'The industrial plan must be constructed not from below but from above'. One year later, at the fifteenth meeting of the party congress between 26 October and 3 November 1926, delegates called for 'the strengthening of the economic hegemony of large-scale socialist industry over the entire economy of the country', and spoke of the necessity of striving to achieve and surpass the most advanced capitalist countries 'in a relatively minimal historical period'.[7] The means to this end, a comprehensive long-term plan for the unified development of the Soviet economy, was conceived shortly thereafter.

On 8 June 1927, the Council of People's Commissars issued a decree calling for the creation of (Nove, 1972, p. 144):

> a united all-union plan, which, being the expression of economic unity of the Soviet Union, would facilitate the maximum development of economic regions on the basis of their speculation...and the maximum utilisation of their resources for the purpose of industrialisation of the country.

The Soviet planning system was far from a unified monolith intellectually. Remember, those charged with managing the economy were largely making it up as they went along because Marx had not provided any direction for the post-revolution organization of the economy. Soviet 'planners' were continuously striving to find a workable compromise between two planning philosophies. The first, known as 'genetic' placed an emphasis on the current economic environment such as the play of market forces, relative scarcities of factors, rates of return, and profitability.

The second, called 'teleological' was based on a forward looking vision that would alter the proportions and size of the economy, maximize growth and emphasize a strategy of development rather than adaptation to circumstances. The more balanced 'genetic' approach was condemned by the Stalin administration as a 'right wing' heresy in the years following 1928. At the same time, economists Popov and Groman were working on a new innovation in Soviet planning; the idea that it was necessary to study the complex inter-relationship between the factors of production and end-use products across all sectors of the economy. Data from 1923–24 was used to construct the first 'input–output' table. Initially condemned as heretic, the tables were destined to form an integral part of the Soviet planning process in later years (Nove, 1972).

Leading experts, generally of the 'teleological' persuasion, gathered to complete a tedious analytical exercise for which there was no known precedent, and only limited statistical foundation. Pressure to adopt targets that would be commensurate with the stature and ability of the Bolsheviks mounted steadily throughout the planning process. The targets became more and more politicized and less and less based on the available resources and technical capacity.

As a result of the sheer magnitude of the planning procedure, six months after it had been promised, in April 1929 a variety of options for the first five-year plan were submitted to the Sixteenth Party Congress for approval. The alternatives, which would govern the evolution of the Soviet oil industry from October 1928 to September 1933, included: (1) a base scenario requiring a 62 per cent increase in the production of crude oil from the 11.7 million tonnes (mt) reported in 1927 to 19.0 mt in 1932; and (2) an 'optimal' scenario specifying an 85 per cent increase in the level of Soviet oil production to 21.7 mt by the end of the planning period (see Table 3.3). The base or 'initial' variant, 'suspected' of being a product of right wing bourgeoisie specialists, was dismissed in the 'pathos of achievement' (Nove, 1972, p 145). While 'genetic' or balanced planning procedures were for the most part studiously avoided in the enthusiasm for growth, an acknowledgement of the crucial importance of geological research was contained in a special resolution appended to the official documentation. According to Hassmann (1953, p. 38):

> The guarantee of further development of the economy makes it absolutely necessary to expedite geological work so that it will surpass the tempo of industrial development and ensure the timely collection of a supply of mineral raw materials.

An NEP inspired increase in large scale industrial investment, which predated the first five-year plan by well over 12 months, lent an aura of credibility to 'optimal' production targets. By year-end 1929, Soviet crude oil production had reached 13.86 mt, an 18 per cent increase over the 11.76 mt reported in 1928. The success of heavy industry spurred on the 'teleological' experts, resulting in a series of unscheduled revisions to the five-year plan. On 1 December 1929, the Soviets issued a decree formally amending (that is increasing) official production targets. The document, which illustrates a clear mid-plan shift in Soviet emphasis towards the rapid development of all heavy or large-scale industry, foreshadowed increased central control over all aspects of the Soviet economy including: (1) the allocation of scarce resources; (2) physical production from all sectors of the economy (GDP); (3) credit; (4) private domestic consumption; and (5) foreign and domestic investment.

Table 3.3: Soviet crude oil production: scheduled and actual
(millions of tonnes)

The First Five-Year Plan	Actual 1927	'Optimal' 1932/1933[a]	'Amended' 1932	Actual 1932
1928–1932	11.7	21.7	40–55	21.4
The Second Five-Year Plan	Actual 1932	Original Version 1937	Planned	Actual 1937
1933–1937	22.3	80-90	46.8	28.5
The Third Five-Year Plan (Interrupted by WWII) 1938–1942	Actual 1938 28.2	Planned 1942 47.4	Actual 1940 31.1	Actual 1942 22.0
The Fourth Five-Year Plan 1946–1950	Actual 1945 19.4	Planned 1950 35.4		Actual 1950 37.9
The Fifth Five-Year Plan 1951–1955	Actual 1950 37.9	Planned 1955 70.9		Actual 1955 70.8
The Sixth Five-Year Plan (Abandoned in 1958) 1955–1960	Actual 1955 70.8	Planned 1960 135		Actual 1960 147.9
The Seventh Seven-Year Plan 1958–1965	Actual 1958 113	Planned 1965 230-240		Actual 1965 242.9
The Eighth Five-Year Plan 1965–1970	Actual 1965 243	Planned 1970 350		Actual 1970 353
The Ninth Five-Year Plan 1970–1975	Actual 1970 353	Planned 1975 505		Actual 1975 491
The Tenth Five-Year Plan 1975–1980	Actual 1975 491	Planned 1980 640		Actual 1980 604

Sources: Nove (1972); Ebel (1961)

The oil target, which stipulated crude oil flows in the 40–55 mt range by year-end 1932, was clearly impracticable. According to Nove (1972, p. 190):

> One old oil expert, given what he regarded as an absurd order to increase production, is said to have written to the central committee as follows: 'I cease to be responsible for the planning department. The (plan) figure of 40 million tonnes I consider to be purely arbitrary. Over a third of the oil must come from unexplored areas, which is like cutting up the skin of a bear before it is caught or even located.

Still, tremendous economic achievement – and the unique Stalin model for rapid economic development – sprang from an attempt to attain the impossible. Approximately 1.6 billion roubles were invested in the Soviet oil industry in a reduced four-year planning period, 1928–32.[8] Key projects included the construction of new pipelines, the promotion of exports, the construction of new refining facilities, increased exploration and development efforts, and a general upgrading of existing technical equipment. Both crude oil flows, and official estimates of Soviet GDP, rose steadily throughout the forecast period. By year-end 1932, Soviet crude production had reached 21.6 mt, a mere 100 000 tonnes short of the 'optimal' 21.7 mt annual production target. The amended variant, approximately 40–55 mt, would not be attained until 1951, 19 years after the 'official' due date.

The original version of the second five-year plan, drafted at the apex of the Great Leap Forward – again in the enthusiasm for 'construction' – was similarly unrealistic. On 30 January 1932 the Seventeenth Party Congress adopted the following proposals as a 'preliminary' basis for the new 1933–37 planning horizon: (1) coal – a 291 per cent increase from 64 mt in 1932, to 250 mt in 1937; (2) pig iron – a 255 per cent increase from 6.2 mt in 1932, to 22 mt in 1937; and (3) oil – a four-fold increase from 21.4 mt in 1932, to 80–90 mt in 1937. The 'financial' and theoretical justification for an emphasis on the already over-strained oil industry was compelling. While the production of crude oil had fallen short of ambitious (amended) planning targets, petroleum exports and associated hard currency receipts had exceeded all expectations. Ample volumes, readily available for loading at Black Sea ports, commanded a considerable geographic advantage in the European market over competition from the US and the Persian Gulf. Low-priced, high-quality Soviet oil exports reached 6 mt in 1932, an incredible 4 mt improvement over the 2 mt reported in 1927. Petroleum export receipts, rarely accounting for more than 7 per cent of 'hard currency' earnings in the pre-revolutionary Russian Empire, mounted steadily throughout the planning period. By year-end 1932 petroleum exports were responsible for a significant 18 per cent of Russia's total hard currency export receipts. Still, the petroleum industry's foreign exchange earnings were dwarfed by timber exports and, in many years, those of grain as well (Goldman, 1980).

The torrent of Soviet crude oil exports, exacerbated by stagnant global demand and increased global competition for limited markets, reduced world oil prices, and Western industry profits, significantly during the early 1930s. By 1932 the posted price of oil had reached levels as low as US$0.69 per barrel, a reduction of over US$1.20 per barrel from the US$1.90 recorded in 1926. The shortfall, primarily attributed to increased Russian exports, led the major international oil companies to explore a number of options to 'cartelize' the international oil industry. The first proposal, formulated by I. B. Aug. Kessler (managing director of Royal Dutch Shell), envisioned global production reductions, restricted drilling programmes and the imposition of export quotas in the United States, Venezuela and Romania. It was abandoned on the grounds that it would violate US anti-trust law.

An alternative plan was floated where it was proposed that the Russians be bought off. Anxious representatives from the major international oil companies met with Soviet officials in New York at the World Oil Conference in the spring of 1932. According to Ebel (1970, pp. 17–18):

> At that time the British–American group proposed that it take over all Soviet exports for the next ten years, at the 1931 level of about 5.2 million tons. As an adjunct, the Soviet Union would be required to dispose of all of its distributing facilities held in foreign countries, to ensure the effective elimination of the Soviet Union from the world oil market. The Russian counter-offer, while not rejecting this proposal out of hand, expressed a firm desire to retain the distribution facilities in question. . . . Although the Soviet delegation indicated the USSR was anxious to co-operate in any manner which gave promise of improving the conditions then prevailing, these differences were too great to be reconciled and the conference dissolved without agreement of any sort having been reached.

In the four short years that defined Stalin's 'first' Great Leap Forward, the Soviet oil industry had gained notoriety as a lucrative, and highly dependable, source of foreign exchange. Its 'political' value as a favourable 'yardstick' for comparison with the US was not lost on the Party. Optimistic expectations, fuelled by the timely announcement of a significant 'new' discovery, a second Baku, in the Eastern regions of the USSR, led to fantastic projections of Soviet oil exports and the fairy tale targets envisioned in the first draft of the second five-year plan.

The discovery of 'second Baku' oil deposits located on the vast landmass between the Volga river and the Ural mountains represents an important milestone in the history of the Russian petroleum industry. It is in the heart of the Soviet Union, between the industrial district of Central Russia around Moscow and the industrial region of the Urals (Hassmann, 1953).

Despite the best intentions of a determined Bolshevik work-force, the year 1933 would be a great disappointment to aspiring 'teleological' theoreticians. That the Soviet economy had been taxed to the limits of practical reality was clear throughout the harsh winter of 1932. Collectivization, and the mass deportation of skilled and ambitious Kulacs, increasing police brutality, and unparalleled economic sacrifice in a race to exceed all reasonable expectations, had ruinous implications for the critical agricultural sector. In 1933, the Soviet Union was plagued by famine, transportation crises and shortages so severe it was necessary to call a halt to the headlong race for economic growth. The economy was too far out of equilibrium for further progress to be made. A significant, and unexpected 14.3 per cent reduction in the volume of capital investment delayed advancement in all heavy industry (Nove, 1972). As progress slowed, eventually to a standstill, it became necessary, no matter how politically unpalatable, to review the optimistic five-year planning targets.

The revisions adopted by the Seventeenth Party Congress in February 1934, and upgraded repeatedly throughout the five-year planning period, reflected an unusual degree of 'human' sensitivity in the Stalinist era. The urgency of 'remedial action', and some improvement in living standards, was apparent even to the Central Committee. Stalin's new slogan 'Life has become better, comrades, life has become more joyous' would carry the nation through years of police terror and brutality, in 1934 there was a respite in the drive for economic growth. In the 'new' world of the 1930s collectivization would be completed, but the peasants on the collective farms, it was forecast, would live much better, with a very much larger number of private (as well as collective) livestock. For urban workers, it was suggested that wages would double as the result of an increase in money wages and a simultaneous fall in retail prices (Nove, 1972). The oil target was reduced by a recognition of the 'structural' decline in flows from ageing Baku oil fields to a slightly more realistic 46.8 mt.

As essential as these revisions were to the internal cohesion of a demoralized, and demographically crippled Soviet society, they would soon be abandoned in an accelerating international arms race. In 1933, the Nazis were elected in Germany, and Hitler, already an established rabble-rouser, made no secret of his hostility to communism. Defence expenditures, inadequate to the needs of the existing, and poorly outfitted, armed forces were increased tenfold, from 1421 million roubles in 1933 to 14 883 million roubles (16.1 per cent of the total Soviet budget) in 1936 (Nove, 1972). The fledgling heavy industrial sector, an ongoing favourite of the Party, was transformed into a leading global producer of armaments. Broad development issues, and the day-to-day needs of ordinary citizens, were forgotten in a supreme effort to revitalize an exhausted Soviet military machine.

By all accounts, and despite considerable discrepancies between 'official' statistics and reality, the second five-year plan was an extraordinary success. To cite only a few examples: national income reached 96.3 billion roubles in 1937, a 112 per cent increase over the 45.5 billion roubles reported in 1932 and only 4.0 billion roubles below the official planning target. The grain harvest reached an impressive 96 mt, a mere 9 mt short of elevated Soviet expectations. The armed forces doubled in size. Electricity generation exhibited an average annual growth rate of 26 per cent and the average annual productivity of a typical coal miner grew from 16.2 tonnes in 1932 to 26.9 tonnes in 1937.

The oil industry was not listed, however, among the rising stars of the 'great Soviet energy initiative'. It fell short of its target by an unacceptable 18.3 mt. While the defence budget had exceeded 14.8 billion roubles in just one year (1936), a cumulative 2.5 billion roubles had been allotted to the oil industry throughout the 1933–37 planning horizon. The allocation of investment funds was plagued by difficult choices between the construction of critical new refineries and cracking plants, and the development of new fields in the Volga-Urals. With no means to make rational choices (von Mises, 1981), the planners constantly struggled with allocation decisions. The emphasis, however, was on new projects. As a result, little was spent to stem the decline of the increasingly decrepit Baku oil fields. Foreign concessions, and investment, had also largely evaporated.

In the years following the NEP, Stalin's acceptance and tolerance of the oil concessionaires was erratic and highly discriminatory. Concessions to 'most' foreign investors ended gradually and by December 1930, the majority of the contracts had been terminated. Standard Oil, however, retained its concession at the kerosene refinery in Batum until at least 1935 and the Japanese retained a presence in Sakhalin until 1944. While the Soviets were ending Standard Oil's concession, they inexplicably issued a new series of contracts to companies such as Badger and Universal Oil Products. Lummus was called back to rebuild and reconstruct refineries. Some of this work continued until 1945 (Goldman, 1980).

To further aggravate matters, flows from the rich, and highly promising Volga-Urals region were below those expected due to the poor quality of the technical equipment in the area, a very large labour turnover due to poor housing conditions, and the backwardness of geological surveying and drilling work (Nove, 1972).

While the number of geophysical crews engaged in exploration increased significantly in the first years of the second five-year plan, from 51 in 1932 to 98 in 1935, their productivity was limited by a heavy reliance on electrical methods most suited to the measurement of surface rock strata. In 1935, the use of the much more effective Seismic method was restricted to nine crews operating in Emba, West Asia, Baku, the Caspian Sea and Volga-Urals. Extensive systematic exploration efforts were not undertaken until the third five-year plan, which placed a significant emphasis on the recovery of oil. The backwardness in Soviet drilling techniques is indicated by the depth of an average Soviet well. According to Hassman (1953, p. 43):

> In most cases Soviet drilling reached only comparatively shallow depths until the Second World War. . . . During World War II some wells in the Second Baku fields reached 10,800 feet, e.g., in Tuymazy. However, the average depth of Russian wells was then but 2,000 to 2,300 feet. The fourth five-year plan endeavored to increase the average depth of the wells in the Second Baku region from 5,700 to 6,560 feet.

The failure of the Soviet oil industry to meet its official production targets, and a substantial increase in domestic demand, which can be attributed to the tremendous investment in the production of fuel guzzling capital goods, had profound implications for the level of petroleum exports. By 1937, Soviet oil exports had fallen to 1.93 mt per annum, the lowest level to be recorded in over a decade (Ebel, 1970). Thus, at the end of the 1930s the Soviet oil industry was committed to production targets that exceeded the ability of the exploration division's ability to find new oil. As a result, the emphasis was on obtaining as much production in the short-run as possible, with disastrous consequences for the management of the reserves. These difficulties, however, were to pale next to the challenges brought by the war with Germany.

3.3 FUELLING THE WHEELS OF WAR

> Many factors shaped Germany's decision to go to war with the Soviet Union:
> Hitler's deep-seated hatred of Bolshevism (its eradication, he said, was his
> 'life's mission'); his personal enmity for Stalin; his contempt for the Slavs,
> whom he regarded as 'little worms'; his desire to dominate completely the
> Eurasian land mass; and his drive for glory. In addition, when he looked East, he
> saw lebensraum ('living space') for the Thousand-Year Reich, his new German
> Empire...From the very start, the capture of Baku and the other Caucasian oil
> fields was central to Hitler's concept of his Russian campaign. 'In the economic
> field', one historian has written, 'Hitler's obsession was oil'. To Hitler, it was
> the vital commodity of the industrial age and for economic power. . . . If the oil
> of the Caucasus – along with the 'black earth', the farmlands of the Ukraine –
> could be brought into the German empire, then Hitler's New Order would have
> within its borders the resources to make it invulnerable (Yergin, 1991, p. 334).

The third five-year plan, drafted in the years 1937–38, and adopted by the
Eighteenth Party Congress in 1939, was doomed from the moment of its
conception. Targets, cautiously optimistic but not lacking the usual
'teleological' imagination, included: a 92 per cent increase in industrial
output, a 58 per cent increase in steel production, a 129 per cent increase in
engineering and the production of machinery, and a 69 per cent increase in
the production of oil. At the same time, Soviet planners were well aware of
the potential 'economic' repercussions of a continued structural decline in
the flows from Baku and Groznyy. The plan placed an urgent and
accelerating emphasis on the development of the strategic Volga-Urals
producing region. The new oil fields were to be connected by an ambitious
pipeline system to the industrial areas of the Urals and Central Russia. The
plan also included the construction of refineries and increased use of rotary
drilling. The geological and geophysical exploration of Central Asia was
given particular emphasis (Hassmann, 1953).

While the Soviet oil industry depended heavily on the rotary drilling
method, the continuous rotation of poor quality drilling pipe led to tool-
joint failures, cracked pipe, and frequent breakdowns. The inability of the
Soviet planning system to provide high-grade, domestically produced pipe
to the oil industry led to 'early' experimentation with substitute drilling
methods. A promising alternative – the hydraulic or turbo system – used a
steady flow of water combined with drilling mud to lubricate the drill bit,
thereby enabling the pipe to remain fixed in place throughout the drilling
process. According to Goldman (1980, p. 40):

Soviet engineers began their first efforts with such a turbo-drill in the 1920s. Some progress was made in 1934, but not enough to solve the basic engineering challenges that were involved. But since the only other alternative seemed to be continuous reliance on imported Western pipe, development work continued. During the war, the experiments were moved from Baku to the greater safety of the Volga-Ural region. In retrospect this proved to be a stroke of luck, since the turbo-drill, which was about to become a reality, was not suited for Baku but was ideally suited for the Volga-Ural region. The hard rock in the Volga-Ural region lent itself nicely to penetration by the turbo-drill, at least up to 2,000 meters in depth. This was fortunate since the Volga-Ural rock all but blocked penetration when the rotary drill process was used with Soviet pipe...[Once] the turbo-drill had proven itself, its use spread rapidly. By 1956, it accounted for 86 percent of all Soviet drilling.

The economic disruption caused by the great purge of 1937–38 along with the transportation bottlenecks and coal shortages that accompanied the Finnish war of 1939–40 meant that progress was tedious and uneven at best. Mass arrests, executions, and the deportation of millions of highly skilled engineers, managers, technicians, and civil servants to distant concentration camps inspired a wave of psychological hysteria that would sweep through the entire economic administration, literally 'paralysing thought and action' (Nove, 1972, p. 37). A severe shortage of labour, and the diversion of goods and resources to the arms industry, resulted in countless economic dislocations and shortages. Investment, construction and initiative plummeted in 1937, forestalling the contemporary and future development of most heavy industry. According to Nove (1972, p. 257) 'output of oil, a vital and strategic item (one would have thought), expanded exceedingly slowly, contributing to a fuel crisis.'

Official targets called for concentrated development of the new Volga-Urals producing regions and specified a steady increase in flows from the 1.6 million tonnes recorded in 1938 to 7 million tonnes by 1942. Initial results were, however, far from satisfactory. According to Ebel (1970, pp. 26–27): '[t]he inadequacy of equipment . . . coupled with a 'no-show' during the first test of the Devonian at Tuymazy in 1940, argued against the expenditure in any significant quantity of investment funds for this purpose.' Production flowed slowly, in limited increments, from the Tuymazy and surrounding fields. By 1940, only two years short of the official planning deadline, the Second-Baku was producing only a disappointing 1.85 million tonnes of crude oil. Soviet production, crippled by the unanticipated shortfall of valuable Volga-Urals crude oil flows, reached only 31.1 million tonnes in 1940 – a disheartening 16.3 million tonnes below target. To the dismay of the Bolsheviks, a full 87 per cent of these volumes, approximately 27.05 million tonnes, was concentrated in the Caucasus around the aged, and militarily exposed, Baku, Maykop and Groznyy producing regions (see Table 3.4 and Appendix A).

In Hitler's mind, the precarious relations between himself and his arch rival Stalin had reached an all time low. One central issue – the coveted Ploesti oil fields of Romania – struck at the heart of German vulnerability; a heavy and strategically unacceptable dependence on imported oil. Feeling protected by the Nazi–Soviet Pact of August 1939, Russia was supplying over 30 per cent of German oil imports (approximately 675 000 tonnes). The bulk of the remainder – some 58 per cent – was provided by the fields

Table 3.4: The regional concentration of the Soviet oil industry

Area	1938 Actual Production	1942 Scheduled Production
(millions of tonnes)		
Caucasus		
Baku	20.7	27.0
Groznyy	2.3	4.1
Maykop	2.0	3.7
Dagestan	0.2	0.6
Russian Platform		
Volga-Urals	1.6	7.0
Emba	0.5	2.0
Ukhta-Pechora	0.1	N.A.
Central Asia	0.3	1.7
Far East		
Sakhalin	0.3	1.3
Total	28.0	47.4

Source: Hassmann (1953).

at Ploesti. The prolific Romanian oil fields, well within range of an invigorated Bolshevik army, would prove irresistible to Stalin. According to Yergin (1991, p. 335):

> In June 1940, the Soviet Union used the terms of the Nazi–Soviet pact as justification to seize a significant part of northeastern Rumania, which put Russian troops all too close to the Ploesti oil fields for Hitler's taste.

Retribution was swift and unmerciful. In December 1940, Hitler initiated Operation Barbarossa – a Directive demanding immediate preparation for a large scale invasion of the Soviet Union. Six months later, on the morning of 22 June 1941, the German army, three million men strong, with 600 000 motor vehicles and 625 000 horses, struck along a wide front. The German onslaught caught the Soviet Union completely off guard and put Stalin into a nervous collapse that lasted several days. To further aggravate matters, the Soviet military–industrial complex had been disrupted by a massive internal purge in 1937–38. The imprisonment, and subsequent execution, of thousands of highly competent generals, officers, and managers, delayed the mass production of modern weaponry and high quality tanks and aircraft, leaving the Soviet 'defence' team to the mercy of an advanced German arsenal. By November 1941 the Soviets had abandoned vast territories – including the Western Ukraine – in a retreat to the perimeter of Moscow.

While the Soviets appear to have been surprised at the timing of the Nazi invasion, they were aware of the threat. There is considerable evidence that the Soviet government was stockpiling oil supplies – withholding them from world oil markets (and Germany) – throughout 1940. Soviet oil exports fell significantly in the latter half of the 1930s – from the 3.4 million tonnes recorded in 1935 to only 0.9 million tonnes in 1940. According to Ebel (1970, p. 24):

> Granted that consumption of oil increased tremendously in the Soviet Union during the late 1930s, owing to . . . the gigantic investment in the production of capital goods . . . It became quite clear that the Soviet government during these years was stockpiling oil for war. . . . In 1938, allocations of more than 2 million tons of products to stocks exceeded exports by 600 thousand tons, and represented the equivalent of about 33 days' supply. Exports continued to decline to only 500,000 tons in 1939, but improved slightly in 1940, increasing to 900,000 tons. Nevertheless, in 1940 allocations to stocks [approximately 1.63

million tons] still were far in excess of exports, if representing only 24 days' supply that year.

The second German offensive – Operation Blau – set its sights on the 'black gold' of the Caucasus. According to Yergin (1991, p. 336):

> With considerable confidence, Germany assembled a Technical Oil Brigade, eventually fifteen thousand men strong, with the charge of rehabilitating and running the Russian oil industry. The only thing standing in the way of Germany's exploitation of Russian oil was the requirement to capture it.

The German forces were beefed-up, but were increasingly isolated from critical domestic supply centres as they made a determined drive for Baku. Rostov, home of the strategic Baku–Moscow oil pipeline, was occupied in July 1942 and the valuable Maykop producing area on 9 August. Less than ten days later, in mid-August, the swastika was planted on the summit of Mount Elbrus, the highest point in the Caucasian Mountains (Yergin, 1991).

Baku, Hitler's primary objective, and in his mind[9] the key to German victory, would prove illusive. The German offensive, stalled by a severe shortage of fuel supplies, was unable to penetrate the narrow, and easily defensible, mountain passes to Baku. Maykop, destroyed by the Russians in a 'hasty' premeditated retreat, provided only 70 barrels per day of 'raw' unrefined crude. Captured Soviet supplies – primarily diesel for use in Russian tanks – were useless to the German panzer division, which required large quantities of gasoline. The ultimate insult, a shortage of fuel in the 'land of plenty', was the deciding factor in the last German assault on the oil fields of Baku in November 1942. By January 1943, German soldiers in the Caucasus had been given the order to retreat.

The military tide turned decisively in February 1943 with the German defeat at Stalingrad. After Stalingrad the war on Germany's eastern front became one of attrition with the key elements being quantities of manpower, equipment and strategic resources, including oil. Despite some local reverses, Soviet material superiority meant that the Germans were unceasingly pushed back until they were driven out of the Soviet Union and ultimately defeated. The Germans were often handicapped by fuel shortages that limited their mobility, constrained their re-supply abilities and reduced their overall fighting effectiveness.

Economic recovery, inspired by periodic celebrations of victory and the gradual rehabilitation of reoccupied territory, would be slow and painful. The Germans, cheated of the fruits of conquest, engaged in a massive 'wrecking' campaign throughout their retreat. By the end of 1943, the gross industrial production of the Ukraine had fallen to 1.2 per cent of 1940 levels, despite the fact that the Red Army had recaptured vast territories including Khartov and Kiev. The consumer goods industry was decimated by German occupation of the principal areas of light industry production and the harsh economic priorities of the Great Patriotic War. In 1942, at the apex of the German offensive, agricultural production fell to 38 per cent of 1940 levels; textiles to 33 per cent; meat and dairy production to 50 per cent; and sugar to 5 per cent (Nove, 1972).

The crude oil industry was laid to ruin by the destruction of fields in Maykop, the Ukraine, Estonia and parts of Groznyy. Stagnation in Baku, and the diversion of exploration and drilling crews to the east, contributed to the decline. By 1943 crude oil flows had fallen to 18 million tonnes per annum, an incredible 15.1 million tonne reduction from the 33 million tonnes recorded in 1941 (see Table 3.5). Crude oil exports, which had been falling steadily since 1932, vanished with the first German offensive. Vital transportation links were sabotaged, and ultimately, destroyed. According to Nove (1972, p. 276): '. . . in the winter of 1942–1943 it was necessary to transport Baku oil to central Russia via Kazakhstan and Siberia by rail, since both the Volga water route and the North Caucasus pipeline were cut'.

Table 3.5: Soviet crude oil production and imports from the United States

(metric tonnes)		
Year	Production	Imports from US
1941	33,000,000	310,342
1942	22,000,000	149,037
1943	18,000,000	362,067
1944	18,300,000	609,300
1945	19,400,000	538,608

Source: Ebel (1970).

The Red Army and the military–industrial complex engaged in massive offensives, the development of a substantial and highly efficient arms industry and the hasty evacuation of 1523 'threatened' industrial enterprises, nearly 1360 of which had been defined as large scale (Nove,

1972). These activities required vast quantities of petroleum products. As a direct result, Russian imports of petroleum products, which had remained constant at approximately 100 000 tonnes per annum from 1937 to 1940 (Goldman, 1980), increased significantly throughout the war years. To cite only one example: the United States committed to providing 'full and complete support to the Soviet war effort' and provided the Soviet petroleum industry with approximately 2.1 million tonnes of fuels and lubricants through a number of Lend-Lease agreements.[10]

Still, by late 1944 victory was plainly within the USSR's grasp. Stalin's new slogan 'Everything for the front' took precedence in all areas of economy and administration. By this time the entire Soviet economy had been revitalized under the watchful guidance of the peremptory State Committee of Defence. The committee, chaired by none other than Stalin himself, displaced the Council of People's Commissars, enacting a series of increasingly successful 'economic–military' plans. In recognition of the strategic importance of oil (and steel), a longer-term plan covering the years 1943–47 was devised for Urals region.

The rapid deployment of troops and resources was essential to the progress, and ultimate victory, of the Soviet Army. According to Nove (1972, pp. 273–74) an established command economy, and the

> experience of centralised planning in the previous years was a great help. In the process of tightening control over resources the government resorted to quarterly and even monthly plans, in far greater detail than in peacetime. The practice of material balances was used successfully to allocate the materials and fuel available between alternate uses in accordance with the decisions of the all-powerful State Committee on Defense. . . . Centralisation was essential to mobilise resources, and the U.S.S.R., after suffering what could have been crippling losses in the first months of war, carried out centralisation very effectively.

The results, particularly in the armaments industry, were impressive by any standard. By 1945, the Soviets had manufactured a grand total of 489 900 pieces of artillery, 136 800 planes, 102 500 tanks and self-propelled guns, and untold quantities of uniforms, boots, and ammunition. In the strategic, and highly prosperous, Urals industrial sector: (1) steel production reached 5.1 million tonnes, an 89 per cent increase over the 2.7 million tonnes recorded in 1940; (2) electricity generation rose by 200 per cent over 1940 levels; and (3) coal production reached 257 tonnes, 245 tonnes higher than the 12 tonnes reported in 1940. An impressive 3500 new enterprises were established on Soviet territory and 7500 of the firms damaged by fighting were restored.

The oil industry would prove slightly less resilient. Still, fuel remained a top government priority and despite a number of unscheduled delays, often attributed to heated differences of opinion among the various groups of geologists, exploratory efforts continued in the second Baku producing region. The first major discovery, well number 100, yielded a substantial 250 tonnes a day from the Devonian at Tuymazy. Activity increased with the success, sending Devonian crude oil flows soaring to 680 000 tonnes (1862 tonnes per day) in 1945 (Ebel, 1961). Soviet production began a slow, albeit accelerating, period of recovery; from 18 million tonnes in 1943, to 18.3 million tonnes in 1944, and an impressive 19.4 million tonnes in 1945.

3.4 STALIN THE GREAT WAR LEADER: RECOVERY AND CONSOLIDATION

> On 9 May 1945 fighting with Germany was over, and the Red flag had been flying over the Reichstag in Berlin for a week. It is impossible to overstate the effect of this dearly bought victory on the morale and consciousness of the Russian people. Stalin was now the great war leader, who had led them to victory (Nove, 1972, p. 287).

The Soviet Union emerged from the horrors of war triumphant, but sorely debilitated. The 'Great Patriotic' battles had been fought with courage, conviction and as if there were an inexhaustible supply of 'dispensable' physical and human resources. Relentless in its pursuit of victory, the Red Army had driven Nazi forces from the Russian 'motherland', Poland, Romania, Bulgaria, Czechoslovakia, Hungary and a significant portion of Germany. Its success provided the allies with an unforgettable demonstration of the power, and potential, of the USSR. By 1945: (1) Russian forces occupied the majority of the Eastern European nations; (2) native Communist resistance parties ruled over Yugoslavia and Albania; and (3) strong Communist political movements dominated the struggle for preeminence in China, France and Italy.

Never in his 20 years of political leadership had Stalin possessed more domestic and international credibility. Civilian hopes ran high. At long last, the average Soviet citizen might be offered relief, some slight reward for the years of war and extreme economic austerity. Stalin, intent on the maintenance of power, had no such predilection. According to Nove (1972, p. 290):

> On 19 August 1945, while the Soviet army was completing its advance in Manchuria against the Japanese in that brief campaign, Gosplan was instructed to draft a five-year plan covering the period 1946–50. Its guiding light: to exceed pre-war output by 1950.

Censorship and oppression intensified in an attempt to discredit Western ideologies.

On 9 February 1946, Stalin spoke to the nation. In this well-documented 'election speech' he recognized, and extolled, the tremendous achievements of the armaments industry. Praise and congratulations were followed by a stern note of caution. While the Germans had been defeated, the world was still full of enemies. The traditional emphasis on arms and strategic heavy industry was to continue, without pause, throughout the decade of the 1950s. By 1960, approximately three five-year planning periods, he projected that: (1) steel production would reach 60 million tonnes, a 390 per cent improvement over the 12.25 reported in 1945; (2) coal production would rise to 500 million tonnes; and (3) crude oil flows would reach 60 million tonnes per annum, a 209 per cent improvement over 1945 levels.

The implied five-year targets, fantastic as they might have seemed to an exhausted civilian work-force, reflected a surprising degree of realism. The fairy-tale quotas, and the outlandish expectations conceived in the early years of central planning had long since been discredited. Nove (1972, p. 292) provides the following insight:

> To quote a Soviet textbook on the period, 'since the possibilities of financing and supplying capital construction at this period were limited, the major part of resources were concentrated on the most important sectors of the national economy – on the restoration and development of heavy industry and rail transport.'

The fourth five-year plan, announced to the Soviet public in March 1946, mirrored the sentiments, and practical ambitions, of the Stalin 'election' decree (see Table 3.6). Gross National Product was to be restored to pre-war levels and then to exceed the 1940 estimate by a plausible 38 per cent by the year 1950. A substantial 87.9 per cent of all Soviet industrial investments was devoted to the reconstruction and development of heavy industry. The residual, 12.1 per cent, was allocated to food and light

Table 3.6: The fourth five-year plan

National Aggregates

	1945	Plan 1950	Actual 1950
National Income (1940=100)	83.0	138.0	164.0
Gross Industrial Production (1940=100)	92.0	148.0	173.0
Gross Agricultural Production (1940=100)	60.0	127.0	99.0
Coal (million tones)	149.3	250.0	261.1
Electricity (million Kwhs)	43.2	82.0	91.2
Steel (million tonnes)	12.3	25.4	27.3
Wool Fabrics (million metres)	53.6	159.0	155.2
Oil Production (million tonnes)	19.4	35.4	37.9

The Crude Oil Industry

(Thousands of Tonnes)

	Plan 1950	Actual 1950	Alternative Estimate 1950
Azerbaijan SSR (Baku)	17,000	17,000	14,822
Russian SSR (Groznyy, Maykop, Dagestan, Molotov, Ufa, Kuybyshev, Ukhta-Pechora, and Sakhalin)	14,500	16,400	18,231
Kazakh SSR (Emba)	1,200	1,300	1,059
Turkmen SSR (Turkmenistan)	1,100	1,250	2,021
Uzbek SSR (Central Asia)	1,066	1,200	1,342
Ukrainian SSR (District of Western and Central Ukraine)	325	330	293
Georgian SSR (Georgian area)	110	120	43
Kirghiz SSR	80	na.	47
Tadzhik SSR	60	na.	20
Total	35,411	37,600	37,878

Note: The Volga-Urals contained the prolific Molotov, Ufa, and Kuybyshev producing areas.

Sources: Ebel (1961); Hassmann (1953); Nove (1972).

(consumer) industries. Factories were given priority over houses; steel and railways over sugar and cotton fabrics.

The oil industry, a critical component of most heavy industry targets, was to be fully revitalized with average annual production reaching a more realistic 35 441 mt by the year 1950. The planners devising the Gosplan were now thoroughly convinced of the potential of the Volga-Urals Devonian reserves and specified the successful completion of 5500 new wells, 3500 of which were to be located in the Second Baku producing region (Hassmann, 1953).

Precise economic directives were followed by an immediate, and uncompromising, call to action. The newly reinstated Council of Peoples Commissars (CPK), burdened with the delicate transition from war to peace time, redoubled its efforts to control and direct the complex Soviet economy.[11] As reported by Nove (1972, p. 296):

> There was . . . a great wave, in 1946 especially, of creating new people's commissariats by sub-division. On 15 March 1946 the designation 'ministry' was substituted for 'people's commissariat', but this had no significance beyond restoring a word formerly regarded as bourgeois.

Glavki (production departments) were promoted to the status of Ministry. The Ministries, in turn, were sub-divided according to geographical and practical considerations.

In 1946, the Ministry of Oil was divided into two separate Ministries: (1) the Ministry of the Petroleum Industry for the Southern and Western Regions – instructed to restore production in the Southern and Western regions to pre-war levels; and (2) the Ministry of the Petroleum Industry for the Eastern Regions – instructed to create a new centre for the production of crude oil in the USSR. This centralization of responsibility with the central government meant that the union republics such as Azerbaijan had little power over their own economic development. The reorganization, and loss of regional control soon proved unworkable and was subsequently abandoned. On 28 December 1948 the two ministries and the 'department of oil supplies' were united as one unit: The Ministry of the Oil Industry.

Despite these setbacks, the centralization of authority continued. Party members, previously responsible for day-to-day industrial operations, were promoted to 'overlords'. Their assignment, to supervise and direct entire industrial sectors – groups of Ministries – enhanced Stalin's control over an intricate, and expanding Soviet economy. In short, while tedious administrative duties and trouble-shooting initiatives were somewhat decentralized (that is, left to 'specialist' Ministers), central control and the unique Stalin command economy was maintained, refined and consolidated.

The results paid tribute to the energy and tenacity of the average Soviet citizen. Discounting a brief period of post-war (transitional) recession, production and achievement increased steadily throughout the planning horizon. By 1950: (1) the broad economic aggregates, Gross National Income and Industrial Production, exceeded pre-war (1940) levels by an impressive 17 per cent ; (2) the production of key commodities – coal, electricity, oil, steel, tractors, sugar, and railway goods and traffic – surpassed official production quotas; (3) the Red Army had conducted a successful test of its first atomic bomb (1949); (4) the Council of Mutual Economic Assistance (COMECON), which defined the trading relationship with countries within the Soviet sphere, was firmly established; and last, but by no means least, (5) the prospects for the consumer goods industry showed clear signs of improvement.[12]

The crude oil industry, inspired by an explosion of exploratory drilling activity, exceeded Gosplan directives by a significant 2.5 million tonne margin. Crude oil production reached 37.878 million tonnes in 1950, a 19.878 mt increase over the 18 mt reported in 1943.

Estimates of Soviet crude oil production, however, vary widely by source and date of publication. To cite only one example: while Ebel (1961) estimates annual average production at 37.878 mt (from approximately 11 600 active wells) in 1950, Hassmann (1953) proposes a slightly lower figure of 37.6 mt (from approximately 18 000-20 000 active wells) (see Table 3.6). The discrepancy is most noticeable in the two estimates of crude oil flows from Baku. In this case, the figure presented by Ebel (14.882 mt) is considerably lower than the 17 mt presented by Hassmann. The Hassmann estimate, published in 1953 (the date of Stalin's death) purports to 'refute a number of incorrect statements heard over and over again about Soviet oil production and policy.' In Hassmann's (1953, p. 49) opinion: 'it is not true that the southern and western regions [Baku, Groznyy, Maykop, Dagestan, Georgia, the Ukraine, and Turkmenistan] fell behind the eastern regions [Molotov, Ufa, Kuybyshev, Ukhta-Pechora, Emba, Central Asia, and Sakhalin] in oil production'. Evidence and statistics uncovered at a later date (1961) would appear to support the estimate presented by Ebel (1961, p. 75); that is, in 1959, 'the greatest concentration of low-productivity wells [was] found in Azerbaijan'. The proposition that the southern and western oil districts suffered from lack of investment and general neglect is supported by the directives of the fourth five-year plan; that is, 3500 out of a total of 5500 new wells were to be located in the promising Second Baku producing region.

The bulk of the increased production has been attributed to increased flows from the immense Second Baku producing region. Extensive exploration efforts, conducted under the strict supervision of the all powerful CPK, resulted in the discovery of the Romashkino oil field in the Tartar Autonomous Republic in 1948. This particular oil field was described by Soviet authorities as being, in terms of reserves, not only the largest in the USSR, but in the world. Supported by this discovery and by others of lesser extent, the yield of crude oil from the Volga-Urals began to increase at an extremely rapid pace.

The super-giant Romashkino oil field eventually boasted crude oil reserves of approximately 14 billion barrels. Production from this field rose steadily for 22 years, to a peak flow rate of 1.65 MMb/d in 1970. The light and medium grade oil reserves found in the Volga-Urals were typically composed of sandstone or limestone layers with good permeability. They were located at a convenient depth, no more than 1500–2000 metres (Reinsch et al., 1992). According to Ebel (1961, p. 65):

> As a portion of total national output, that from the [Volga-Urals] increased to 29 percent in 1950 and to 58.7 percent by 1955. Further increases were registered during 1956–60, to about 70 percent of the U.S.S.R. total. It is estimated that by 1965, almost 80 percent of the Soviet production of crude oil in that year will come from fields in the [Volga-Urals].

By 1950, Second Baku production averaged 10.985 mt per annum, an incredible 8.152 mt higher than the 2.833 estimated in 1945.

The resurgence of crude oil supplies, tempered to some extent by a marked decline in flows from Azerbaijan, would prove inadequate to the task of reconstruction. Despite optimistic 'pre-planned' expectations, the Soviet Union remained dependent on imported oil supplies, sometimes euphemistically referred to as foreign interference, for the duration of the fourth five-year plan. Severe shortages, aggravated by the lack of sophisticated secondary refining capacity, sent planners scrounging for particular qualities of kerosene and diesel. At the same time, Stalin was determined to minimize Soviet exposure to hostile Western influences. Trading, and dependency, were limited; imports restricted solely to sympathetic (neutral) and submissive (communist) nations.

Non-Communist imports were minimized throughout the fourth five-year planning period. The meagre volumes reported in Table 3.7 (55 500 tonnes in 1949 and 263 400 in 1950) were imported from Austria. Non-Communist exports included: (1) substantial shipments to Finland – 197 100 and 147 300 tonnes in 1949 and 1950 respectively; and (2) minor quantities to Great Britain, Italy, France, Sweden, Yugoslavia, Greece and Denmark. It should be noted, however, that Table 3.7 does not include the 89.5 thousand tonnes of motor oil shipped to the USSR through a number of US Lend–Lease agreements.

Still, the rapid expansion of heavy industry demanded fuel. Imports from Romania and East Germany rose steadily throughout the planning period, from the 547 900 tonnes reported in 1947, to 2 636 600 tonnes in 1950 (an average annual growth rate of approximately 48 per cent). Exports, primarily low grade petroleum products supplied to friendly 'communist' nations, increased at a slower pace – from 500 000 tonnes in 1946, to 1.1 million tonnes in 1950 (an average annual growth rate of only 21.8 per cent). According to Ebel (1970, p. 30) '[T]he weak position in crude oil, combined with the continuing need to import high-quality products, kept the Soviet Union as a deficit trader in oil through 1953'.

The years 1950–1953 were somewhat scattered and disoriented. This was the apex of suppression: the Zhadanovschina.[13] Stalin had celebrated his 70th birthday in 1949, and was starting to show the signs of old age and the years of strain. His mental condition began to deteriorate markedly under the impact of hardening of the arteries and growing paranoia. Milovan Djilas, the Yugoslavian communist, commented that in 1948 the formerly quick-witted Stalin began to act 'in the manner of old men'. Khruschev reported that in Stalin's last years, 'He trusted no one and none of us could trust him'. Once renowned for public oration, Stalin remained in the Kremlin, condemning himself – and his closest subordinates – to a sedentary, administrative existence. His last publication, *Economic Problems of Socialism in the USSR* was released in 1952. The treatise, essentially a textbook on political economy, warned Socialist economists 'not to meddle in the affairs of the Politburo. The rational organisation of the production forces, economic planning, etc., are not problems of political economy, but of the economic policy of the directing bodies. They are two provinces which must not be confused'.[14]

In the meantime, the country waited patiently for an announcement from the Politburo of the targets to be included in the fifth five-year plan. Days stretched into months, months into years. Finally, in October 1952, a plan covering the years 1951–55 was announced at the long overdue Nineteenth Party Congress.[15] To the dismay of Nikita Krushchev, the principal report to Congress was delivered by Malenkov, a promising young upstart (only

Table 3.7: Soviet crude oil production and international trade statistics, 1946–55

(thousand metric tonnes)

Year	Production	Exports	Imports	Net Exports	Non-Communist Exports	Non-Communist Imports*	Net Non-Communist Exports
1946	21,746	500	909.1	(409.1)	58.6	0	58.6
1947	26,022	800	574.9	225.1	356.7	0	356.7
1948	29,249	700	874.0	(174.0)	200.5	0	200.5
1949	33,444	900	1,831.9	(931.9)	251.3	55.5	195.8
1950	37,878	1,100	2,636.6	(1,536.6)	164.4	263.4	(99.0)
1951	42,253	2,500	2,659.0	(159.0)	344.8	0	344.8
1952	47,311	3,100	3,797.6	(697.6)	783.9	121.6	667.7
1953	52,777	4,200	4,704.6	(504.6)	1,304.7	23.9	1,280.8
1954	59,218	6,500	3,993.0	2,507.0	2,550.2	0	2,550.2
1955	70,793	8,000	4,374.8	3,625.2	4,039.2	467.1	3,572.1

* All crude oil imports are from Austria.

Source: Ebel (1970).

50 years of age) who held a prominent position in the Secretariat. The requirements, all fairly straightforward, suggested 'business as usual at the Kremlin': (1) Gross National Income was to rise by 60 per cent; (2) industrial production would increase at a slightly faster pace, 70 per cent; (3) real wages would show a 35 per cent improvement; and (4) crude oil production would reach 70.9 million tonnes in 1995, a 33 million tonne increase over the 37.9 million tonnes reported in 1950.

Stalin did not survive the winter. The great dictator would die from complications following a 'minor' stroke suffered on 1 March 1953. His personal physician, the best man to treat the condition, had been arrested in Stalin's last purge: the 'Doctor's Plot' of late 1952. The news evoked extreme emotions from a demoralized Soviet population. A sense of relief swept across the countryside. The random episodes of extreme terror, oppression and paranoia might finally come to an end. At the same time, Stalin's status and popularity had been cemented by World War II and a massive, and highly effective, propaganda machine. Millions mourned the death of a comrade. Many more feared reform. Paranoid and secretive to the end, Stalin had refused to rally the political forces necessary to sustain his apparent successor Malenkov. Further, administrative deficiencies and the gross misallocation of energy and resources were becoming more and more obvious in the unique Stalin planning model.

The petroleum sector, while it had achieved much, was riddled with waste and ad hoc solutions. While some of this is understandable given the pressures put on the system during World War II, much of it had to do with the obsession with economic growth, and particularly the expansion of heavy industry. The planning process was driven by these priorities with the petroleum industry expected to come up with the fuel required to sustain an ever larger economy as well as growth itself. The planners had great difficulty with unknowns such as 'drilling success' rates and well head production. As a result, they had no real idea about the resource requirements of the industry. The system lacked any incentive for industrial enterprises to save on fuel use so energy was wasted on a large scale. All this combined to force those in charge of oil production to focus on short-term output rather than maximizing the long-run production from the fields.

The Soviet oil industry at the time of Stalin's death was not the one that was needed by a country with aspirations to be one of two equal superpowers. Certainly, it was now clear that the Soviet Union was endowed with sufficient petroleum resources to support its superpower goals, in fact it was much better endowed than its rival, the United States.

With the war in the past and the Communist Party firmly in power, the urgency that characterized the entire Stalinist era was gone. While the 'Cold War' was a considerable drain on resources, it did bring a long period of calm. The question to be answered in the next chapter is whether the Soviet oil industry was, under these very different conditions, able to become a supporting pillar of the new superpower's economy.

NOTES

1.   As cited in Braudel (1995), p.554.
2.   As cited in Yergin (1991), p. 239.
3.   The Supreme Council of the National Economy (VSNKh) was established in 1917, shortly after the revolution. The VSNKh, which derived its power from the Council of People's Commissars (CPK), was given full control over the oil industry. On 17 May 1918 the CPK created the Chief Oil Committee (*Glavny Neftianoi Komitet*) to supervise the development of the Soviet oil industry. The Chief Oil Committee served under the jurisdiction of VSNKh until 5 January 5 1932, when the VSNKh was formally abolished.
4.   As cited in Yergin (1991), p. 241.
5.   There appears to be some confusion surrounding the exact timing of the collapse of the *Front Uni*. In the opinion of Marshall Goldman (1980, p. 24): 'the *Front Uni's* embargo was broken even before it began to operate. Shell, itself a leader of the boycott, made a purchase of Russian oil in February 1923 and the French followed soon after.' On the other hand, according to Robert E. Ebel (1970, p. 15), 'The united front against the Soviet Union lasted roughly five months after its formalization and came to an end with the purchase of certain quantities of Soviet oil products by Royal Dutch.' Daniel Yergin (1991) suggests that a Standard Oil–Shell alliance lasted well into 1927.
6.   In January 1928 the Communist Party was discredited by a significant shortage in grain procurements. The shortfall was particularly evident in the West Siberia and Volga Urals region – where the harvest had been plentiful. The discrepancy, which was primarily attributed to the large gap between official and free grain prices, threatened industrial crops (cotton in Uzbekistan), the supply of grain to the cities, and ultimately the credibility of the Central Committee of the Communist Party. 'Ignoring the proposals of Bukharin and others to increase grain prices, Stalin decided to launch a direct attack, which revived memories of the excesses of war communism' (Nove, 1972, p. 150). Article 107 of the criminal code was used to provide legal justification for the seizure of grain 'surpluses' in the Urals and West Siberia. This arbitrary procedure for the procurement of grain for state purposes was known as the 'Urals–Siberian method', and repeated in many regions in the years 1928–29. According to Nove (1972, p. 153):

> In retrospect this must be regarded as a great turning-point in Russian history. It upset once and for all the delicate psychological balance upon which the relations between the party and peasants rested, and it was also the first time that a major policy departure was undertaken by Stalin personally, without even the pretence of a central committee or Politbureau decision.

7.   As cited in Nove, (1972, pp. 143–4).

8. In 1930, the central planning procedure was enhanced by a decision to synchronize the Russian 'economic' (fiscal) and calendar years. The Soviet fiscal year was defined as beginning on 1 October and ending on 30 September (Hassmann, 1953, p. 46).

9. According to Yergin (1991, p. 336) in a midnight phone call, Field Marshal Erich von Manstein begged Hitler to transfer the German forces in the Caucasus to his command in order to help the embattled Sixth Army at Stalingrad. Hitler refused. 'Its a question of the possession of Baku...Unless we get the Baku oil field, the war is lost.'

10. According to Ebel (1970, p. 28) 'These shipments continued during 1946 and 1947 for an additional 89.5 thousand tons, all motor fuel...Soviet trade statistics for these latter two years omit, purposely or otherwise, mention of these imports from the United States.'

11. The State Committee on Defence was abolished, and power returned to the Council of People's Commissars, on 4 September 1945.

12. The production of wool, cotton fabrics and sugar had been restored to pre-war (1940) levels. While the textiles and footwear industries showed promise, both failed to fulfil official production targets.

13. Zhadanovschina was a vicious campaign against Western influence named after Andrei Zhadanov, 'the saviour of Leningrad' who had been the Party boss of Leningrad during the siege of World War II.

14. As reported in Nove (1972, p. 320).

15. This was the first Party Congress to be held in thirteen years. The delay, which occurred despite the fact that Party regulations clearly stipulated the convention of a major Congress every three years, has been attributed to World War II and the rising international tensions accompanying the Cold War.

4. An Industry Fit for a Superpower

4.1 KRUSHCHEV'S SCHEMES

> Stalin's death added a tremendous problem to the many others left unsolved
> while he was still alive. The dead tyrant's successors knew that they could not
> govern as he had. Their personal security could not be achieved simply by the
> end of terror from above; it also required additional reforms to avoid a possible
> threat from the abused and oppressed masses. The dilemma that Stalin's death
> presented was to determine what kind of reform the system could stand without
> being undermined. At what point would the reforms that were necessary to
> sustain the system begin to threaten it. (Kort, 1993, p. 231)

Georgi Malenkov, the youngest of Stalin's inner circle and a devout student
of political intrigue, had prepared carefully for the struggle for personal
power and domination that would consume all post-Stalin administrators.
He moved quickly to assert control, claiming the top government and party
positions, prime minister and senior party secretary, for himself. A close
friend and co-conspirator, Lavrenti Beria (the chief of the secret police
under Stalin) received two prominent, and complementary, appointments;
Minister of the Interior and first deputy prime minister. Viacheslav
Molotov, a renowned Bolshevik and distinguished member of the Politburo,
assumed the position of foreign minister. The three formed an uneasy
alliance, a political triumvirate, which would succeed Stalin in the days
immediately following his death.

The new order was quickly discredited. According to Kort (1993,
p. 233):

> On March 14, only nine days after Stalin's death, Malenkov was compelled by
> his 'colleagues' to give up his post on the Secretariat, probably to prevent him,
> or anyone else, from accumulating too much power and following in Stalin's
> footsteps. That important job now went to Khrushchev, possibly because the
> three senior men did not consider him a serious candidate for power. On June
> 26 the crack widened when Beria, the member of the ruling group most feared
> by his colleagues because he controlled the secret police, was secretly arrested

in his Kremlin office while army tanks surrounded the secret police
headquarters.

Denounced as a 'capitalist agent', Beria was 'tried', convicted, and
executed in December 1953. Malenkov, apparently tainted by a close
'personal' association with Beria, clung desperately to his sole remaining
position: Chairman of the Council of Ministers.

The winds of change were blowing in Russia, and 'political' success
now demanded substantial 'consumer-oriented' reform. The first measure,
a 10 per cent reduction in the level of retail prices, was announced on
1 April 1953. Ten days later on 11 April 1953, Malenkov issued a decree
increasing the authority of ministries. Nove, (1972, p. 325) states: 'They
could within stated limits, alter staff establishment of their own enterprises,
redistribute equipment, materials and resources, and approve plans for
small- and medium-scale investment.' These changes eventually filtered
through the petroleum industry. In 1954, the Ministry of the Petroleum
Industry USSR was transformed from an 'all-union' Ministry, to an 'all-
union-republic' Ministry. The subtle change in status, and responsibility,
was facilitated by the creation of a new Ministry – the Ministry of the
Petroleum Industry Azerbaijan SSR – to oversee production in the republic
of Azerbaijan.

The general reorganization of the economy was touted as a 'New
Course' in Soviet economic policy – the development of a 'new' consumer
incentive programme. By August 1953, compulsory–voluntary bond sales
had been sharply reduced, average wages had been increased by 3 per cent,
a number of debts and tax arrears had been cancelled, and 4000 'political
prisoners', a small fraction of the total, had been released. On 28 October
1953, *Pravda* carried the following announcement: 'In the light of the
remarkable post-World War II achievements (primarily in the realm of
heavy industry), it was now possible to accelerate the production of
consumer goods'. The fifth five-year plan was revised accordingly. To cite
only a few examples: (1) the production of cotton textiles would reach 6267
million metres by 1955, a 24 per cent improvement over the 5044 recorded
in 1952; (2) the production of radios and TVs would be expanded to 4527
(million) units, 3195 (million) more than the 1332 (million) units available
in 1952; and (3) 3445 million units of bicycles would be produced, a
substantial 109 per cent increase over 1952 levels. Targets for the crude oil
industry, not yet a vital component of the 'consumer goods sector',
remained constant – 70.9 million tonnes in 1955 (see Table 4.1).

Table 4.1: The fifth and sixth five-year plans

	1950	Plan 1955	Actual 1955	Plan 1960
National Income	100.0	160.0	171.0	160.0
Gross Industrial Production	100.0	170.0	185.0	165.0
Producers' goods	100.0	180.0	191.0	170.0
Consumers' goods	100.0	165.0	176.0	160.0
Coal (million tones)	261.1	373.4	389.9	592.0
Electricity (million Kwhs)	91.2	164.2	170.2	320.0
Steel (million tonnes)	27.3	44.2	45.3	68.3
Cotton Textiles (million metres)	3,899.0	6,267.0	5,905.0	n.a.
Oil Production (million tonnes)	37.9	70.9	70.8	135.0

Source: Nove (1972).

The 'New Course' – higher incomes, the expansion of the 'consumer goods' industry, the maintenance of 'steady growth' in the heavy industrial sector, and an ambitious programme to achieve superiority in the US/USSR 'arms race' – led to severe economic dislocations, and overstrain. Resistance, doubt, and hesitation (the traditional assassins of reform) were bolstered by the failure to fulfil ambitious production targets. Valuable time and resources were also diverted by periodic episodes of political instability. In the immediate post-Stalin period, the official attitude of liberalization combined with a degree of apparent timidity among the new leaders led to anti-Soviet violence in the Soviet satellite countries, particularly Czechoslovakia and East Germany. In East Berlin tanks were used to restore order. In the Soviet Union itself there were several rebellions in the Gulag, and at the Vorkuta coal mine an uprising was only suppressed after considerable violence and bloodshed (Kort, 1993).

In the final analysis the regime championed by Malenkov was overturned because he was never strong enough to solve decisively conflicts over demands for resources. Of course, Khrushchev constantly worked to weaken the regime politically. When he was forced to resign in February 1955, Malenkov stood accused of not giving sufficient weight to heavy industry (and too much weight to consumer goods) in his priorities.

The means by which Malenkov had been outmanoeuvred and forced into political oblivion belong strictly to the realm of Kremlinology. Suffice it to say that by the end of 1954 (early 1955), Khrushchev's trusted allies controlled Leningrad, Moscow, the secret police and the Komsomol.

Bulganin succeeded Malenkov as prime minister. Khrushchev, who had achieved pre-eminence in his position of First Secretary of the Communist Party, simply retained the position. While Khrushchev held the real position of power, he never was able to use it to dominate in the way Stalin had, nor maybe did he want to. The 'New Course', and all Malenkov's amendments to the fifth five-year plan, were, however, forgotten in the struggle for primacy.

The crude oil industry was unaffected by the debate and progressed as scheduled throughout the five-year planning period. Crude oil production reached 70.79 million tonnes per annum in 1955, a 32.90 million tonne increase over the 37.89 million tonnes reported in 1950. As predicted, the bulk of the increment (approximately 30.57 million tonnes) was due to increased flows from the prolific Second Baku producing region (see Table 4.2). Azerbaijan registered a modest 500 000 tonne improvement, from the 14.8 million tonnes recorded in 1950 to 15.3 million tonnes in 1955. The drilling programme did, however, exceed Gosplan expectations. By 1955, there were 27 600 active wells in the USSR – 16 000 more than the 11 600 recorded in 1946. Not all of the growth in the number of active wells during these years represented new effort. A sizeable number were successful restorations to production of wells which, for various reasons, had been shut down. Attention was also given to raising the coefficient of utilization of active wells. During the period 1940–55, the coefficient of utilization increased sharply from 0.907 to 0.955. This was a result of the lengthening of the time interval between repairs and a reduction in the length of time that wells were shut down for repairs (see Table 4.3).

Prior to the Second World War, due to a lack of either manpower or materials, it was not possible to bring a number of fields back into production, those wells that had been shut down because of, for example, mechanical breakdowns or excessive water flooding. If a well was considered economically non-viable it was taken off the active list.

The torrent of oil production from the Volga-Urals fields, accompanied by a deliberate policy to 'cut costs' (whatever that meant in an economy with prices established by fiat) through the elimination of residual fuel oil as an input to Soviet heavy industry, led to a substantial increase in the volume of petroleum that was readily available for export. This represented a fundamental change from the years 1950–55 (the fifth five-year plan) when Soviet energy policy had been geared towards the attainment of regional self-sufficiency in energy. The primary motivation for Stalin's strategy was a reduction in waste and inefficiencies associated with the

Table 4.2: USSR crude oil production, 1950–65

Republic	1950	1955	1958	1959	1960	1965 Estimate
			(thousand metric tonnes)			
RSFSR	18,231	49,263	87,978	102,792	118,900	220,360
Inc. Volga-Urals	10,985	41,555	76,000	90,000	104,000	210,000
Ukraine	293	531	1,236	1,627	2,159	6,000
Uzbek	1,342	996	1,297	1,465	1,601	3,000
Kazakh	1,059	1,397	1,511	1,544	1,601	2,000
Georgia	43	43	35	35	35	80
Azerbaijan	14,822	15,305	16,497	17,076	17,800	22,000
Kirgiz	47	115	490	424	464	464
Tadzhik	20	17	18	17	17	60
Turkmen	2,021	3,126	4,154	4,577	5,278	10,000
Total	37,878	70,793	113,216	129,557	147,864	265,000

Source: Ebel (1961).

61

Table 4.3: Production, productivity and idle wells, 1946–60

(million metric tonnes)

Year	Crude Oil Production	Production Per Active Well (metric tonnes)	Number of Active Wells	Active Wells as a Percent of Total Wells	Total Wells	Idle Wells
1946	21.7	1,872	11,600	65.0	17,800	6,200
1950	37.7	2,016	18,800	na	na	na
1951	42.3	1,968	21,500	na	na	na
1955	70.8	2,568	27,600	89.5	30,800	3,200
1958	113.2	3,708	30,500	87.8	34,700	4,200
1959	129.6	4,050	32,000	na	na	na
1960	147.9	4,368	33,900	na	na	na

Source: Ebel (1961).

transportation of natural fuels, particularly petroleum, to remote and inaccessible regions of the Soviet Union. Emphasis had been given to the following initiatives: (1) the 'mineralization' of the fuel balance which meant a pronounced and sustained increase in the share of hard coal in the fuel and energy balance of the USSR, and (2) an increase in the production of synthetic fuels. The former, in particular, showed the inability of the planning system to make resources tradeoffs based on relative scarcity, particularly in the face of technological change. It led the Soviet Union, with its vast petroleum resources to emphasize coal production long after Western economies had begun the major shift away from coal to oil. It left the Soviet Union with a high opportunity cost energy system when it was attempting to close the gap between market and command economies.

The change in policy away from self-sufficiency combined with rising production meant that the Soviets had 8 million tonnes of oil available for export in 1955 – a remarkable 6.9 million tonne improvement over the 1.1 million tonnes made available in 1950. It is interesting to note that a significant, and growing, volume of these shipments were destined for non-communist (European) nations. Soviet international oil trade with the market economies was expanding despite the fact that East–West relations had been stretched to the limits of mutual tolerance by the tensions of the Cold War.

While trade with non-communist nations had resumed, at least in minor quantities, in 1946, these volumes were strictly limited by the requirements of the 'Plan'. Net non-communist exports had fallen steadily from 356.7 thousand tonnes in 1947, to a deficit of 99 thousand metric tonnes in 1950. According to Ebel (1970, p. 33) the

> turning-point appears to have been made in 1951, the first year of the fifth Five-Year Plan. In that year oil sales to the West showed a net of 344.8 thousand tonnes. . . . Thereafter net trade in oil with non-Communist countries began to increase rapidly, buoyed by concomitant high rates of growth in [the] production of crude oil.

The bulk of the exports were delivered to 'established' Soviet trading partners – Finland, Sweden and Italy – all nations that had stuck by the USSR at a time when Western trade with communist nations was almost imperceptible. With new 'deals' with Ireland, France, the UAR, Turkey and Yugoslavia, Soviet exports to non-communist nations reached 2.55 million tonnes in 1954, a startling 95 per cent improvement over the 1.30 million tonnes reported in 1953 (see Table 4.4).

Table 4.4: Soviet oil trade with non-communist countries,
       1950–54

Non-Communist Countries	(thousand metric tonnes)				
	1950	1951	1952	1953	1954
Exports					
Great Britain	5.5	7.9	0.7	1.4	11.2
Italy	5.7	110.4	382.4	138.2	183.2
West Germany	0.0	0.0	0.0	59.5	12.1
Finland	147.3	221.6	380.7	716.6	897.7
France	0.0	0.0	0.6	28.4	123.4
Sweden	5.3	2.7	15.0	269.0	743.9
Yugoslavia	0.0	0.0	0.0	0.0	48.2
India	0.0	0.0	0.0	4.6	0.0
United Arab Republic	0.0	0.0	0.0	0.0	207.9
Greece	0.1	1.5	0.9	0.5	74.8
Denmark	0.5	0.1	1.1	20.3	0.2
Ireland	0.0	0.0	0.0	65.5	231.9
Turkey	0.0	0.6	2.5	0.7	15.7
Total	164.4	344.8	783.9	1,304.7	2,550.2
Imports					
Austria	263.4	0.0	121.6	23.9	0.0
Total	263.4	0.0	121.6	23.9	0.0
Net Trade	-99.0	344.8	667.7	1,280.8	2,550.2

Source: Ebel (1970).

By May 1955, the Soviet Union was, in many ways, well on its way to the completion of its ambitious post-war reconstruction plans. Malenkov had been 'removed' from the office of prime minister, and despite the 'gross misallocation of resources' that had accompanied his 'absurd' 'New Course' in Soviet economic policy (at least according to the official Party line) – the key industrial targets of the 'original' fifth five-year plan would be fulfilled on schedule. Regardless of these considerable quantitative achievements, however, the development and production of heavy industry was still woefully inadequate to the task it had been set. Soviet enterprises simply lacked the facilities and the technology to meet the high, exacting standards of their Western counterparts. In the final analysis Soviet-manufactured items could not compete in foreign markets with those produced in market economies. Shoddy in appearance, poor in quality, overpriced, and of questionable durability, they could scarcely hope to attract the eye of the potential buyer. At the same time, the Soviet Union had lost ground in the race with the West, and was in no position to waste valuable time and resources in an attempt to duplicate (re-invent) existing equipment and technology. If Soviet enterprise was to acquire the required modern technology, the only alternative was to find a way to finance large-scale imports (Ebel, 1970).

The solution lay in the Soviet Union's abundance of natural resources. In particular, the extremely rich deposits of oil that had been discovered in the Volga-Urals region in the late 1940s and early 1950s were just beginning to show promise. With only 'limited' additional capital investments, the promising Volga-Urals production basins could be completely developed, refining capacity expanded, and the pipeline network extended to accommodate both domestic and international requirements. If Soviet geologists were correct, the resulting flood of crude oil would be more than sufficient to satiate domestic consumers, while simultaneously facilitating a steady, and sustained, increase in the volume of crude oil exports. The implications for the hard currency reserves available to the government were unmistakable. In May 1955, *Pravda* published an article stressing the importance of increasing the share of oil production in the fuel and energy balance of the Soviet Union. Gosplan's decision to take action to increase capital investment in the oil industry, and thereby to disrupt and reorient the entire Soviet energy balance, was reached shortly thereafter.

As with the change in emphasis that was shaping the broader Soviet economy, the decision indicated a clear change in the direction of established Soviet energy policy. As suggested above, in the years 1932–50, the production and consumption of hard coal had commanded a large, and rising share of the Soviet energy balance – from 50.8 per cent in 1932 to 64.6 per cent in 1950. Over the same time frame, the share of crude oil had fallen significantly from 28.7 per cent to a mere 17 per cent. The focus on coal to the exclusion of residual fuel oil intensified in the years 1950–55 through the 'mineralization' of the fuel balance specified by Stalin's fifth five-year plan. The May 1955 *Pravda* article provided the Western world with the first indication that the Soviet government was dissatisfied with the status of petroleum in the Soviet energy balance. In the words of Robert E. Ebel (1961, p. 7):

> it is not clear what prompted the decision for a shift in priority from coal to crude oil and natural gas. Surely a number of economic factors – reduced capital investment and reduced cost per unit of output and a higher index of

labour productivity – weighed heavily, but strategic and political decisions must also have supported this shift.

It should be re-emphasized that this somewhat radical revision in Soviet energy policy reflects one of the major deficiencies in the unique Stalin planning model. In the command economy, demand, and the allocation of scarce resources towards competing ends were determined by command from the centre. Prices, and any costs based on them, as well as rates of return were largely irrelevant, their roles officially having been reduced to a minimum by the supremacy of Gosplan. The reality was that the conditions prophesised by von Mises in the 1920s had come to pass and the planners had no way to make efficient decisions regarding the allocation of resources.[1] The result was that prices were changed only infrequently, and bore little relationship to opportunity costs. For example, it is hardly surprising given the absence of a means for determining relative costs that residual fuel oil – deemed prohibitively expensive in the years 1950–54 – should suddenly have been discovered to be 'cost' efficient, and highly desirable only one year later.

Economic reform – and an effort to revive the ailing manufacturing sector – continued under the supervision of the dynamic duo, Khrushchev and Bulganin. Political credibility had become a scarce and highly coveted commodity, and no expense could be spared in the preparation of the sixth five-year plan. First and foremost were problems associated with extreme centralization and the confinement of draft planning to a small group of senior officials.

Another endemic problem with the formal system of planning was the difficulty that the planning system had in incorporating new technology. Unlike western market economies where new technologies are adopted and adapted on a trial-and-error basis with viable technologies separated from non-viable technologies through the pressure of market forces, the planning system has no mechanism to incorporate technological experiments. Further, given the absence of opportunity cost pricing, there is no direct means of determining the relative efficiency of a technology that has been developed and that has experienced widespread adoption in market economies. While scientific research could be funded, there was no mechanism to encourage the movement of a technological idea to one with economic application. The planning process, with its emphasis on material balances and set technical relationships between inputs and outputs, has difficulty dealing with change of any kind. In short, incorporating technological change created much extra work for planners and an inherent bias against incorporation. The problems with incorporation of new technology are similar to the difficulties that planners had in dealing with the incorporation of international trade, given the volatility of international markets (Henderson and Kerr, 1984/85).

While Soviet authorities could recognize the problem, they were never able to see (or more likely admit) the central problems with the formal planning system. Hence, their solution was usually to add more layers of bureaucracy.

In recognition of a conspicuous absence of technical progress and innovation throughout the Soviet Union, the State Committee on New Technique (Gostekhnika) was established on 28 May 1955. Its mission: to investigate foreign achievement and, somehow, incorporate new technology into the formal planning procedure. On 9 May 1955 union-republics were given increased control over their enterprises, the allocation of raw materials, and a limited number of small-scale investment decisions. The reorganization of the union-republic Ministries was accompanied in some sectors by the creation of new Ministries to handle excessive work schedules. In recognition of the large number of construction projects to be undertaken by the oil industry, a new Ministry of Construction of Enterprises was formed with the following mandate: To co-ordinate – and facilitate – the activities of the construction enterprises of the union-republic Ministry of the Petroleum Industry.

Gosplan was revitalized and, in June 1955, divided into two complementary and highly specialized divisions: (1) The State Committee on Long-Term Planning (Gosplan); and (2) The State Economic Commission for Current Planning (Gosekonomkomissiya). As part of the reform, the sole remaining barrier to reform, Stalin and his iron-clad reputation for political and economic brilliance, would be discredited at every opportunity.

The sixth five-year plan was approved by the Twentieth Party Congress on 24 February 1956. Its major provisions may be summarised as follows:

1. a substantial commitment to heavy industry – with gross industrial production increasing by 65 per cent over the traditional five-year planning horizon;
2. increased investment in consumption and agriculture – specifying a 60 per cent increase in consumers' goods by 1960;
3. the development of a 'third' metallurgical base in Kazakhstan and Siberia – with the production of pig-iron rising to 53 million tonnes in 1960;

4. a substantial 64.2 per cent increase in crude oil flows – with production rising steadily from the 70.8 million tonnes recorded in 1955 to 135 million tonnes in 1960 (Nove, 1972).

With the business of planning and co-ordination having been settled quietly and without conflict, Khrushchev moved on to address more pressing considerations. Delegates at the 20th Party Congress were requested to remain seated for a special closed session. In the early hours of the morning of 24–25 February 1956 Khrushchev delivered his infamous secret speech 'On the Cult of Personality and its Consequences'. In a tirade lasting four and a half hours, he officially recounted for the first time ever the crimes committed by Stalin and, by close association, the vast majority of attendant party delegates. According to Kort (1993, pp. 239–40)

> Khrushchev accused the late *Vozhd* of being a brutal dictator. Stalin, the first secretary revealed, had ravaged the party by murdering thousands of its best people, including over 70 per cent of the Central Committee at his own 1934 'Congress of Victors'.

In short, Khrushchev suggested that Stalin was little more than a tyrant, whose 'annihilation of many military commanders . . . because of his suspiciousness and through slanderous accusations'[2] had been directly responsible for the early military disasters, and incalculable human casualties, of Barbarossa (World War II). According to Kort (1993, p. 240) Khrushchev added insult to injury by suggesting 'Stalin had promoted a "personality cult" that glorified him beyond recognition and gave him credit for what so many others had done.'

The speech, a blatant attempt to discredit those closest to Stalin – and not coincidentally, Khrushchev's most feared political opponents Malenkov, Molotov, and Kaganovich[3] – left the reputation of the Communist Party untarnished. Lenin, the Party and Khrushchev were beyond reproach: the 'avenging' angels of a harsh, and heartless, Stalinist bureaucracy. At the same time, there were 'minor hazards' and 'bottlenecks' associated with the 'Cult of Personality' which would soon be redressed. Khruschev wanted to streamline the bureaucracy and to reduce its authoritarian character that he realized was, in part, responsible for the lack of responsiveness in the economy. In the wake of the speech a number of decrees were issued that attempted to reduce bureaucratic malaise and the overstaffing of government offices (Nove, 1972).

The dilemma – to compromise Stalin (and Stalinism) while simultaneously maintaining the pristine appearance of the Communist Party and, hence, Khrushchev – would prove insoluble. Stalin had dominated Party proceedings for over 25 years without even a whisper of opposition. According to Kort (1993, p. 240): 'Once Khrushchev cast doubts on Stalin, the party's symbol of truth for so long, he inevitably cast doubt not only on the party's future policy, but on its very legitimacy to rule.'

News of the 20th Party Congress, and Khrushchev's Secret Speech, spread quickly across the Soviet Union, to Eastern Europe, and to the 'West' where the contents of Khrushchev's 'secret speech' were revealed in a document published by the US State Department. The confession exposed a rift in the invincible Party monolith, providing new impetus, a gateway, for independent inquiry and reform. Approximately 8 million people would was by horrific tales of unendurable suffering and personal tragedy – inspired a dangerous episode of overt public inquiry. Who would be held accountable for these flagrant crimes against humanity? Tensions and frustration mounted, leading to violent anti-Soviet demonstrations in Poland and Hungary in 1956.

Nikita Khrushchev, treading the fine line between heroism and 'guilt by be released from the Gulag in 1956 and 1957. Their reunion with an already disenchanted Soviet population – accompanied as it political association', could offer only token gestures such as the chastisement of Stalin era officials to appease 'confused' Soviet civilians. In Khrushchev's view, however, the reputation and authority of the Communist Party had to be preserved at all costs. Valuable time and resources were diverted to 'wayward' frontiers. In the early hours of Sunday morning, 4 November 1956, the Red Army reached Hungary. According to Soviet reports thousands of 'treacherous' Hungarian citizens were killed in a blaze of 'small arms fire and flaming Molotov cocktails' (Moynahan, 1994, p. 198). In the final analysis, it was the people and an exhausted Soviet economy that would pay for the chaos of reform.

Despite the best efforts of a new and improved Gosplan, the sixth five-year plan was riddled with inconsistencies. To cite only one example: gross industrial production was scheduled for a 65 per cent increase. Its principal components – producer goods and consumer goods – were to rise by 70 and 60 per cent, respectively. A discrepancy – the mathematical impossibility of even this basic assumption – was noted by Alec Nove (1972, p. 342): 'I am aware that, since 70 per cent of industrial output consisted of producers' goods, the 1960 plan indices for gross industrial output are inconsistent. But that is how they appeared in the plan'. The 'problems' were attributed to flaws in Khrushchev's reform programme – administrative confusion and internal inconsistencies resulting from the reckless reorganization of Gosplan. It is also true that the plans had become political documents in the Cold War side-show of competing economies, more propaganda than scientific planning. While plans had always had a value as propaganda, the sixth had a particularly weak internal structure.

The sixth five-year plan became the first long-term plan to be formally abandoned during peacetime. Khrushchev's credibility had been stretched to its limits by the uprisings in Poland and Hungary – events attributed solely to the de-Stalinization exercise. His opponents, sensing a moment of weakness, moved quickly to complete his demise. A meeting of the Central Committee was called for 20 December 1956. Delegates, taking great pains to examine the provisions of the entire sixth five-year plan in minute detail, concluded 'that the plan was out of balance', upset by 'excessive tautness' and blatant 'defects in planning' that suggested a 'lack of adequate coordinating powers.' (Nove, 1972, p. 344). The document was returned to Gosplan to be revised, rewritten and, if at all possible, resubmitted. A number of new proposals were suggested to enhance the efficiency, and co-ordination, of the ministries.

The revised version of the sixth five-year plan was submitted to the Central Committee on 9 April 1957. The members' unanimous rejection of (in their words) the 'half-hearted' blueprint – and by implication the first in a long list of what would come to be known as Nikita Khrushchev's 'hare-brained schemes' – was swift and unmerciful. Severe economic inefficiencies, shortages, had arisen from a lack of central control (co-ordination) and Ministerial 'empire building'. Regional and local planning activities were frustrated by the fact that most enterprises were still subordinate to the whims of Moscow. While the role of sub-national governments had been enlarged, their powers were strictly limited. Even minor decisions were subject to the approval of distant, and overworked, ministries. Ministerial 'empire-building' arose not so much from the aggrandisement instincts of the bureaucrats but, rather, from the divergence between the plans and what was actually delivered. Faced with chronic shortages due to the failure of their designated suppliers, prudent bureaucrats, charged with implementing the plans, attempted to compensate by having as much as possible produced within their own production units. As they were given targets, and would suffer if the targets were not achieved, they also attempted to stockpile as much materiel as possible and to keep as large a labour force as they could acquire to deal with the inevitable 'end of the month rushes' that were required to meet planning targets in the face of late deliveries of raw materials. A new vision would be required to correct the gross misallocation of valuable resources, and the wasteful duplication of effort. In response, and to maintain some semblance of consistency, the next plan would be required to oversee the development of the Soviet economy over an 'artificially expanded' seven-year period, 1959–65.

Khrushchev, now thoroughly convinced of the errors of his reform package – and undoubtedly fully apprised of an impending Malenkov/Gosplan *coup d'état* – took drastic measures to regain control. In May 1957, the Supreme Soviet adopted the following legislative initiatives:

1. the Industrial Ministry system – consisting of 140 ministries at the national and union-republic levels – was abolished, and replaced by a regional system to be co ordinated by Gosplan;
2. all enterprises of more than local significance were placed under the authority of Suvnarkhozy: the 105 regional economic councils responsible for all aspects of the economy within a specified geographical jurisdiction;
3. the Suvnarkhozy were to be appointed by, and answerable to, the republican Councils of Ministers;
4. Gosplan retained its responsibility for the 'drafting' and co-ordination of plans;
5. the all-powerful, all-union Council of Ministers was given full executive authority over all Gosplan decisions; and
6. The State Committee on Current Planning (Gosekonomkomissiya) was abolished after only two years of operation.

The Ministry of the Petroleum Industry USSR and the Ministry of Construction of Enterprises of the Petroleum Industry were abolished, and their responsibilities handed over to the local Sovnarkhoz. The Ministry of the Petroleum Industry Azerbaijan SSR remained intact until July 1959, when it too was abolished.

Less than a month later, on 19 June 1957, an enraged anti-Party coalition – Malenkov, Molotov, Kaganovich – secured a majority (7 to 4) in the Presidium to have Khrushchev, the 'estranged' first secretary, removed from office. The prime minister, Nikolai Bulganin, voted for Malenkov. Despite the personal insult, Khrushchev fought back bravely, insisting that the dispute be settled accurately – according to 'official' Party procedure – by the 'entire' Central Committee. Victory was guaranteed. Favourable weather conditions, and the abundant harvests of 1957 had enhanced

Khrushchev's support base throughout the provinces. Marshal Zhukov, the minister of defence, was persuaded to employ military planes, and personnel, to 'escort' sympathetic (provincial) Committee members to the meeting. The triumph was completed by a massive party 'purge' and reorganization. One year later, on 28 March 1958, Nikita Khrushchev consolidated his position, relieving 'Judas' Bulganin of all duties and responsibilities, and claiming the title of Prime Minister for himself. I. Kuzmin, a minor, but loyal, party official was appointed to the head of Gosplan, on 1 April 1958.

The oil industry suffered like the rest of the economy as a result of the lack of realism in the sixth five-year plan, particularly shortages of equipment and spare parts. Thus, it was ill-prepared for the demands placed on it by the Soviet Union's status as an international superpower.

4.2   THE GREAT SOVIET SUPERPOWER

In 1962, Soviet life was dominated by the hope of rapidly advancing towards the final stages of the Industrial Revolution. The Khrushchev revolution seemed to open the way to such progress, since the seven-year plan of 1958 had stressed the new industries of a 'sophisticated' consumer society – electronics, electro-mechanics, nuclear energy, plastics, chemicals. All of these were industries which, even before they called forth a new generation of consumers, required and would have to train 'a new type of working class' – white-coated technicians, technologists, scientific and industrial research workers, and so on. The pressure of these new social forces would sooner or later make the democratisation of the USSR inevitable and irreversible, concluded the sociologist from whom we gleaned these details. But that pressure, of course, had to make its way through both the live and the inert counter-pressures of the Communist society and the Party itself. It was logical, moreover, for the Party to try to control and apportion any new prosperity and comfort, so as to make the success its own (Braudel, 1995, p. 571).

By all accounts, and despite problems inherent in the 'bizarre' Suvnarkhozy panacea, Nikita Khrushchev had reached the pinnacle of his success. The year 1958 again brought excellent weather conditions, and 'bumper' harvests contributed to both civil satisfaction and the rapid expansion of State farms. At the same time, the substantial financial investments made in the fifth five-year plan were beginning to pay off. Healthy industrial growth rates were ensured for the remainder of the decade. While in the cold light of history it may seem incredible, the achievements of the Soviet Union, a rising international superpower, would capture the imagination of the world. In October 1957, the Soviets launched Sputnik, the first artificial satellite. Soviet science and technology seemed poised to overtake the USA, if it hadn't already.

In the realm of international relations, careful Soviet diplomacy and the strategic application of foreign aid – and 'friendly' joint-venture capital – won promising political allies in India, Asia, and the Middle East. Relations with the West had thawed considerably in the years 1955–58, resulting in a significant 30 per cent increase in the level of Soviet exports.

The Twenty-First Party Congress met, as scheduled, in January 1959. Khrushchev, riding high on the success of Sputnik, was full of promises. Suvnarkhoz reform – aided by the recent discovery of new mineral resources – would revitalize the nation, permitting significant advances in all aspects of economic growth and development. With these advantages – and under the discriminating guidance of Gosplan – the seventh 'seven-year' plan could accelerate investment in both heavy industry and consumer goods, enabling per-capita production in the Soviet Union to surpass that of the United States by 1970. The following areas were found worthy of special attention: (1) the chemical industry, held back by a shortage of technical expertise and capital, would be invigorated – with the added benefit that the production of synthetic fibres would reach 666 thousand tonnes in 1965, a 300 per cent improvement over the 166 thousand tonnes reported in 1958; (2) labour – adversely affected by low World War II birth rates – would be supplemented by a reduction in official Red Army personnel; and (3) a gross fuel imbalance 'which was too heavily oriented to coal and neglected oil (which was abundantly and cheaply available in the Volga-Urals fields) and above all natural gas, which was available in large quantities and very little used' (Nove, 1972, p. 355).

New energy resources were also coming on stream. One of the major new sources was in Western Siberia. According to Goldman (1980, p. 35) exploration for liquid energy in this region began before the World War II but the first find occurred by accident in West Siberia in September 1953:

> A drilling team was delayed while sailing up the Ob river near the town of Berezov. On the spur of the moment it made a test boring and found gas in what it came to call the Berezovskoe gas field. It was seven more years, in 1960, however, before the first oil was discovered in a Jurassic zone near Shaim on the River Konda, a tributary of the Ob and Irtysh.

According to the plan, with new production in Western Siberia, zealous use of water flooding and an expanded drilling and exploration programme,

crude oil production was to reach 230–240 million tonnes by 1965, a 117–
127 million tonne improvement over the 113 million tonnes reported in
1958 (see Table 4.5).

Table 4.5: The seventh seven year-plan

	1958	Plan 1965	Actual 1965
National Income (1958=100)	100.0	162-5	158
Gross Industrial Production (1958=100)	100.0	180	184
Producers' goods (1958=100)	100.0	185-8	196
Consumers' goods (1958=100)	100.0	162–5	160
Coal (million tonnes)	493.0	600-612	578
Electricity (million Kwhs)	235.0	500-520	507
Steel (million tonnes)	54.9	86–91	91
Synthetic Fibres (thousand tonnes)	166.0	666	407
Oil Production (million tonnes)	113.2	230-40	242.9

Source: Nove (1972)

The seven-year plan called for a significant increase in the level of
exploratory and development drilling. In the years 1959–65, approximately
9155 thousand metres of new exploratory and development holes would be
drilled in areas deemed appropriate by the Suvnarkhoz. The emphasis on
exploration was unmistakable. For the first time in the history of the USSR,
the volume of exploratory drilling for both oil and natural gas (6743
thousand metres) was scheduled to exceed that of development drilling
(2412 thousand metres). In terms of field development, the production from
free-flowing wells was to be maximized through an accelerated process of
water injection. The use of water injection to enhance oil production in the
Soviet Union had been rising steadily throughout the post-war
reconstruction era – from 42 755 cubic metres a day (cm/d) in 1951, to 348
818 cm/d in 1957. According to Ebel (1970, p. 103) by 1960

> the deposits which [were] included in the water flooding program [accounted]
> for 63 per cent of the national crude production. Furthermore it [was] claimed
> that up to 70 per cent of the oil in place [was] being recovered. By 1965, 81 per

cent of the national crude output [was] to be provided by deposits where water flooding [was] employed.

The Gosplan specified a daily water injection 'quota' of 850 000 cubic metres in 1965.

An insight into the extent of Gosplan's commitment to the new energy balance can be gleaned from a brief inspection of the investment strategies, and capital expenditures, envisioned for the seven-year plan. In the years 1946–58 inclusive, approximately 100 317 million roubles (over US$25 billion) had been invested in the oil and gas industries (see Table 4.6). These funds represented a significant portion of total Soviet capital investments – approximately 5.6 per cent in the fourth five-year plan (1946–50) and a substantial 6.7 per cent during the fifth and sixth five-year plans (1952–58). The bulk of the 100 000 million roubles (89 000 million roubles) as to be devoted to oil, leaving a mere 11 000 million roubles for the entire natural gas industry – extraction, transportation, and storage facilities. While similar figures are not available for the coal industry, an increase in the share of coal in the primary energy balance – from 58.6 per cent in 1940 to 64.6 per cent in 1950 – and the mineralization strategy envisioned in Stalin's fifth five-year plan, suggest that coal was given clear fiscal precedence over oil and gas in the years 1946–55.

Soviet oil investments were heavily concentrated in two regions throughout the early post-war reconstruction era. These were the Caucasus and the Volga-Urals. The Caucasus received 34.2 billion roubles or 38.4 per cent over the 1946–58 period. Approximately 34 billion (1955) roubles, 38.2 per cent of total Soviet oil investments, were allocated to the Volga-Urals region. The investments in the Volga-Urals paid off immediately, generating an impressive 79 per cent of all incremental Soviet oil flows in the years 1946–58. As production per rouble of investment continued to exceed Gosplan expectations, the portion of funds earmarked for the prolific Volga-Urals production area was gradually increased. By 1958, 45.8 per cent of Soviet oil and gas investments, approximately 33 068 million roubles, was committed to the development of the Volga-Urals region. While little information is available on the disaggregation of capital expenditures envisioned for the seven-year plan, a substantial portion of Soviet oil investments is believed to have been scheduled for the continued, extensive development of the Volga-Urals region (Ebel, 1970).

The de-mineralization of the energy balance called for a large, and sustained, increase in the portion of energy investments devoted to the oil and gas industries. While some progress in this direction had been achieved in the sixth five-year plan (1955–58), coal was still grossly over emphasized. In the years 1952–58, the fifth and sixth five-year plans, a total of 208 500 million (1955) roubles was invested in the coal, oil, natural gas, and electric power industries. Approximately 29 per cent of these funds – 61.2 billion roubles – was devoted to the coal industry. The oil industry – aided by de-mineralization in the years 1956–58 – received 30 per cent, or 62.7 billion roubles. The electric power industry absorbed 36 per cent, or 75.1 billion roubles, and the natural gas industry only 4.5 per cent, or 9.5 billion roubles.

In 1959, Gosplan created a 'special commission' to oversee the delicate – and highly complex – transition to a rational domestic fuel balance. It is, of course, ironic that the only way that the planners knew that their mix of energy sources was not appropriate for a modern economy was by observing the transition to oil-based economies that had already taken place in market economies. In a similar vein, the planners were very slow to de-mineralize other aspects of the Soviet economy. For example, in a period when market economies were increasingly reliant on petroleum-based plastics, whether and eventually how to incorporate a plastics industry in the plan became a major stumbling block to modernization of the Soviet economy. It was often commented on by visitors to the Soviet Union as late as the 1970s how many products were still produced using (costly) metals that had been replaced with plastics in the West.

The Grosplan's guidelines for the seven-year plan (1959–65) may be summarized as follows: (1) Capital investment in the oil and natural gas industries would exceed 171 500 million roubles – a significant 137.5 per cent increase over the 72 200 roubles invested in the years 1952–58, and a substantial 46 per cent of Soviet energy investments; and (2) Capital investments in the coal industry would be restricted to only 75 000 million roubles – a mere 20 per cent of total Soviet energy investments. A significant portion of these funds – approximately 130 200 million (1955) roubles – was devoted to the oil industry. By 1965, the entire Soviet energy balance was to be completely transformed, with the production of crude oil and natural gas comprising a significant 49.4 per cent of total primary energy production (see Table 4.7). While the exact derivation of the shares specified by the new fuel balance remains something of a mystery, the USSR appeared to be 'striving to achieve an energy balance resembling that presently [1961] in the US, wherein fuel supplies are furnished in part by each of the major sources of energy (crude oil, natural gas and coal)' (Ebel, 1961, p. 12).

Table 4.6: Capital investment in the Soviet energy economy, 1946–65

(millions of rubles in July 1, 1955 prices)

Oil Industry	1946	1950	1946–50	1955	1951–55	1956	1957	1958	1952–58	1959–65 (Planned)
Extraction Phase Exploratory and Development Drilling	643	2,754	8,172	3,667	17,133	3,357	3,670	4,590	25,700	43,000
Production	795	1,977	6,395	2,300	11,834	2,465	2,830	2,930	17,500	29,000
Total	1,438	4,731	14,567	5,967	28,967	5,822	6,500	7,520	43,200	72,000
Refining, Transportation and Storage	345	1,325	3,778	2,886	13,806	2,545	2,600	3,040	19,500	58,000
Total Oil Industry	1,783	6,056	18,345	8,853	42,773	8,367	9,100	10,560	62,700	130,200
Total Natural Gas Industry	103	113	629	983	3,922	1,581	2,200	2,840	9,500	41,300
Oil and Natural Gas	1,886	6,169	18,974	9,836	46,695	9,948	11,320	13,400	72,200	171,500[a]
Coal									61,200	75,000[b]
Electric Power									75,100	125,000[c]
Total									208,500	371,500

Note: The above analysis assumes an official rouble exchange rate of 4.0 roubles to 1.0 US dollar, which was enforced from the year 1950 to 1 January 1961.

a: The seven-year plan states a range of 170–173 billion roubles (approximately $42.50–$43.25 billion US). The midpoint of the range has been selected.

b: The seven-year plan states a range of 75–80 billion roubles. The low side of the range has been selected.

c: A range of 125–129 billion roubles has been given for the years 1959–65. The low side of the range has been used in this computation.

Source: Ebel (1961).

Table 4.7: The production of primary energy sources in the USSR, 1955, 1958 and 1965

(converted to a standard fuel – millions of metric tonnes)

Source of Energy	1955		1958		1965	
	Standard Fuel	% of Total	Standard Fuel	% of Total	Standard Fuel	% of Total
Coal	310.8	63.2	362.1	56.9	475.0	42.0
Crude Oil	101.2	20.6	161.9	25.4	379.0	33.5
Natural Gas	11.4	2.3	33.9	5.3	180.0	15.9
Peat	20.8	4.2	21.1	3.3	26.6	2.4
Shale	3.3	0.7	4.5	0.7	6.0	0.5
Fuelwood	32.4	6.6	32.9	5.2	25.0	2.2
Total Mineral Fuels	479.9	97.6	616.4	96.8	1,091.6	96.5
Hydroelectric Power	11.6	2.4	20.4	3.2	40.0	3.5
Total	491.5	100.0	636.8	100.0	1,131.6	100.0

Source: Ebel (1961).

The 'oil and gas' campaign initiated by Khrushchev having originated 'in a country where every decision taken has overriding political motivations' (Ebel 1970, p. 38), was greeted with growing apprehension on the part of the major players in the international oil industry. The implications for world oil prices were grim. By 1959, Soviet net oil exports had already reached levels in excess of 21.8 million tonnes per annum – a 489 per cent improvement over the 3.7 million tonnes reported in 1955. The deluge, aggravated by new sources of supply from countries outside the Soviet Union, would soon surpass both international expectations, and import requirements (see Table 4.8). In particular, the sudden resurgence of Soviet oil exports had penetrated world oil markets at the worst possible moment for beleaguered US producers. The Suez Canal crisis had been resolved in 1957, facilitating a flood of new oil exports from the Middle East, and reducing the demand for US oil shipments by a substantial 1.2 million barrels per day. At the same time, US domestic requirements had been sharply reduced by a severe industrial recession. The US was rapidly developing a substantial, and extremely bothersome, crude oil overhang, and the new quantities expected from Russia were staggering. Markets in Western Europe were quickly saturated, leading to price wars and a significant US$0.10 reduction in the benchmark West Texas Intermediate price, from US$3.07 per barrel in 1958 to only US$2.97 per barrel in 1960.

As expected, the major oil companies responded immediately to Soviet competition by cutting prices. The reaction to the Soviet-induced oil glut was to have long-lasting consequences for the international oil industry and the global economy in the 1970s and 1980s. According to Riva (1994, p. 15):

> However, the companies, being caught in the middle, could not absorb the entire price cuts themselves. In 1959 they cut posted prices as well, forcing the producing companies to share the burden of competing with the Soviets. The result of this unilateral decision was diminished national revenues for the oil-producing countries. The oil exporters reacted quickly and at the Arab Oil Congress in Cairo in April took the first steps in creating a common front against the oil companies, an alliance that would develop into the Organisation of Petroleum Exporting Countries, better known as OPEC.

To add insult to injury, US producers were locked in fierce competition with an aggressive, and highly discriminatory and sometimes politically motivated, Soviet marketing campaign. A newcomer to an established global energy market, the Soviet Union was eager to offer prices – and a host of additional incentives – that would be sufficiently agreeable to tempt

Table 4.8: Soviet crude oil production and international trade statistics, 1955–68

(million metric tonnes)

Year	Production	Total Exports	Total Imports	Net Exports	Non-Communist Exports	Non-Communist Imports	Exports to Communist Countries[a]
1955	70.8	8.4	4.7	3.7	3.8	0.467	4.2
1956	83.8	10.6	5.6	5.0	4.9	1.375	5.2
1957	98.3	16.3	4.5	11.8	5.9	1.124	7.6
1958	113.2	18.9	4.6	14.3	9.2	1.005	8.9
1959	129.6	26.5	4.7	21.8	14.1	1.012	11.1
1960	147.8	34.5	4.7	29.8	18.1	1.003	15.2
1961	166.1	42.7	3.8	38.9	22.8	0.773	18.4
1962	186.2	47.1	3.0	44.0	24.7	0.499	20.7
1963	206.1	53.1	3.0	50.1	28.4	0.548	23.0
1964	223.6	58.3	2.3	56.0	31.3	0.036	25.3
1965	242.9	66.2	2.1	64.1	35.5	0.046	28.9
1966	265.0	75.6	1.8	73.8	41.4	0.267	32.3
1967	288.0	81.0	1.5	79.5	43.5	0.248	35.3
1968	309.0	88.5	2.7	85.8	44.0	na	42.2

Note: [a] Defined as the communist countries of Eastern Europe (Bulgaria, Czechoslovakia, East Germany, Hungary, Poland, and Rumania) plus Cuba, Yugoslavia, North Korea, North Vietnam, Mongolia, Communist China, and Albania.

Source: Ebel (1970).

even 'hesitant' consumers to switch suppliers. Soviet oil could be purchased at bargain basement prices in local 'soft' currencies, or bartered for agricultural and/or manufactured commodities. By 1959, discriminatory Soviet pricing policies had become a matter of major international dispute. According to Ebel (1970, p. 37) this Soviet approach to marketing:

> paid off handsomely and, so far as is known, the Soviet Union was able to find buyers for all of the oil it had for sale in any given period. A place for Soviet oil was found in Italy in exchange for large-diameter steel pipe and associated equipment, in Brazil in return for coffee, in Cuba in return for sugar, and in Egypt in return for cotton.

This quasi-barter approach was also simpler for planners to deal with than selling on the volatile international market. As fluctuating international prices made it difficult to predict foreign exchange earnings, it was difficult to plan the quantities of imports that would be available to be incorporated into the plan. Oil for commodity swaps provided some assurance as to the quantities that would be available for import, facilitating the planning process (Henderson and Kerr, 1984/85). To gain this degree of 'surety', however, probably meant that Soviet oil was exchanged at even lower effective prices than if they had simply been sold. This further undercut the market.

On the other hand, in a path-breaking study by Robert E. Ebel, official Soviet statistics were 'applied to demonstrate that the USSR [was] taking advantage of the captive markets in other communist countries to charge a much higher price for crude oil compared to prices obtained in the West' (Ebel, 1970, p. 57). The magnitude of this discriminatory injustice is illustrated in Table 4.9. It is important to notice that a significant portion of the discrepancy between average prices paid by communist and non-communist customers – approximately US$1.13 per barrel in 1959 – can be attributed to differences in transportation costs to various locations along the Soviet border. As the exact magnitude of this 'transportation' factor was virtually impossible to determine, discriminatory pricing policies were identified on a country-by-country basis. To cite only one uncomfortably conspicuous example: East German customers paid approximately US$2.70 for a barrel of Soviet oil in 1959 while West German consumers were charged only US$1.69 (Ebel, 1970, p. 63).

The resentment of US-based international oil companies rose steadily with Khrushchev's financial commitment to the oil industry. Given this was taking place at the height of the Cold War, many in the West believed that the aggressive Soviet marketing campaign was politically motivated, the purpose of which was to create energy dependence in Western Europe, thus weakening the unity of NATO. Further, influence of the West in the Middle East would be weakened. In the words of American Senator Kenneth Keating: 'Khrushchev has threatened to bury us on more than one occasion. It is now becoming increasingly evident that he would also like to drown us in a sea of oil if we let him get away with it'.[4]

Table 4.9: Average prices paid for Soviet crude oil, 1955–67

	(US$ per barrel)		
Year	Other Communist Countries	Non-Communist Countries	WTI Posted Price
1955	3.38	2.16	2.82
1956	3.30	2.17	2.82
1957	3.28	2.55	3.04
1958	2.97	2.08	3.06
1959	3.01	1.88	2.98
1960	3.01	1.57	2.97
1961	2.54	1.26	2.97
1962	2.52	1.26	2.97
1963	2.55	1.43	2.97
1964	2.57	1.41	2.95
1965	2.42	1.40	2.92
1966	2.18	1.39	2.94
1967	2.10	1.50	3.03

Note: All average prices are calculated from apparent annual prices paid f.o.b. at the Soviet border. Conversions are based on 7.3 barrels per metric tonne, and 1 rouble = US$1.11.

Sources: Ebel (1970); *International Petroleum Encyclopedia* (1994).

The euphoria generated by the USSR apparently closing both the military and economic gap with the US, justified by continued over achievement in the realm of heavy industry, would last well into 1961. At the same time, Khrushchev and his closest associates were becoming increasingly erratic, and irresponsible. While cracks – deficiencies – had appeared in both Soviet foreign policy and the unique Suvnarkhozy planning system, these were greeted with dismissal and outright defiance. Calls for retrenchment – a return to Stalinism – were ignored; the accomplishments of the seventh seven-year plan were discovered to be modest, too humble by far for the great Soviet superpower. A new party programme – only the third to be adopted since 1903 – and a series of upward amendments to the seventh five-year plan were announced to the Twenty-Second Congress in October 1961. The new programme 'solemnly' assured Soviet citizens that 'a state of whole people' had risen to replace 'the dictatorship of the proletariat', and that the 'present generation [would] live under communism'. By 1980, the 'foundations of communism would be built'. The production of everything would be doubled, tripled, quadrupled, etc. by then, and 'the high road to complete communism would be opened' (Kort, 1993, p. 246). The program was an obvious attempt to inspire mass enthusiasm and a sense of nationalistic idealism in a disgruntled Soviet population. It was accompanied by another crushing denunciation of Stalin.

The days of the 'secret speech' were over. A plethora of orations from Khrushchev, and his distinguished associates, condemned Malenkov, Kaganovich, and Molotov as Stalin's most evil accomplices. Khrushchev rose to the occasion, agreeing to erect a monument to Stalin's victims, and permitting an obscure delegate, D. Lazurkina, 'to make a speech . . . claiming that the ghost of Lenin had appeared to her, to say: "It is unpleasant for me to lie next to Stalin in the Red Square mausoleum" (Moynahan, 1994, p. 206). A resolution was passed immediately. Stalin would be removed from the mausoleum that night, and his remains placed near the Kremlin wall, an area generally reserved for only 'second-class' dignitaries. Clearly, all was not well in the supposed bastion of 'scientific socialism'.

With the affairs of the Marxist 'afterworld' having been resolved to the satisfaction of the Twenty-Second Congress, Khrushchev turned his attention to more tangible matters; state and the economy. By 1962, the deficiencies in the Suvnarkhoz structure had become apparent even to devout Khruschev admirers. As the Soviet economy had grown, so had a multitude of minute economic decisions, which in the command economy could only be solved by overworked senior planning administrators. As a result, the majority of plan 'blue-prints' specified only gross aggregate requirements – criteria that could be met in a variety of convenient, and often highly imaginable, ways. With the passing of time and increasing complexity in the economy, von Mises' calculation problem was becoming more and more evident, and transparent. The planners simply did not have sufficient information to match supply and demand over the myriad of transactions required to make even the simplest goods and had no means upon which to evaluate trade-offs. Not surprisingly, production was often entirely incompatible with demand, and in the case of producer goods, critical user requirements. To cite only a few examples provided by Nove (1972, p. 357): 'Steel sheet was made too heavy because the plan was in tons, and acceptance of orders from customers for thin sheet threatened plan fulfilment. Road transport vehicles made useless journeys to fulfil plans in ton-kilometers'. Enormous sofas and heavy chandeliers had no place – and indeed could not fit – in the average Soviet apartment. At the direct request (order) of Nikita Khrushchev, 85 million acres of maize were planted in virgin territories in which the soil and climate was unsuitable for cultivation – a mere 16 million acres were harvested. While nothing had really changed since the beginning of the formal era of planning, it was now obvious that market allocation was not replaced by 'scientific' planning in command economies but rather administrative fiat based on the 'enthusiasms' of the political leadership.

Despite the best efforts, and direction, of Gostekhnika, all innovation – new technologies – were studiously avoided. Any disruption of established procedure, however promising, threatened the timely fulfilment of strict quantitative production targets. According to Goldman (1980, p. 37):

> For the oil drillers, this meant that the steel manufacturers they depended on had no incentive to produce or even contemplate producing high-grade qualities of steel. . . . Khrushchev put it vividly when he complained: 'The production of steel is like a well travelled road with deep ruts; here even blind horses will not turn off, because the wheels will break. Similarly, some officials have put on steel blinkers; they do everything as they were taught in their day. A material appears which is superior to steel that is cheaper, but they keep on shouting 'steel, steel, steel!'

To further aggravate matters, the abolition of the national ministries had removed a vital link in the chain of an established, but ailing, command economy. The Suvnarkhozy had only one clear criterion – the requirements (and preservation) of their own geographical region. The needs of enterprises in the other 104 Suvnarkhoz were secondary, and in most instances, entirely irrelevant. Scarce economic resources, and investment funds, were hoarded – diverted to local enterprises – to the detriment of competing regions. The shortages facilitated numerous cases of embezzlement and theft, leaving central authorities with little alternative but to increase the number of items to be monitored (and controlled) by

Gosplan. Draconian measures were taken to re-establish workable supply chain links.

On 7 May 1961, Khrushchev introduced the death penalty for a range of economic offences, including speculation and the direction of valuable state resources to private ends. One year later, in 1962, the Communist Party was split in half; one side to preside over agriculture, the other over industry. Suddenly there were two 'first' secretaries, neither – or both – of whom could claim ultimate responsibility for neutral services such as education, police protection and health care (Kort, 1993). Within a matter of months – early 1963 – Khrushchev would order another major administrative reform; a reduction in the number of Suvnarkhozy from 105 to 47. As Nove (1972, p. 362) indicates: 'By 1963 no one knew quite where they were, or who was responsible for what. Pungent criticisms were appearing in the specialised press. Planning was being disrupted'.

In early 1963, the bureaucratic community was alarmed by Khrushchev's declaration that Gosplan's failures could only be attributed to 'excessive conservatism'. Further reforms, panaceas, were sure to follow. The last campaign, an overly aggressive chemical investments programme, would threaten the industrial balance of the entire Soviet economy. While few doubted the necessity for a substantial expansion of the chemistry industry, the target – a 300 per cent increase in the production of chemicals in less than seven years – was simply unattainable. It simply reflected a new enthusiasm from the leadership. A new annual plan (1964–65) specified the reduction of lesser targets, such as steel and housing, to make room for chemicals. Nove (1972, p. 363) attempts to suggest that this endemic problem could have been overcome with better planning: 'It was typical that sensible changes – such as the substitution of non-solid fuels for coal, or prefabricated concrete for bricks – became much too drastic . . . causing grave shortages'. A dramatic increase in the armaments quota – including quantities specified by the space and missile programme – and soaring military expenditures placed a tremendous strain on scarce domestic resources. By 1963, official Soviet statistics placed industrial growth rates at levels well below 8 per cent per annum, the lowest peacetime figures to be recorded since 1933.

Khrushchev was dismissed in October 1964. Suddenly called back to Moscow from vacation, the 79 year-old leader was informed of his request to be relieved of all duties, and responsibilities, on the grounds of ill health. With no time, or ability, to counter-manoeuvre, he retired gracefully, accepting an apartment in Moscow, a country house and a private car and chauffeur. He is reported to have said:

> I am old and tired, [and] I've done the main thing. Relations among us, the style
> of leadership, has changed drastically. Could anyone have dreamt of telling
> Stalin that he didn't suit us anymore, and suggesting that he retire? Not even a
> wet spot would have remained where he had been standing. . . . His own epitaph
> was short and fitting: 'The fear's gone. That's my contribution' (Moynahan,
> p. 210).

The successes in nuclear weapons technology and the mass production of
military hardware (as well as the space programme) had given the Soviet
Union superpower status. This 'status', however, masked serious
deficiencies in economic organization. It had become clear that improving
the planning process could not solve the problem. As a result the economy,
including the oil industry, could not really support the needs of a
superpower. The Party, however, had 'hung its hat' on an alternative
allocation mechanism to markets. They were too inflexible, and fearful of
losing their legitimacy as society's leader, to admit their mistake. Unlike
their more pragmatic Chinese counterparts 25 years later, they could not
begin to take the road back to market allocation through creative and
flexible concepts such as 'market socialism' (Kerr and MacKay, 1997). The
heady experiment with rapid forced growth was clearly over but it would
take decades for the myth to die. No matter what it was called, conservatism
became the watchword and patching the creaking economy to keep it going
(and the Party in power) became the objective.

NOTES

1. See the discussion of von Mises (1981) in Chapter 1.
2. Nikita Khrushchev, 'On the Cult of Personality and its Consequences', translation by
    Brian Moynahan in Moynahan (1994, p. 197).
3. Lazar Kaganovich was Stalin's chief 'collectivizer' and a close personal associate
    throughout the Great Leap Forward.
4. As reported in Yergin (1991, p. 517).

5. Malaise at the End of the Command Era

5.1 BREZHNEV'S SEARCH FOR STABILITY STIFLES GROWTH

> The country wanted no further alarms and excursions and it yearned rather for stability and perks; cheap food, cheap rents, jobs for life and three-ruble vodka. They were provided by Leonid Ilyich Brezhnev, the Ukrainian steel-workers son from Dneprodzerzhinsk. . . . He was vain; vanity, the importance of appearance, was key to his era. He had a modest, noncombatant war as a political advisor, but loved uniforms and medals, which he awarded himself. He won Lenin Prizes for Peace and for Literature; he became a Marshal of the Soviet Union. . . It was the form, not the substance, that was important to him; on the surface, at least, the country prospered under him (Moynahan,1994, pp. 211–12).

Khrushchev's dismissal put an end to the 'hare-brained schemes', and endless reorganizations that had undermined the security of both state and party bureaucracies. The Soviet ruling class, the Communist Party, had at long last come of age and was willing to fight tooth and claw for the simple privilege of stability. No errors could be made in the next transition. With powerful vested interests, and job security, now clearly on the line, the new leader must be ideally suited to the primary directive: the preservation of the Party, status, and appearances. The appointment of Leonid Brezhnev – a tough Stalin enthusiast – to the position of First Secretary would guarantee this result for many years to come. His newly instated Soviet security team included such trusted old-line henchmen as Alexi Kosygin (Prime Minister), Nikolai Podgorny (appointed president of the Soviet Union in 1965) and Mikhail Suslov (chief party ideologist and media director).

The first order of business, the 'dekrushchevization' of the Soviet economy, required the immediate, and absolute, reversal of all ill-conceived reforms. The most offensive violation of the unique Stalin planning model, the heretic division of the Communist Party, was corrected, and the two sides reunited in November 1964. Less than a year later, in September 1965, the Suvnarkhozy were abolished and replaced by the familiar system of industrial ministries.

The Ministry of the Petroleum Industry USSR was in due course reinstated but, in the mid-1960s, split into two independent divisions – The Ministry of the Petroleum Industry USSR (MNP) and the Ministry of the Gas Industry USSR (MGP). The two fuel ministries were given full responsibility for the development, transmission, and wholesale distribution of oil and gas supplies. A separate entity, the Ministry of Geology (Mingeo), was reinstated to oversee the preliminary stages of exploration – geophysical mapping, prospecting and so on – for all mineral resources located within the boundaries of the USSR. In newly discovered oil and gas regions (West Siberia), Mingeo was assigned additional responsibilities including oil field delineation and establishing the volumes available in proved reserves (Gustafson, 1989).

Gosplan was restored to full stature, and subsequently given full authority over all aspects of both long-term and short-term planning activities. An extremely unpopular Khrushchev 'reform' – the rule requiring the regular turnover of all party officials – was dropped in 1966. Re-Stalinization was accompanied by a massive front-line offensive against the newly obtained fledgling freedom of intellectual and cultural expression. A relentless campaign to repress all anti-Soviet literature, and simultaneously to restore Stalin's battered reputation, was launched in 1965.

With economic growth rates at the lowest levels to be recorded in over 30 years, the faltering Soviet economy remained a top government priority. Policies to improve per-capita productivity – to minimize the waste and inefficiencies inherent in both the agricultural and industrial sectors – were announced in 1966. The effect on the agriculture sector was to turn it from a source of surplus that could be used to finance industrial expansion into a subsidized sector. This was accomplished by raising agricultural prices paid to farmers but not raising prices to consumers.

Alexi Kosygin, acting Prime Minister and a proven economic administrator, seized the first opportunity to assert control, undertaking the personal supervision of all industrial reform. Ironically, his plan for Soviet industry reflected an economic theory that had been developed under the 'liberated' academic atmosphere of the Khrushchev administration. Evsie Liberman, a leading Soviet economist, had long advocated a system in which a manager's performance might be measured in terms of a firm's profitability. Under the new rules of engagement, managers would be given greater control over all facets of business transactions, including decisions related to hiring, wages, raw materials, and the quality of production. To secure an acceptable level of profitability, the enterprise would be forced to produce goods that consumers might actually be willing to purchase. The revision, a radical departure from the simple fulfilment of aggregate production targets, was targeted at eliminating the waste and inefficiencies that had plagued, and too often embarrassed, Soviet administrators throughout the Khrushchev years; shirts with no buttons, heavy sheets of low quality steel, enormous chandeliers and so on (Kort, 1993). The central problem remained, however, and with all prices arbitrarily set by the state, not reflecting their product's opportunity cost, managers were profit maximizing to perverse incentives. Thus, while some local maximizing could take place, this could not contribute to solving the global maximization problem of the economy.

Kosygin, inspired by visions of an effective industrial renaissance – and the accompanying political distinction – seized the initiative, and in September 1965 adopted a decree that had the 'declared intention' of increasing the powers of managers, 'reducing considerably the number of compulsory indicators "passed down" from the centre' (Nove, 1972, pp. 374–5). The 'Liberman' reforms were highly sophisticated – involving a plethora of detailed economic calculations – and unnecessarily complex (see Appendix D). Intricate accounting attempted to provide managers with bonuses tied to performance.

Despite the apparent devolution of managerial authority, Stalin-like centralization reigned paramount and industrial reform, paralleling difficulties in the Party, suffered a form of intense schizophrenia – severe internal resistance, and blatant inconsistencies. According to Davies (1979, p. 231): 'Soviet planning after the 1965 reform became much more complicated as well as much more concerned with economic efficiency. But it . . . remained administrative planning'. To cite only a few examples: (1) the majority of the new ministries were all-union (not all-union-republic), thereby re-centralizing most planning activities and ministerial powers in Moscow; (2) the ministries retained full authority over the activities of enterprises and, as a result, often issued orders which nullified (superseded) the newly gained powers of managers; (3) the success of managers, and bonuses, were primarily determined by the fulfilment of 'plans' (quotas), which, According to Nove (1972, p. 375), 'meant that it always "paid" to have a modest, "fulfilable" plan'; (4) severe shortages of critical producers goods, inputs, led to hoarding and unscheduled delays in the fulfilment of plans; (5) the profit motive was undermined by the fact that 'excessive' profits (the free remainder) were too often seized and returned to the budget; and (6) central planners simply refused to allow prices to respond to the forces of supply and demand (Kort, 1993).

To aggravate matters further, the reforms had been implemented in a piecemeal fashion – only a few enterprises were granted the privilege of altering production and prices in order to maximize profits, at any given time. Reform, and indeed all new products and technologies were greeted with suspicion, and outright resistance by managers and senior ministry authorities. The few enterprises which supported an innovation soon found it to be counter-productive. According to Goldman (1980, pp. 38–9):

> Inevitably they found that their new freedom was negated when they could not locate suppliers who were willing to match their willingness to innovate...As for the ministers, they found that they were still being held responsible for increasing the quantity of goods being produced even though they had been instructed to sit back and turn over some of the powers to the managers.

The experiment was terminated, and planning restored to the traditional Stalin method, in 1968. One result was that the Soviet oil industry suffered from the absence of domestically produced high-quality steel and pipe. This reduced its productivity relative to the west and limited the exploitation of some reserves.

The targets of the eighth, ninth and tenth five-year plans are illustrated in Table 5.1. At first glance it would appear that the Soviet economy had thrived under the direction of Brezhnev and Kosygin. Aggregate growth rates, while similar in some respects to those obtained in the previous five-year plan, revealed a clear, and sorely needed, improvement in a number of key manufacturing sectors. To cite only two examples: (1) the production of consumer goods rose by 49 per cent in the years 1966–70, a 13 per cent improvement over the 36 per cent reported in 1961–65; and (2) the production of fertilizer reached 55 million tonnes in 1970, a five-year growth rate of well over 77 per cent. Much of this was simply reaping the benefits of previous investments.

The crude oil industry had been designated only a minor priority, following armaments, agriculture and the implementation of the Kosygin

Table 5.1: The eighth, ninth and tenth five-year plans, 1965–80

Index Numbers	(1965=100)			(1970=100)		(1975=100)	
	1965 actual	1970 plan	1970 actual	1975 plan	1975 actual	1980 plan	1980 actual
National Income	100.0	139.5	141.0	138.6	128.0	126.0	120.0
Industrial Production	100.0	148.5	150.0	147.0	143.0	137.0	124.0
Producer Goods	100.0	150.5	151.0	146.3	146.0	140.0	126.0
Consumer Goods	100.0	144.5	149.0	148.6	137.0	131.0	121.0
Electricity (milliard kwhs)	507.0	840.0	740.0	1,065.0	1,039.0	1,380.0	1,290.0
Oil (million tonnes)	243.0	350.0	353.0	505.0	491.0	640.0	604.0
Gas (milliard cubic metres)	129.0	233.0	200.0	320.0	289.0	435.0	435.0
Coal (million tonnes)	578.0	670.0	624.0	694.9	701.0	800.0	719.0
Steel (million tonnes)	91.0	126.0	116.0	146.4	141.0	168.0	155.0
Motor Vehicles (million tonnes)*	616.0	1,385.0	916.0	2,100.0	1,964.0	2,296.0	2,199.0
Fabrics (million square metres)	7,500.0	9,650.0	8,852.0	11,100.0	9,956.0	12,800.0	10,700.0
Fertilizer (million tonnes)*	31.0	63.5	55.0	90.0	90.2	143.0	104.0

Note: 1980 'actual' figures are estimates.
* Gross Weight

Source: Nove (1972).

reforms. Years of steadily increasing crude oil flows – and the crucial ability to generate hard-currency receipts – had fostered a large measure of complacency, a sense of easy optimism that would influence Gosplan decisions throughout the latter half of the 1960s. Indeed, the Soviet Union's financial commitment to the oil industry had been waning steadily since 1959. A total of only 13.1 billion roubles was invested in the oil industry over the years 1966 to 1970. The implied fiscal allocation – a mere 4.27 per cent of total Soviet capital investments – reflects a sizeable retrenchment from the 6.7 per cent reported during the fifth and sixth five-year plans, 1952–58 (see Table 5.2).

At least a portion of this reduction appears to have been spontaneous, that is unplanned. According to detailed annual investment figures, capital investment in the oil and gas industry fell short of Gosplan targets by a substantial 6.9 and 4.3 per cent in the years 1959 and 1960, respectively. Robert Ebel (1961, p. 24) attributes this to:

> the failure to construct new refining capacity and, to a lesser extent, to the lags in petroleum pipeline installation and in construction of new storage facilities. Although construction and installation trusts had the necessary capital, failure of industry to produce the required material and equipment prevented the investment of these sums. . . . It was also apparent that in an oil field, the planners would allocate as much investment as necessary to secure the maximum production (consistent, of course, with reasonable conservation) and no evaluation of capital investment versus volume would be made.

It is important to notice that these failures were consistent with the two most widely publicized deficiencies of the unique Stalin planning model: (1) Gosplan's inability to provide an adequate blueprint of the detailed production and allocation decisions required to motivate and sustain the complex Soviet economy: the von Mises problem. As mentioned above, frequent miscalculations led to severe shortages of critical producers' goods, hoarding, and the inability to fulfil plan targets. To cite only one example, the production of steel, a critical input to the oil industry, fell short of its target by 10 million tonnes in the eighth five-year planning period (1965–70); and (2) The race to fulfil aggregate production quotas, which led to an excessive and reckless concentration on volume at the expense of quality. In the case of the crude oil industry, the focus fell clearly on the maximization of extraction and recovery with a minimum of time and effort. In other words, excessive water flooding and the rapid depletion of promising producing regions.

Table 5.2: Capital investment in the Soviet energy economy, 1965–80

(billions of roubles in 1 January 1969 prices)

Five-Year Plan	1961–65	1966–70	1971–75	1976–80
Total Soviet Investments	213.3	306.3	437.2	567.2*
Total investment in Industry	87.7	119.7	168.6	219.3*
Energy	30.0	42.0	61.0	88.6*
Oil	9.3	13.1	20.5	32.2*
Gas	3.9	7.3	13.8	23.7*

Note: * 1969 prices corrected with consideration for the 'new' norms for construction introduced in January 1976.

Source: Anon (1995).

As witnessed in the seventh seven-year planning period, the results of even this restrained capital investment programme would satisfy Gosplan expectations. Crude oil production reached 353 million tonnes in 1970, a figure which lay safely within the boundaries of the 345–355 million tonne range specified by the final version – 'revised' provisions – of the eighth five-year plan (1966–70). The sustained concentration of capital investments in the North Caucasus and Volga-Urals producing regions had, once again, paid dividends, and production from these areas rose to 242.5 million tonnes in the year 1970 – a 48.9 million tonne increase over the 193.6 million tonnes reported in 1965 (see Tables 5.3 and 5.4). At the same time, exciting new prospects in West Siberia had been cultivated and were, at long last, beginning to bear fruit.

Forecasting crude oil production was always a crude science at best. This can be illustrated by The Twenty Year Plan (1961–80) released in October 1961. It forecast Soviet crude oil production of 390 million tonnes by 1970. Crude oil production stayed on schedule, exceeding Gosplan expectations, right up to 1965. From then on, the ambitious 1970 estimate was periodically revised downwards to reflect changing trends and conditions in 'all' sectors of the Soviet oil industry. By November 1968, the (1970) forecast had been reduced to 348 million tonnes. One month later on 11 December 1968, *Pravda* issued the following retraction. The revised oil production figures had been 'subsequently complicated by the severe winter of 1968-1969, the effects of which may hold 1969 output to 325 million tons. If so, the 1970 estimate may be an overstatement'.[1] Neither forecast would prove to be correct, and crude oil production averaged 353 tonnes in 1970 – well within the 345-355 million tonne range specified by the final, 'revised' provisions of the eighth five-year plan (1966–70). The noticeable departure from the 'ambitious' 20-year planning schedule was the first clear indication of potential problems, deficiencies in the mighty Soviet petroleum sector. In the words of Robert Ebel (1970, p. 86):

> While in retrospect it is easy to find correlative evidence to justify the sharp dip in crude oil growth rates, clear-cut indicators of an immediate slowdown were absent. It is apparent that after enjoying a succession of years of prosperity and continued growth in increments to production, the oil extraction industry of the Soviet Union has been forced to pause temporarily at least, to allow other sectors of the oil industry which had begun to lag badly, to catch up. This rather abrupt slowdown reflects admitted failures in pipeline construction and in the development of new oil finding and producing equipment responsive to the more demanding conditions of greater depths, higher formation temperatures

and pressures, and exacting climate and terrain now being encountered in the expanding search for oil and gas in the Soviet Union.

The consistent problem of not being able to plan in sufficient detail to match requirements and production caused bottlenecks that plagued production, but particularly expansion in the oil sector.

The first successful exploitation of West Siberia was achieved in 1960, and included the development of a number of modest sized fields in a Jurassic zone near the Shaim district of the Konda River Valley – a tributary of the Ob' and Irtysh rivers. On the advice of Mingeo, the Soviet exploration effort migrated to the Middle Ob' River Valley. Less than one year later, in 1961, the giant 2.3 billion barrel Ust'balykskoe field was discovered in Tyumen region to the southwest of Surgut (Riva, 1994). The tip of the Siberian iceberg had been exposed, unleashing an intensive search for oil in the Middle Ob' valley.

By 1969, a total of 59 new oil fields – nine of which were classified as 'giants'[2] – had been discovered in the harsh climate, swamps, and permanently frozen terrain of the West Siberian lowlands. The most important of these, the super-giant Samotlorskoe, was discovered in the Nizhnevartovsk region of the Tyumen Province. Samotlorskoe – the largest oil field ever to be discovered in the Soviet Union was estimated to have recoverable reserves of more than 14.6 billion barrels at the date of its discovery in 1961 (Hewett, 1984). These estimates have been revised upward to a total of 20.53 billion barrels. The second largest discovery in West Siberia, the 7.19 billion barrel Fedorovoskoe oil field, was discovered less than twelve months later in 1962 near Surgut. In the same year, the 4.2 billion barrel Sovetskoe oil field was discovered in the Nizhnevartovsk region of Tyumen (Riva, 1994).

The new discoveries were by no means, nor by any subsequent exaggeration of Gosplan's limited capacity to seize the initiative in 'unfamiliar' territory, restricted to West Siberia. As in the past, the exploration effort was concentrated in the prolific – albeit admittedly aging – Volga-Urals, and North Caucasus producing regions (see Table 5.5). In short, while test drilling in these two regions had fallen significantly, from 60 per cent in the years 1951–60 to only 49 per cent in the 1960s, they still commanded the bulk of the Soviet exploration budget. Still, some experimentation was necessary for the fulfilment of ambitious Gosplan 'reserve' quotas, and the number of 'test wells' (wildcats) was cautiously expanded in the Ukraine, Siberia, Komi and Central Asia. As a result, during the eighth five-year plan (1966–70) proved reserves were expanded

Table 5.3: Capital investment in the oil industry by region, 1965–85

(expressed as a % of the total RFSFR)

	1965	1970	1975	1980	1985
Russia	63.58	68.50	71.32	83.53	88.15
North Caucasus	13.90	14.80	9.90	5.60	3.20
Volga-Urals	46.90	45.40	43.10	23.90	24.50
Komi	1.50	3.10	5.90	5.30	3.80
West Siberia	1.30	5.20	12.50	48.70	56.70
Ukraine	6.05	5.96	5.57	2.71	1.52
Byelorussia	0.14	2.31	2.77	1.09	0.61
Georgia	0.44	0.55	0.54	0.66	0.67
Azerbaijan	13.42	6.36	4.99	2.58	1.91
Kazakh	4.90	5.83	7.10	5.54	4.45
Turkmen	7.88	6.82	4.77	2.39	1.63
Uzbek	1.93	1.91	1.46	0.72	0.53
Tadzhik	0.74	0.96	0.95	0.50	0.35
Kirgiz	0.93	0.80	0.53	0.28	0.19
Total RFSFR	100.00	100.00	100.00	100.00	100.00

Source: Anon (1991).

Table 5.4: USSR crude oil production, 1965–80

(thousand metric tonnes)

Region	1965	1970	1975	1980	1985	1986	1987	1988
Russia	199,191	282,010	405,345	na	534,375	550,590	557,513	552,740
Kaliningrad	0	0	0	0	1,523	1,511	1,421	1,300
North Caucasus	19,971	34,156	23,037	18,412	10,521	10,116	9,835	9,380
Volga–Urals	173,634	208,357	224,905	185,869	135,544	128,394	122,853	115,831
Komi	2,223	5,609	7,120	18,075	18,215	18,269	17,344	15,600
West Siberia	953 *	31,416 *	154,679	322,459	365,805	389,665	403,403	407,845
Sakhalin	2,410	2,472	2,244	na	2,589	2,452	2,410	2,400
Others	na	na	na	25	178	183	247	384
Ukraine	7,339	13,501	11,544	6,383	4,853	4,756	4,652	4,487
Byelorussia	39	4,234	8,244	2,551	2,019	2,028	2,041	2,010
Georgia	30	24	261	3,186	552	179	183	120
Azerbaijan	21,500	7,287	5,840	5,053	3,909	3,902	3,734	3,700
Kazakh	1,688	13,161	23,890	18,836	21,493	21,688	21,914	21,925
Turkmen	9,636	14,430	15,307	7,150	5,423	5,359	5,241	5,080
Uzbek	1,563	1,016	947	909	939	956	981	960
Tadzhik	47	181	274	391	387	367	322	300
Kirgiz	305	298	230	219	190	190	186	180
Unspecified	1,550	3,997	7,555	42,864	11,918	15,318	17,340	17,979
Total RFSFR	242,888	353,039	490,801	603,207	595,291	614,752	624,177	619,401

Note: * Estimate

Source: Anon (1991).

in West Siberia, the North Caucasus, Komi ASSR, Udmurt ASSR, Perm Oblast, Orenaburg Oblast, the Ukraine, and Turkmenia (Elliot, 1974).

According to statistics provided by the United Nations Yearbook, the volume of proven oil reserves in the USSR reached 7987 million tonnes in the year 1969, approximately 10 per cent of the world's total (73 062 million tonnes). As always, however, a measure of caution is advised in the simple interpretation of Soviet statistics. According to Elliot (1974, pp. 18–19), the United Nations estimate 'is unlikely to be an accurate indication of the present position, since Soviet oil reserves have been a state secret since 1947'. To further complicate matters in the case of the oil industry the customary assumption of 'errors' in the estimation of Soviet statistics was validated by the existence of a unique (that is, a system for which there was 'no exact equivalent in Western terminology' (Elliot, 1974, p. 80)) Soviet classification system for mineral resources. It is interesting to note the fact that – despite the tremendous influence of perestroika and reform in the early 1990s – the State Secret Act of 1947 was still in force and has not yet been repealed.

Table 5.5: The distribution of test drilling in the USSR, 1920–70

(percentage of the total for the USSR)

Region	1920–40	1941–50	1951–60	1961–70
Volga-Urals	8	26	40	34
North Caucasus	29	23	20	15
Komi ASSR	2	3	2	5
Siberia			2	10
Azerbaijan	42	25	11	5
Ukraine	1	5	7	11
Kazakstan	7	6	4	4
Central Asia	4	8	9	10
Others	7	4	5	6

Source: Elliot (1974).

Remarkably, amidst a decade of tremendous geological achievement, Gosplan found cause for complaint. In the words of Elliot (1974, p. 81):

A representative of the State Planning Commission (GOSPLAN SSR) nonetheless had to criticise prospecting organisations for their inefficient geological exploration over this period [1966–70]. The planned expansion of reserves in the three categories $A+B+C_1$ [proven, probable, and some possible] was underfulfilled by 10.9 per cent, and the two top categories alone (A+B) were 16.2 per cent below target.

A tentative programme to develop the promising new Siberian oil fields was initiated in 1964. By the end of the year a total of 15 development wells had been completed, yielding crude oil flows of approximately 209 000 tonnes per annum (see Table 5.6). Progress continued slowly throughout the eighth five-year plan, so that by the end of the planning cycle, December 1970, only 10 out of 59 new fields had started production. They were:

1. Trekhozernoe in the Shaim region (1964);
2. Ust-Balyskoe in the Surgut region (1964);
3. Megionskoe in the Nizhnevartovsk region (1964);
4. Zapadno-Surgutskoe in the Surgut region (1965);
5. Vatinskoe in the Nizhnevartovsk region (1965);
6. Teterevo-Mortyminskoe in the Shaim region (1966);
7. Sovetskoe in the Nizhnevartkvosk region (1966);
8. Pravdinskoe in the Surgut region (1968);
9. Samotlorskoe in the Nizhnevartovsk region (1969); and
10. Mamontovskoe in the Surgut region (1970).

Table 5.6: West Siberian crude oil production, 1964–70

(thousand metric tonnes)			
Region	1964	1967	1970
Surgut	120	2,561	15,191
Nizhnevartovsk	73	945	11,588
Shiam	16	2,287	4,637
Total West Siberia	209	5,793	31,416

Source: Elliot (1974).

To the amazement of Gosplan, the acceleration in Western Siberian production, which had the advantage of more modern technology and excellent availability of thermal mineral water for flushing, would soon surpass that of any other region in the Soviet Union. The crude oil flows from 1230 new development wells reached 31 416 million tonnes in the year 1970 – a 447 per cent increase in only three years (Elliot, 1974).

The aura of 'easy' optimism which would determine the future of the West Siberian oil industry was inspired by the realization of an instantaneous, and nearly effortless, prosperity. Indeed, the discovery and initial development of an abundance of giant, and super-giant oil fields had pared the implied capital cost – investment per tonne of increased production capacity – to the bone. Over the period 1964 to 1970, Western Siberia received 20 per cent of the total capital investment in the oil industry. It is estimated that the greatly increased capacity for production attained over this period (1964–70) cost 32 per cent less per ton than the average for the industry as a whole, and it is expected that returns for investment will continue to improve in this new oil centre, the 'third Baku' (Elliot, 1974).

Progress on the international oil market was equally impressive. Oil exports both to communist countries and the market economies had been rising steadily since the Khruschev revolution and the USSR's tentative emergence as a 'reluctant' participant on world oil markets in 1954. The volume of crude oil and petroleum product exports reached 88.5 million tonnes in 1968, a 54 million tonne increase over the 34.5 million tonnes reported in 1960. The substantial increment suggests an average annual growth rate of approximately 12 per cent over the period 1960–68 (see Table 5.7). Approximately one half of the total, 44 million tonnes, was delivered to non-communist countries. According to Ebel (1970, p. 50) remarkably, this achievement had been accomplished 'with virtually no investment in foreign marketing facilities and with a tanker fleet which, although greatly enlarged, [ranked] only eighth in the world, with 3.3 per cent of the ship tank fleet deadweight tonnage (dwt)'.

As mentioned above, Gosplan had adopted an unusual 'competitive' approach to the marketing of oil and petroleum products, offering low prices to western consumers in an effort to wrestle market share from the firmly established – and comfortable – international oil companies. To cite only one example, over the period 1957 to 1962, the average price for Soviet crude oil exports to the non-communist nations was reduced from US$2.55 to US$1.26 per barrel – a 50 per cent price cut in only five years. Over an identical time frame Soviet oil exports to non-communist countries

Table 5.7: Hard currency earnings from Soviet oil exports to non-communist nations, 1955–67

(nominal US$)

Year	Foreign Exchange Earnings (US$)	Annual Growth Rate %	Volume of Exports Million Tonnes	Annual Growth Rate %
1955	$81,600		3.8	
1956	$112,800	38.2	4.9	28.9
1957	$157,400	39.5	5.9	20.4
1958	$178,700	13.5	9.2	55.9
1959	$244,200	36.7	14.1	53.3
1960	$263,200	7.8	18.1	28.4
1961	$284,400	8.1	22.8	26.0
1962	$313,700	10.3	24.7	8.3
1963	$365,000	16.4	28.4	15.0
1964	$383,000	4.9	31.3	10.2
1965	$421,000	9.9	35.5	13.4
1966	$485,100	15.2	41.4	16.6
1967	$539,000	11.1	43.5	5.1

Note: Conversions are based on the official exchange rate (1Rbl=US$1.11).

Source: Ebel (1970).

101

reached 24.7 million tonnes, a 319 per cent increase (see Tables 4.8 and 5.7). Of course, ideologically multinational oil companies represented the Marxist 'arch-villain' of rapacious capitalism and the proof of Lenin's theory of international capitalism's exploitative nature in developing countries. Hence, anything that contributed to their demise was viewed as a means to shortening the time to the triumph of 'workers' around the world.

The benefits of oil sales to the 'decadent' west, in terms of hard currency receipts, were irrefutable. Assuming the official Soviet exchange rate, hard currency earnings from the sale of crude oil and petroleum products to non-communist nations were valued at US$313.7 million in 1962 – a 99 per cent increase over the US$157.4 million achieved in 1957.

Facing a torrent of low-priced Soviet oil exports, the major international oil companies had little alternative but to cut prices. In 1959 the posted price of Saudi Arabia Light crude was reduced by 9.5 per cent from US$2.08 to US$1.90 per barrel. The reduction accelerated an already aggressive Soviet marketing campaign. As hard currency receipts were necessary for the purchase of industrial technology, as well as food imports, the flood of Soviet oil exports continued at an even faster pace. According to Riva (1994, p. 15): 'It became possible to load Soviet oil at Black Sea ports for about one-half the posted price of Middle East oil'. Less than one year later, in August 1960, the posted price of Saudi Arabia Light crude was cut again, this time to a mere US$1.80 per barrel.

As in the past, the Middle Eastern exporting countries were outraged. The Iraqis seized the opportunity to organize the petroleum exporting countries. Representatives of the exporting countries were invited to Baghdad in September. On 14 September Iraq, Saudi Arabia, Kuwait, Iran, and Venezuela formed the Organisation of Petroleum Exporting Countries (OPEC) with the goal of defending the world price of oil (Riva, 1994). Unperturbed by the potential implications of a powerful new oil cartel, Soviet exports continued as scheduled by the five-year planning process. Of course, for the Soviets, any increase in the international price of oil would represent a financial windfall. In the meantime, OPEC apparently took some, limited, action to rein in 'runaway' oil supplies, and world oil prices stabilized at the US$1.80 per barrel mark.

Shortly after 1962 – a year distinguished by the US victory in the Cuban Missile crisis – the benign and politically motivated 'non-communist pricing' policy was gradually, and deliberately, reversed. According to Ebel (1970, p. 63): 'Once a place in the [West European] market has been secured, effort has been made to secure the greatest return for the oil sold, which means price increases if at all possible'. By 1963 the average crude oil price charged to non-communist nations had risen to US$1.43 per barrel; a 17 per cent annual increase. Four years later, in 1967, the Arab oil embargo provided the Soviet government with another opportunity to raise both the volume and prices of crude oil exports to Western European markets. It is important to note the fact that the price increases were offered, and accepted, at a time when world oil prices had been remarkably stable (that is the posted price of Saudi Arabia light crude was frozen at US$1.80 per barrel over the period 1960 to1967, inclusive). In retrospect, the Soviet marketing scheme was extremely successful. By 1967, the hard currency earnings from oil exports to non-communist nations had reached US$539 million, a healthy 72 per cent increase in only five years. The success of the Soviet oil marketing campaign is most evident in terms of total revenue – or hard currency receipts. The physical goal, as measured in terms of market share, would appear to have been the repossession of 14 per cent of the European oil market, which by 1967, had yet to be reclaimed. According to Ebel (1970, p. 75; p. 83):

> The Soviet Union on several occasions argued that it was just as much entitled to a share in the export market as any of the Western countries, that its 'historic' share of European oil trade was 14 per cent, and that intentions were to recover this position. . . . In 1960, Communist oil represented an estimated 8 per cent of the demand for oil in West Europe, and 4.5 per cent of the demand in the Free World outside the United States, in the succeeding 7 years, no important changes in these shares have taken place.

While the crude oil prices charged to non-communist nations had been rising steadily since 1962, the prices paid by 'friendly' satellite nations were, clearly, on the decline. As mentioned in Chapter 4, official Soviet statistics offer considerable evidence in support of the proposition that the USSR was practising a 'unique' form of price discrimination (that is charging a higher oil price to captive satellite markets than to the 'free world' Western European consumers). Table 5.8 illustrates the magnitude of the discrimination in East versus West Germany as implied by data provided by the Ministry of Foreign Trade of the USSR. To cite only one example: in 1962, East Germans paid an average price of US$2.66 per barrel of Soviet crude (f.o.b. the Soviet border), more than twice as much as the US$1.30 per barrel charged to West Germans.

It is important to note the fact that all allegations of 'price discrimination' were vehemently denied by the Soviet government at every possible opportunity. On closer inspection, however, it would appear that public statements protesting the fact that 'satellite' nations were being 'unjustly' overcharged for their oil were, for the most part, deliberately exaggerated. As always, the issue has been hopelessly confused by the pre- eminence of politics, and the administrative role of prices in the unique Stalin command economy.

Table 5.8: Annual prices for Soviet crude oil in East and West Germany, 1959–67

Year	West Germany	East Germany
1959	$1.69	$2.70
1962	$1.30	$2.66
1965	$1.40	$2.33
1967	$1.40	$2.04

Note: All prices are F.O.B. the Soviet border. Currency conversions are based on official Soviet exchange rates   (1 rouble = $1.11 US)

Source: Ebel (1970).

Under strict guidelines imposed by the Council for Mutual Economic Assistance (CMEA), at least 90 per cent of all transactions with the socialist countries (satellites) were dictated by the terms of detailed bilateral trade agreements. The Soviet government's standard answer to accusations that they were 'exploiting their comrades' was to reply that higher than necessary prices were paid for Soviet imports from their satellites and, further, that they provided markets for low quality or obsolete products that would not find a market in market economies. While this backhanded condemnation of the economic system imposed by the Soviet Union on its satellites is reflective of the way CMEA trade was organized, few doubt that the Soviet Union was the major beneficiary (Hobbs et al. 1997).

These complex 'barter' agreements were negotiated on an annual basis, and dovetailed with the five-year planning horizon in a loose way by five-year trade agreements. Transactions negotiated under these agreements [were] denominated in transferable roubles [TR], an accounting currency with the same official exchange rate as the Soviet domestic rouble.[3] Given that internal prices could not be used to represent value, in terms of opportunity cost (Hobbs et al. 1997) – the von Mises problem – the prices for all goods to be included in the CMEA agreements were set, and/or 'negotiated at five-year intervals on the basis of world market prices that prevailed in some previous period, usually also a five-year period' (Hewett, 1984, p. 161). To cite but one example, the base price for crude oil was fixed according to the level of world oil prices in 1957–58, 1960–64, 1966–70, and 1971–75; that is the crude oil prices for bilateral trade agreements negotiated in the seventh five-year plan (1958–1965) were fixed according to the 1957–1958 base price (Ebel, 1970).

Still, the glaring example of disparate prices charged to East and West Germany was hard for the Soviet leadership to refute, as was the inevitable temptation to profit from the delicate political status of captive East European satellites. To cite an 'early' example: There is clear evidence of 'discrimination' in the USSR–Poland bilateral trade agreements negotiated in the fourth and fifth five-year plans (1946–55). In the words of Alec Nove (1972, p.351):

> The Soviet leadership even felt it necessary, in its agreements with Poland in November 1956, to cancel Polish indebtedness of past credits as compensation for the 'full value of the coal supplies to the USSR from Poland in the years 1946–53,' a clear and public admission of past underpayments.

As always, the use of official Soviet exchange rates (TRs) complicates any sensible economic evaluation. Over the period 1961–72, the official Soviet exchange rate was fixed at US$1.11 per barrel – a figure which, even in the late 1960s and early 1970s, was assumed to be grossly overvalued. According to Goldman (1980, p. 133):

> Since the Soviets do not allow for free convertibility of the rouble, the rate of exchange set by the Soviets between the rouble and other currencies is an arbitrary matter. Moreover, it is widely accepted that the rouble is overvalued. . . For example, even though it cost the Soviets only in 1971 about 11.5 to 12.5 roubles to produce and ship a tonne of petroleum which was then sold at only about 13 roubles, it may not have been a foolish transaction.

Further discussion, and/or the presentation of a definitive solution to the 'multi dimensional' puzzle of price discrimination, is beyond the scope of this analysis. However, at least one point is clear: statistical estimates of the 'apparent' average crude oil prices charged to the Communist allies did fall significantly from 1960 to 1967. While a portion of the reduction may indeed have been due to a 'political' reversal of discriminatory pricing policies, this is an unanswered question – riddle – that has yet to be resolved.

Despite delays in the construction of both pipelines and storage facilities, a number of important new projects were completed in the seventh and eighth five-year plans. To cite only a few examples: (1) the Friendship (Druzhba) pipeline system was completed, and operational, by 1964. The complex system included 3004 kilometers of 426–1020 millimetre pipe in the USSR, 675 kilometres in Poland, 27 kilometres in East Germany, 836 kilometres in Czechoslovakia, and 123 kilometres in Hungary; (2) the final stage of the 3682 Trans-Siberian pipeline from Ufa to Irkutsk was completed in 1964; (3) the 410 kilometre West Siberian pipeline from Shaim to Tyumen was completed in 1965; and (4) the 1000 kilometre West Siberian pipeline from Ust'Balyk to Omsk was completed in the year 1967.

By the end of the eighth five-year plan (1970) a total of 37 400 kilometres of pipeline had been laid across the vast territories, and often inhospitable landmass of the USSR (see Table 5.9). Still, the Soviet Union's progress in the modernization of transportation networks for crude oil and petroleum products was slow at best, and by the late 1960s pipeline construction in the USSR had only recently begun to keep pace with the increase in oil output (Elliot, 1974). To further aggravate matters, a number of critical producer goods such as steel were in short supply, so that the pace of construction fell significantly from 1965 to 1970 (the eighth five-year plan). Better quality steel production and the ability to domestically produce large diameter pipe reduced the need for the sector to compete for foreign exchange to acquire these products from the West. By the end of the planning period (1970), however, with pipelines accepted by the planners as the preferred transportation mode, well over 50 per cent of the total volume of crude oil and petroleum products was still being transported by the archaic channels of river, sea and railways.

Railways dominated oil transportation. The volumes moved by rail are staggering. According to Elliot (1974, p. 110): 'oil and oil products accounted for 342 800 million ton-kilometres [in 1969], and the average length of haul per ton was 1204 km., both indicators being significantly higher than those for pipelines'. The disturbing 'economic' reality that pipeline transportation costs had fallen to less than one third (33 per cent) of railway costs by the year 1970 had not passed unnoticed by Gosplan. Again, according to Elliot (1974, p. 110):

It is expected, however, that by 1980 pipelines will have taken over the transportation of oil and oil products almost completely...[Over] 22,000 km. of new trunk lines are planned for the 1971–75 period. The actual construction

Table 5.9: The transportation of crude oil and petroleum products in the USSR, 1940–71

(millions of tonnes)

Method of Transportation	1940	1950	1960	1965	1968	1969	1970	1971
Rail	29.5	43.2	151.0	222.2	275.9	284.7	302.8	322.8
Sea	19.6	15.8	32.5	53.5	70.1	70.5	75.1	79.8
River	9.6	11.8	18.4	25.0	29.2	30.3	33.5	35.2
Pipeline	7.9	15.3	129.9	225.7	301.3	324.0	339.9	352.5
Pipeline Construction and Throughput								
Length at Year-End in Thousands of Km	4.1	5.4	17.3	28.2	34.1	36.9	37.4	42.9
Throughput of Crude Oil and Petroleum Products in Millions of Tonnes	7.9	15.3	129.9	225.7	301.3	324.0	339.9	388.5

Source: Elliot (1974).

107

achieved in 1971 and 1972 [a disappointing 7,700 km.], however, made this target seem improbable.

In retrospect, the completion of the immense 'Friendship' pipeline system would prove to be a mixed blessing for Gosplan, forming as it did, a permanent physical link between the Soviet Union and its East European allies. As early as the mid-1960s, one finds clear evidence that the East European satellites were already becoming a definite fiscal burden on the over-stretched Soviet economy. In 1965, a number of Soviet economists would gain global recognition for the suggestion that the USSR was, in fact, being exploited by a complex 'mercantilist' relationship with Eastern Europe. According to Goldman (1980, p. 61):

> As they saw it, the Soviet Union was not the coloniser but the colonised. While the Soviet Union was busy supplying raw materials to its East European allies, its allies were using the Soviet Union as a dumping ground for their manufactured goods. . . . [As a result], in what clearly was a major change of emphasis, in 1965, the East Europeans were told to redesign their foreign aid program, so that they would be compensated by the developing country recipients with oil and ferrous and non-ferrous metals.

While the exact basis for these demands was not entirely transparent in the mid-1960s, the economic justification having been clouded by TRs and the complex terms and conditions of detailed bilateral trade agreements, the Soviet Union's new East European policy was in fact a subtle request for fiscal or economic relief. It would be repeated frequently, and with a growing sense of urgency, throughout the decade of the 1970s.

THE PROBLEM WITH APPEARANCES

> Things had never looked so good since the revolution. The economy was the world's second largest, after the US. Meat and milk production climbed by a third. Consumption of fish and eggs doubled; that of alcohol quadrupled. Prices were stable. Wages doubled, but the price of meat remained pegged at 1962 levels... Apartments rented at five dollars a month; heating, in cities with six-month winters, was included. Socialism had been a struggle and now here was the pay-off. As distant oil and mineral reserves were exploited, hard currency tumbled into Soviet coffers with the huge increases in world energy and precious metal prices. The country became the world's largest oil producer, and Muruntau in the Uzbek wilds was its largest gold mine, producing eighty tonnes a year. . . . Surgut, the center of a huge oil zone in the middle reaches of the Ob', boomed from a village settlement among frozen marshes and soon outgrew

Anchorage, Alaska; tool pushers on the Samotlor field it serviced could make $18,000.00 a year, phenomenal money. Bratsk, heart of the Siberian power industry, quadrupled its population, whilst that of Yakutsk, deep in the permafrost zone, grew to 100,000, with a university and a branch of the Academy of Sciences (Moynahan, 1994, pp. 212-213).

The unique Brezhnev management style – dekhrushchevization and stability – was flattered as having 'performed miracles' for the estranged Soviet economy. Indeed, the strange mixture of planned 'competition' (the Kosygin reforms), and a characteristic Party preference for heavy industry, paid handsome dividends in select sectors of the USSR. By the late 1970s, the Soviet Union led the world in the production of such basic items as steel, oil, machine tools, and heavy military hardware (Kort, 1993). At least a portion of this newfound prosperity reached the consumer goods sector, and real wages were increased by 50 per cent over the years 1965 to 1977. By the early 1980s, a typical Soviet family had acquired such coveted modern luxuries as a radio, refrigerator, and even a washing machine. When measured in terms of US dollars, the average GNP per capita reached $3130 in 1975 – an 86.5 per cent improvement over the US$1678 reported in 1963, and only US$670 below the US$3800 per capita average for the European NATO countries (Rakowska-Harmstone and Gyorgy, 1979).

On closer inspection, there were cracks, deficiencies, in the 'improved' Stalin planning system. Behind a thin veneer of prosperity, even the official statistics betrayed a faltering national economy. At best, the economic performance of the Brezhnev government can be described as mixed. According to Kort, (1993, pp. 259–60):

> Innovation lagged; much of the Soviet Union's equipment was obsolete. Also, the Soviet Union did not produce many of the finished industrial goods that in Japan, Western Europe, and the United States were the basis for increased productivity and a far higher standard of living than Soviet citizens knew...The housing situation remained unsatisfactory, to say the least; in 1981 the Soviets could not meet minimum standards that had been set by the government in 1928...[Twenty] percent of all urban families still shared kitchen and bathroom facilities.

National income, and a host of broad economic indices fell short of their targets in the ninth and tenth five-year planning periods. While Gosplan had envisioned a 38.6 per cent increase in national income in the years 1971–75, the official statistics suggested only 28 per cent – a significant shortfall. Similarly, the 26 per cent national income growth rate target was under-fulfilled by 6 per cent in the years 1976–80, the tenth five-year plan.

The main culprit was the consumer goods industry, which fell short of its target by 11.6 per cent in the ninth five-year planning period, and 10 per cent in the years 1976–80 (see Table 5.1). According to Nove (1972, pp 372–73)

> the consumers' goods target proved far beyond reach, due partly to the shortfalls in agriculture (and so in the food industry), and partly to increasing delays in construction; the latter factor has become of great importance, with a persistent rise in the volume of uncompleted investments and in the time taken to bring new productive capacity into operation.

Steel, a critical input to most manufacturing industries, fell short of its targets by 10 per cent in the period 1965–70; 5.4 per cent in 1971–75; and rising to 13 per cent in the tenth five-year planning period.

Disruptions and shortfalls spread to every sector of the Soviet economy, so that by 1980, only two industries, natural gas and passenger vehicles, had fulfilled Gosplan expectations. The production of nearly all other commodities – including consumer goods, crude oil, coal, fertilizer, cement, tractors, locomotives, civilian machinery, and construction materials – fell far below quota (Nove, 1972).

The failure to satisfy even the 'modest' targets of the ninth and tenth five-year planning periods has been interpreted as the first clear warning, a premature manifestation, of the severe industrial crisis of the early 1990s (late 1980s). Economic deficiencies and shortfalls were exacerbated by the following 'unplanned' complications:

1. *Low birth rates, and a growing shortage of labour.* Rapid urbanization, and planned improvements in the education system led to a reduction in birth-rates among the European and Slavic citizens of the USSR – approximately 80 per cent of the population at that time. The large reservoir of cheap and abundant labour which had provided the impetus for the rapid industrialization of the post-World War II era, simply stopped growing. The result was an acute shortage of labour despite 'overmanning', and the waste of valuable labour resources in some sectors of the Soviet economy (Nove, 1972). As pointed out previously, managers in Soviet firms had an incentive to hoard labour. The surplus labour was required to allow the achievement of the quotas set out in the plan. As upstream firms in Soviet supply chains were often late in delivering their products, downstream firms were faced with having to complete the target quantities in less than the allotted time. This could

only be accomplished through the use of labour intensive methods (Hobbs et al., 1997).

2. *Massive investments in the agricultural sector.* The succession of Leonid Brezhnev to First Secretary of the Communist Party marked a major turning point for the official role of agriculture in the USSR. At the March 1965 plenum of Central Committee, delegates denounced the Khrushchev government for shoddy agricultural policies, including: (1) a neglect of the non-black-earth regions; (2) the inadequate provision of new machinery; (3) major capital losses in the production of livestock; and (4) sporadic shortages of food supplies (Nove, 1972). To compensate Soviet citizens for past 'errors in judgement', the expansion and modernization of the agricultural sector was designated as a priority for the government. Substantial investments were scheduled for the eighth, ninth, and tenth five-year planning periods. At the same time, the planners sought to improve the profitability of primary agriculture. Procurement prices were increased substantially, especially for livestock (Nove, 1972). The pricing arrangement, however, reflected 'traditional' Soviet values – stability and an absolute intolerance of domestic inflation – so that the retail prices for all agricultural necessities were held constant. A similar pricing scheme was applied in the petroleum industry (see Appendix D). Of course, this was the primary cause of the chronic shortages and long queues that characterized the last decades of the command economy (Hobbs et al., 1997)

The pricing strategy, of course, resulted in massive and expanding subsidies to the agricultural sector. Total investments in agriculture rose steadily throughout the Brezhnev era from 48.6 milliard roubles in 1961–65, to 82.2 milliard roubles in 1966–70. By the tenth five-year planning period the level of agricultural investment exceeded 131.5 milliard roubles – an unsustainable 26.2 per cent of total capital investments for the entire USSR.[4] Instead of contributing to the wealth of the nation, agriculture was rapidly becoming a burden on the already strained Soviet economy.

3. *The failure of the central planning.* The central planning dilemma – the von Mises problem – grew apace as the Soviet economy increased in size and complexity. The stubborn focus on quantity measured by clumsy success indicators provided incentives to produce shoddy products. Innovation, and the introduction of new goods and services were stifled by administrative procedures. All new products and technologies had first to be given the approval of a hesitant Soviet bureaucracy. Only after its 'official' acceptance could a new product,

and the accompanying material inputs, be carefully incorporated in the five-year planning process. Not surprisingly, the pre-planned quantities of producer goods frequently failed to match the precise product requirements required by management (Nove, 1972). Chronic shortages, and bottlenecks encouraged hoarding.

The deficiencies were so pronounced that even the stolid bureaucrats of the Brezhnev era were forced to contemplate reform. In 1973, after years of cautious experimentation, a new system of Associations (*ob'edineniia*) was introduced to replace the complex, and lethargic network of sub-Ministerial entities. To cite only a few examples: (1) the *glavki* (Ministerial production departments) were replaced by 'industrial associations'; (2) factories were regrouped and placed under the direction of 'production associations'; and (3) research and development institutions were linked to factories and placed under the direction of 'science production associations'. In recognition of the need for economic efficiency, the funding for the new industrial associations was to be drawn, not from the State budget, but from the profits of constituent enterprises.

Many constraints were, however, placed on the powers of the new production associations, so that the 1973 reform did not allow the economy to escape from the allocation problems associated with central planning. While industrial associations were granted a greater degree of freedom – and some powers – they were still effectively subordinate to the authority of the Ministries. Further, given the priority given by the Party to stability, the reform – in reality little more than a simple rearrangement of administrative entities – was to be introduced only gradually throughout the ninth and tenth-five year planning periods. In the words of Davies (1979. p. 232): 'it may be safely predicted that the 1973 reform, like the 1965 reform, will not fundamentally change the economic planning mechanism because it does not involve a substantial change in planning indicators or in prices'.

4. *The cost of the military–industrial complex.* The cold war, and with it an obsession with gaining and maintaining military superiority over the US, placed an unsustainable burden on the scarce financial resources of the USSR. Given the preoccupation with secrecy at the time, the exact level of military expenditures still remains a mystery. The drain on Soviet resources has been estimated at levels as high as 25 per cent of GNP throughout the decade of the 1970s (Kort, 1993). When measured in terms of US dollars, Soviet defence expenditures exceeded those of the US by 45 per cent in some years. This could only be accomplished

by 'starving' other sectors of resources. It did, however, provide the Party with the means to influence events around the world – the Vietnam War, the various conflicts in the Middle East, in the chaos that often characterized post-colonial Africa. The cost of that influence was a decaying economy at home.

In retrospect, it is impossible to over-emphasize the ruinous effect of the military on the scarce financial resources of the USSR. According to Moynahan (1994, p. 221):

> The vast military and political effort – Cuba alone was costing $4.5 billion [US] a year by the start of the Eighties – was extracted from a country in which consumer goods were chronically *defitsitny*, in deficit. No recognition came from a geriatric politburo.

The limited attempt to force a mutual accommodation, or détente, provided the sole indication that the arms race was becoming unduly burdensome to an over-stretched Soviet economy (Kort, 1993). The pre-eminence of the military continued and produced some successes. By the early 1980s Solidarity, and 'socialist' insurrection was temporarily silenced in Poland, and pro-Soviet regimes had been established in Angola, Mozambique, Ethiopia, South Yemen and Nicaragua. Fidel Castro survived in Cuba on the doorstep of the US despite the best efforts to overthrow him. Thus in the final analysis, even the Soviet Union's most extravagant military expenditures, could be justified in the minds of the Communist Party leadership by international recognition and conquest (Moynahan, 1994).

The crude oil industry, forgotten in the struggle for international supremacy, along with a host of other industries suffered the debilitating consequences of neglect. Given the tremendous potential of the West Siberian producing region, the central planners expressed considerable confidence in the oil industry at the start of the ninth five-year planning period. The planners and the Party leadership seemed entirely unaware of the impending OPEC–induced energy crisis. According to Gustafson (1989, p. 23): 'In his reports to the Party Central Committee in December 1972 and 1973, Brezhnev gave hardly more than a passing reference to the subject; and Kosygin, in his few published speeches on domestic policy during this period, had equally little to say about energy production or conservation'. 'Minor problems' in the oil industry, if mentioned at all, were dismissed as an inevitable consequence of rapid growth.

Staying the course – stability – was the watchword at the energy ministries. Approximately 20.5 billion roubles were invested in the oil industry in 1970–75. This constituted only 4.69 per cent of total Soviet investment. The major portion of these expenditures, some 16 billion roubles, were devoted to the extraction of crude oil (see Tables 5.10 and 5.11). It is important to note that the basic structure of the Soviet investment strategy remained virtually unchanged through the entire 15-year period from 1961 to 1975. Industrial investment absorbed approximately 40 per cent of total Soviet investment in the eighth and ninth five-year planning periods. Each year a little over one third of these funds was earmarked for energy. Oil investments, which should have been adjusted to reflect variations in exploration and development costs, accounted for a virtually constant share of total Soviet investment – 4.36 per cent for 1961–65, 4.28 per cent in 1966–70, and 4.69 per cent in 1971–75 (see Table 5.10). This exceptional rigidity has been attributed to the inherent difficulties associated with incorporating change in the Stalin planning model and obsession with stability that was characteristic of the Brezhnev era. Hewett (1984, p. 38) is harsher in his criticism: 'Presumably that stability reflects the bureaucratic inertia endemic to the Soviet planning system.'

This unimaginative and inflexible capital investment programme would prove entirely inadequate. Crude oil production rose only lethargically throughout the ninth five-year planning period, falling short of its annual production target in every year (Hewett, 1984). At the end of the planning period in 1975, crude oil flows had reached only 491 million tonnes, a disappointing 14 million tonnes below the 505 million tonne production quota (see Tables 3.3 and 5.4). The shortfall was due to an 'unplanned' acceleration in the rate of decline from the older producing regions. Despite a sustained concentration of capital investments in the North Caucasus and Volga-Urals, production from these areas reached only 247.9 million tonnes in 1975 and a meagre 5.4 million tonne increase over the 1970 total.

While the North Caucasus absorbed a solid 10 per cent of investments in oil during the ninth five-year plan, the region experienced a sharp (33 per cent) reduction in crude oil flows. Annual production fell from 34.2 million tonnes in 1970 to only 23.0 million tonnes in 1975. The Volga-Urals, which consumed over 43 per cent of Soviet oil investments throughout the ninth five-year planning period, experienced a significant deceleration in crude oil flows. Production reached only 224.9 million tonnes in 1975, a disappointing 7.9 per cent increase over the 208.4 million tonnes reported in 1970. The implied rate of deceleration from a five-year growth rate of 20

Table 5.10: Capital investment in the Soviet energy economy, 1966–80

(billions of roubles in 1969 prices)

5 Year Plan	Total Extraction	Coal	Oil		Natural Gas		Electricity
			Extraction	Extraction and Transportation Costs	Extraction	Extraction and Transportation	
1966–70	22.76	7.24	11.06	n.a.	4.46	n.a.	13.63
1971–75	31.61	8.34	15.98	18.98	7.29	13.79	17.00
1976–80	46.25	9.76	26.22	30.22	10.27	28.57	19.39
1980	10.89	2.09	6.63	n.a.	2.17	n.a.	4.19

Source: Hewett (1984).

Table 5.11: Capital investment in the Soviet oil industry, 1966–80

5 Year Plan	Thane Gustafson		Ed A. Hewett			Yesterday, Today and Tomorrow		
	Soviet Oil Investment	Percent of Total Soviet Investments in Industry	Soviet Oil Investment (Billions of pre-1973 Roubles)	Percent of Total Soviet Investments in Industry	Percent of Total Soviet Investment	Soviet Oil Investment (Billions of 1 Jan. 1969 Roubles)	Percent of Total Soviet Investments in Industry	Percent of Total Soviet Investment
1966–70						13.1	10.94	4.28
1971–75	16.0	9.23	16.0	9.28	3.25	20.5	12.16	4.69
1976–80	26.4	11.92	26.2	11.72	4.13	32.2	14.68	5.68
1981–85	45.2	16.06	43.0	15.64	6.14	n.a.	18.79	7.43

Sources: Gustafson (1989); Hewett (1984); *Yesterday, Today and Tomorrow of the Russian Oil Industry* (1995).

116

per cent in the period 1965–70, to less than 8 per cent in the years 1970–75, was, to say the least, disconcerting (see Tables 5.3 and 5.4).

After years of relatively 'effortless prosperity' in the oil industry, the Soviet Union had run into a non-linear planning problem which would temporarily stupefy Gosplan. In most other Soviet industries, consistent levels of production were achieved by the simple maintenance of capital, producer goods, and labour inputs, a formula that could not be applied to the oil industry. According to Hewett (1984, pp. 48–9):

> Extractive industries are different. Even if the capital stock and labour inputs are maintained at a constant level, output will typically fall because reserves are used up and new reserves are of lower quality (more costly to find and exploit) than the old reserves. Consequently, barring technical improvements in the production techniques for an extractive industry or the discovery of new reserves that are cheaper to exploit than those currently under development, maintaining a current output level in an extractive industry will require increasing inputs of labour and capital.

Crude oil flows from ageing producing regions, particularly the Volga-Urals, had declined by far more than anticipated. Soviet planners had envisioned a 2.96 MMb/d (147.3 million tonne per annum [mt/y]) decline in production from older oil fields in the ninth five-year planning period. The actual reduction exceeded 5.22 MMb/d (260 mt/y), a 2.26 MMb/d (112.5 mt/y) discrepancy (Hewett, 1984).

Production from the West Siberian basin had been rising steadily, however, and would mask the decline. Although investment in West Siberia had more than doubled in the ninth five-year planning period, the region received less than 12.5 per cent of total Soviet investments in the oil industry (see Table 5.3). This parsimonious financial commitment would yield a wealth of reserves and a degree of prosperity to those involved in its development. A mere 14.2 per cent of Soviet exploration funds was allocated to West Siberia in the ninth five-year planning period. By 1975, the region would provide more than two thirds of the gross additions to industrial reserves $(A+B+C_1)$ for the entire USSR (Gustafson, 1989).

The rapid rate of discovery was unexpected and according to Gustafson (1989, p. 79): 'in both 1970 and 1975 the planners set aside the cautious preliminary five-year targets recommended by local oil men and substituted sharply higher ones'. The planners were lucky and were able to respond to production declines in the Volga-Urals by artificially elevating the West Siberian production quota (see Table 5.12). In the short term, at least, the decision would pay handsome dividends. Crude oil flows from West Siberia accelerated steadily, exceeding planning expectations throughout the ninth five-year planning period. By 1975, West Siberian production had reached 154 679 million tonnes per annum, a remarkable 392 per cent increase over the 31 416 million tonnes reported in 1970 (see Table 5.4).

Table 5.12: Five-year planning targets for Tyumen, West Siberia

	(millions of tonnes)	
	1975 (Target formulated in 1970–71)	1980 (Target formulated in 1975–76)
Preliminary Targets	70–75	250–260
Five-Year Plan	120–125	300–310
Actual Production	141.4	303.8

Source: Gustafson (1989).

It is important to note that this feat was accomplished through the rapid exploration, and development, of the region's richest oil fields. By April 1973, an impressive 113 oil fields had been discovered in the vast and inhospitable permafrost of the West Siberian Plain (see Appendix A). The majority (75 fields) of these discoveries had been located before 1970. By 1975, however, only 19 of these fields had been brought into production. The explanation for this lethargic development programme can be found in the operating procedures and often perverse incentives of the command economy system.

The most important of these 'incentives' was production targets. The incentive, to maximize development at the expense of the Soviet exploration programme was very strong. As suggested by Hewett (1984, p. 58): 'Any sensible minister of oil and gas, if forced to choose between proving up new reserves at the planned rate and fulfilling this year's output plan, will choose to fulfil the output plan'. While aggregate drilling statistics registered a 64 per cent increase between the eighth and tenth five-year planning periods, from 56.5 million metres in 1966–70 to 92.7 million metres in 1976–80, there is clear evidence of stagnation in exploratory drilling. Approximately 26.1 million metres were drilled for exploratory purposes in the ninth five-year plan (1971–75) representing a four percent reduction from the 26.2 million metres reported in the years 1966–70. This was far less than planned and the shortfall did not pass unnoticed by the central planners. The tenth five-year plan called for a 7 per cent increase in exploration effort to 28 million metres. To the continued dismay of planners, actual exploratory drilling reached only 26.7 million metres in 1976–1980, an unacceptable 1.3 million metres below quota (see Table 5.13).

The Ministry of Petroleum (MNP), which was primarily responsible for advanced exploration efforts (deep drilling) in the established petroleum regions (notably the Volga-Urals), reported a sharp, and prolonged reduction in exploratory drilling activity. Official MNP exploratory drilling statistics fell sharply from 15.4 million metres in 1966–70, to 14.5 million meters in the years 1971–75, and only 11.27 million metres in the tenth five-year planning period. With critical production targets threatened and an accelerating rate of output decline in the ageing oil regions, the Ministry had little alternative but to focus its resources on the rapid and technologically enhanced development of existing oil fields. By the tenth five-year plan, MNP development drilling had reached 65 million metres, a 119.9 per cent increase over the 29.56 million metres reported for 1966–70. Despite this remarkable achievement, however, the Ministry would be reprimanded for a lack of initiative. The stagnation in the exploration effort was so pronounced that MNP fell short of its aggregate (exploration and development) drilling quotas by 3.7 million metres over the period 1971–75, and a full 4 million metres in the tenth five-year plan (see Table 5.13).

This alarming new trend which included stagnation in the exploratory drilling activities of not just the MNP, but Mingeo as well, was exacerbated by a plethora of industrial difficulties including:

1. *Bureaucratic inertia, and a reluctance to shift the focus of the Soviet exploration effort from established producing regions.* The acceleration in Volga-Urals decline rates presented the Gosplan's framers with a unique form of planning dilemma. Something had to be done to offset the 260 mt/y decline from the established producing regions. A debate, complicated by the complexity of even a subtle change in the energy plan, raged throughout the first half of the 1970s. According to Hewett, (1984, p. 59) many:

> including N.K. Baibakov, the chairman of Gosplan, contended that the appropriate response was to maintain relatively high rates of drilling activity

Table 5.13: Exploration and development drilling for oil and gas in the USSR

(million metres)

Five Year Plan	All Ministries			Ministry of Petroleum (MNP)			Planned (MNP) Drilling Quota
	Exploration	Development	Total	Exploration	Development	Total	
1961-65	24.7			16.30			
1966-70	26.2	30.4	56.6	15.40	29.56	44.71	
1971-75	26.1	41.8	67.9	14.50	37.80	52.30	56.0
1976-80	26.7	66.0	92.7	11.27	65.00	71.00	75.0
1981-85	32.1	132.3	161.5	13.70	130.10	143.80	130.0

Sources: Hewett (1984).

in traditional areas, presumably along with the secondary recovery program put in place for FYPX [the tenth five-year plan], in order to slow the decline in the old fields, and therefore opt for slower development in West Siberia.

Of course, this debate stemmed from the central problem of the command planning system – the inability to value alternatives. The planners realised that choices were available but had no method to guide them. There were, on the one hand, clear advantages to be derived from stability. The established producing regions (Volga-Urals) were ideally located; that is in closer proximity to consumers than West Siberia, and fully supported by a comprehensive network of pipelines, production infrastructure, and manpower.

West Siberia, on the other hand, was an unknown entity. As was the case with all 'new products' and innovation in the USSR, the transfer of resources to West Siberia had to be discussed at great length, at all levels of the planning hierarchy. To aggravate matters further, both the Oil and Gas Ministries as well as Mingeo, which all had well established roots in the traditional producing regions, were reluctant to expend valuable financial resources on risky ventures that might fail and thus put them in disfavour with the planners and the Party. According to Gustafson (1989, p. 298):

> The geological establishment long resisted the idea that a major oil province would be found there; and even after it had been proven wrong, it was similarly sceptical about the prospects for gas. The gas industry was equally resistant. As late at the mid-1970's MGP [Gas Ministry] leaders considered Siberian gas too expensive to gamble on, an 'exotic commodity', and even in the Tenth Five Year Plan (1976-1980) they continued to push investment in the older gas producing regions faster than in Siberia.

In the final analysis, the transfer of resources to West Siberia would be delayed, beyond reason, by 'simple bureaucratic inertia' (Gustafson, 1989).

2. *Significant increases in drilling and crude oil production costs.* The natural ageing of the Russian oil industry combined with the massive infrastructure required for the development of West Siberia resulted in a significant increase in drilling and production costs throughout the USSR. In the established producing regions, Soviet exploration and development teams were, according to Gustafson (1989, p. 73): 'obliged

to drill deeper, travel to more remote places, and work in more complex formations'. The average depth of exploratory wells rose steadily throughout the seventh, eighth, ninth and tenth five-year planning periods from the 1845 metres in 1960, to 2788 metres in 1980 (see Table 5.14). As expected, the tendency was most pronounced in the Central Asian and European regions. By 1977, the average depth of exploration wells had reached 4307 million metres in Azerbaijan, and a nearly impossible 4660 million metres in the Ukraine (Gustafson, 1989).

Soviet geologists were, however, aware of the difficulties that would be caused by the failure to fulfil industrial (A+B+C$_1$) reserves targets in the eighth five-year plan, and sought relief in the form of new technology including field computers. The use of rotary drills, and advanced seismic reflection techniques, increased considerably from 1970 to 1980.

The increased use of the rotary drill was necessitated by the increase in average well depths throughout the USSR. According to Hewett (1984, p. 58): 'In many areas the average depth of exploratory wells is exceeding the range within which turbo drills – the technology on which the Soviets rely quite heavily – can operate efficiently'. The inability of the Soviet industrial system to produce the high quality steel pipe necessary for the production of an effective rotary drill rig, often exacerbated by the sporadic shortages of steel supplies, necessitated significant, and costly, imports of steel pipe and pipe casing. Goldman (1980, p. 42) reports that by the late 1970s: 'Soviet pipe imports [for both drilling and pipelines] constituted about 15 per cent of all Soviet hard currency imports. In addition, significant quantities of pipe were imported from some of the East European countries, like Romania and Poland'. Such 'unplanned' imports and extra drain on hard currency reserves were extremely difficult for the planners to deal with (Henderson and Kerr, 1984/85) and would have been resisted at every opportunity.

In West Siberia, where well depths were not yet a problem, exploration costs reflected problems associated with the severe northern winters, and remote, often roadless, territories. The conditions in West Siberia were so formidable that a new research institute, Giprotyumenneftegaz, was created for the sole purpose of solving the problems faced by the oil industry in the north, such as long severe winters and the vast roadless wastes of marshland and taiga (Elliot, 1974). Gustafson (1989, p. 73) reports that the:

average cost per exploratory meter in West Siberia more than doubled between the Eighth and the Tenth Five-Year Plans, while average well depth increased only 24 per cent during the same time period. However, the average of the Tenth Plan includes the three frantic years 1978–1980, during which cost per meter rose sharply.

The effect on the Soviet drilling programme was devastating. According to figures provided by Oil and Gas Ministries, the average cost of 'exploratory' drilling in the USSR reached 401.77 roubles per metre in 1980, a 253 per cent increase over the 113.67 Rbls/m reported in 1960. Development drilling costs rose due to an increase in average well depths and reached 119.56 Rbls/m in 1980, a 137 per cent increase over those of 1960 (see Table 5.14). Soaring costs and expenditures meant that the Oil and Gas ministries as well as Mingeo fell short of their aggregate drilling quota in the ninth five-year planning period. Soviet drilling teams completed only 67.9 million metres in 1971–75, an unacceptable 4.6 million metres below the 72.5 million metre drilling target (see Table 5.13) (Hewett, 1984). As might have been anticipated in the context of a centralized incentive system which granted the highest priority to the production target, the discrepancy was reflected in a sharp reduction in exploration activity. Only 5973 exploratory oil wells were completed in the ninth five-year planning period, a 16 per cent reduction from the 7117 reported in the 1961–65 quinquennium (see Table 5.15).

The natural ageing of established oil fields, and remote and inhospitable location of new discoveries – specifically the West Siberian oil fields – affected costs in all divisions of the Russian oil industry. In the development and extraction phase, the use of enhanced water-injection systems was accelerated significantly, and complex Russian 'PAT' and 'Sputnik' systems were adapted to monitor, and correct, oil flows and water pressure in inhospitable and/or remote locations by remote control (Elliot, 1974). Expenditures were elevated by an improvement in trade relations with the West resulting in a significant increase in the level of technological imports. Despite the complex requirements of the Soviet oil industry, however, technical imports from the United States were restricted by the ongoing tensions of the Cold War. These difficulties were particularly evident in the 1960s and early 1970s. As Elliot (1974, p. 109) suggests:

It is occasionally admitted that American oil-extraction technology is ahead of the Soviet equivalent in many cases, and it is doubtless hoped that the

USSR. will benefit from the improved trade conditions of the 1970s by
importing equipment from the United States.

This hopeful prophecy would subsequently be proven correct. In 1970,
the Soviet Union spent approximately 20.7 million roubles on oil- and
gas-related technical imports. The term 'oil- and gas-related technical
imports', Category 128 in Soviet Foreign Trade Statistics, includes
machinery and equipment for drilling, well development and geological
exploration. Pipe and refinery equipment is not, however, included in
this estimate. The contribution of the United States was extremely
small. To some extent Britain and France were able to accommodate
Soviet needs accounting for 36 per cent of the expenditures on imports.
By 1972, the year of the SALT I agreement to limit nuclear armaments,
which represented one of the first 'thaws' in the Cold War, the value of
technical imports from the US had risen to 13.8 million roubles. Despite
the USSR/US supported conflict in Angola, the calming effect of
détente reigned supreme, and Western imports continued to rise
throughout the ninth five-year planning period. By 1978, the value of
oil- and gas-related technical imports had risen to 251.2 million roubles,
the bulk of which – approximately 185.5 million roubles or 74 per cent
– was imported from the West (Gustafson, 1989).

According to figures provided by the Oil and Gas Ministries of the
USSR, the average production cost for crude oil in the USSR reached
7.21 roubles per metric tonne in 1980, which represented a 148 per cent
increase over the 2.91 roubles per metric tonne reported in 1965 (see
Table 5.14). Of course, these figures, and indeed all Soviet prices
(costs), were maintained primarily for accounting purposes, and as a
result do not provide an accurate valuation of the level, and/or rate, of
increase in Soviet production costs. As always, an extreme measure of
caution is advised in the interpretation of Soviet financial statistics. In
this particular example, the rate of increase in crude oil production costs
has been exaggerated by the Kosygin reforms where there was a limited
initiative by those responsible for developing the Grosplan to improve
both the planning value of prices and the general efficiency of the Stalin
command economy. On 1 July 1967, crude oil producing enterprises
were suddenly, and for the first time ever, required to include two new
charges as basic or essential costs of production: (1) a 6 per cent capital
charge levied on the fixed and working capital stock of an enterprise;
and (2) a 25 per cent finding fee, or geological exploration cost (see
Appendix D). It is important to notice that the production costs reported

in Table 5.14 significantly underestimate the actual costs of production. In short, while an exploration fee may have been included in the basic costs of production as of 1 July 1967, rent payments, royalties (for the raw mineral deposits), and charges for the outright purchase of land, were not. Of course, all this tinkering with the pricing schemes could not overcome the central difficulty identified by von Mises (1981) that without markets to establish the relative worth of products, there can be no way to determine that any allocations are efficient relative to the alternatives.

Putting aside difficulties in the interpretation of the Soviet pricing system, official statistics suggest that crude oil production costs rose by 68 per cent in the 1970s – from 4.28 Rbls/tonne in 1970 to 7.21 Rbls/tonne by the year 1980. To these estimates the following can be added: (1) the substantial capital expenditures required for the exploration programme; and (2) the costs of expanding the pipeline transportation network. According to estimates provided by Hewett (1984), the capital investments required per net increment to oil output (a simple ratio of capital expenditures to net production increments) increased by 98 per cent in the 1970s. The costly new technology and more intensive use of older technologies was able to boost production to some extent (see Table 5.16) but clearly the easy years were over.

3. *Labour shortages.* As mentioned above, rapid urbanization and improvements in the Soviet education system led to acute shortages of labour throughout the USSR. This trend was exacerbated in the oil industry, where a growing number of ageing oil wells demanded frequent repairs and maintenance turnovers. Technological progress, and the use of sophisticated enhanced recovery schemes, underscored the need for an educated and dynamic work-force – a group of Soviet citizens that was understandably reluctant to migrate to West Siberia. As always, the pre-eminence of the production quota diverted valuable time and resources from the development of even basic social amenities.

According to Hewett (1984, p. 73):

> The natural reluctance to live and work in northwestern Siberia is enhanced by the fact that the plans most often and most significantly under-fulfilled in western Siberia are for the construction of housing, stores, schools, recreational facilities, and other elements of infrastructure. The social infrastructure in Siberia suffers the predictable fate of relatively low-priority projects in a high-priority campaign under conditions of labour scarcity. The result is that ... many people are living in portable dormitories and

Table 5.14: Drilling and production costs in the USSR, 1960–85

Year	Average Well Depth (metres)		Estimated Cost of Drilling One Metre (Roubles/Metre)		Costs of Production (Roubles per Metric Tonne)	Capital Investment in New Oil Industry Capacity (Roubles/Tonne of New Capacity)	
	Development	Exploration	Development	Exploration		USSR	Tyumen
1950	1148	1362	44.31	92.81	5.89	n.a.	n.a.
1955	1454	1748	50.22	122.63	4.65	n.a.	n.a.
1960	1582	1845	50.29	113.67	3.14	n.a.	n.a.
1965	1641	2195	58.08	159.27	2.91	n.a.	n.a.
1970	1728	2492	83.69	246.38	4.28	29.30	22.40
1975	1821	2718	105.01	313.72	5.70	23.80	14.80
1980	2003	2788	119.56	401.77	7.21	46.00	33.90
1985	2100	2813	132.71	490.96	13.68	62.40	46.10

Sources: *The Oil and Gas Industry of the USSR.* (various issues); Khartukov (1995); *Soviet Energy: An Insider's Account* (1991).

Table 5.15: Exploration and discovery in the USSR, 1961–75

Five-Year Plan	Number of Gas and Oil Structures Discovered	Number of Gas and Oil Structures Prepared for Drilling	Number of Gas and Oil Structures Drilled	Number of New Fields Identified	Number of Exploratory Oil Wells Completed	Exploratory Wells Completed in West Siberia	Exploratory Drilling in West Siberia (Thousand Metres)
1961–65	2316	1672	1585	n.a.	7117	541	1074
1966–70	2035	1971	1721	453	6074	932	2087
1971–75	2047	1861	1778	388	5973	938	2645

Source: Gustafson (1989).

Table 5.16: Production, productivity and idle wells, 1960–76

Year	Crude Oil Production ('000 mt)	Production Per Active Well (mt)	# of Active Wells	Active Wells as a % of Total Wells	Total Wells	Idle Wells	Coefficient of Utilization of Active Oil Wells
1960	147,859	4368	33,900	n.a.	n.a.	n.a.	n.a.
1965	242,888	5637	43,091	90.4	47,652	4561	0.949
1970	353,039	6581	53,643	94.8	56,568	2925	0.950
1975	490,799	7398	66,340	95.9	69,191	2851	0.954
1976	519,678	7502	69,274	95.8	72,311	3037	0.954

Source: The Oil and Gas Industry of the USSR. (1977).

small huts because apartments are not available, and they cannot avail themselves of even the simplest amenities.

By the mid-1970s, the labour shortage in West Siberia had reached chronic proportions. A unique Gosplan incentive programme offered bonuses of up to 70 per cent of the normal wage simply to attract a permanent work-force to Siberia. The attempt failed as opportunistic workers seized the windfall profits, only to return to 'civilization' in less that two years (Gustafson, 1989). Labour turnover rates of 30 per cent were common and in some locations reached 100 per cent per year. Hence, there was little result from the considerable expense associated with relocating a worker to West Siberia. When the labour shortage threatened the fulfilment of the crude oil production quota, temporary and expensive solutions were devised to salvage the Plan. Prominent among these were: (1) The *Vakhtovyi Metod* whereby workers were housed in permanent West Siberian base camps, and assigned temporary duties in remote field locations; and (2) The *Ekspeditsionnyi Metod* in which workers were flown in from the established oil regions of the USSR to complete a two-week rotation in West Siberia.

Despite difficulties associated with the efficiency, and attitude, of a temporary, and exhausted work-force, the *Metod's* provided Gosplan with a satisfactory solution. Gustafson (1989, p. 174–75) states that:

> By the end of 1983 about 200,000 workers in Tiumen' were employed under one or the other system. This has been the crucial margin of difference in preventing manpower shortages from becoming a bottleneck in both the oil and gas campaigns, and it will remain so for the indefinite future.

To the detriment of the Soviet planning system, these temporary solutions – a complex repertoire of ad hoc targets, economic bandages, and inadequate incentive policies – would be applied to a growing number of economic ailments throughout the 1970s, and beyond (see Appendix D).

With little in the way of increased financial support, and in response to mounting pressure from the central planners, the Western Siberian production associations focused their efforts, and investment, in the prolific Middle Ob' region of the Tyumen oblast. The primary directive of 'increased production' was paramount so that the bulk of these investments were further concentrated on the 'proving up' and development of previously discovered oil reserves. In the eighth five-year plan (1966–70) a disproportionate 72 per cent of all exploratory drilling in West Siberia fell under the category of *razvedochnoe burenie*, exploratory or 'outlining' wells drilled for the purpose of proving up commercial oil reserves for immediate development (Gustafson, 1989).

The deficiency in the exploration programme was, eventually, reflected in the volume of reserve additions. The number of oil and gas discoveries (new fields identified) fell significantly in the ninth five-year planning period from 453 in 1966–70 to only 388 in 1971–75 (see Table 5.15). These meagre 'new' discoveries contained a disappointing volume of industrial oil reserves. According to Gustafson (1989, p. 81):

> The average field discovered in Tiumen' between 1971 and 1975 held 40 per cent as much prospective reserves (that is, $A+B+C_1+C_2$) as the average Tiumen' field discovered before 1966, but only 27 per cent as much industrial reserves (i.e. $A+B+C_1$). An inventory conducted in 1980 showed that the [oil] fields discovered in 1971–1975 represented only 20 per cent of total industrial reserves in West Siberia, while fields discovered before 1966 still held over 58 per cent.

In response to numerous and clear warnings from respected Siberian geologists a few tentative measures were taken to improve the exploration effort. In the ninth five-year plan, the share of *razvedochnoe* drilling was cautiously reduced, and the share of *poiskovoe burenie* (prospecting wells) was simultaneously increased to 42 per cent of the total for West Siberia. The improved exploration targets were, however, fulfilled with only a minimum of effort and by taking few risks. By 1975, 75 per cent of deep well exploratory drilling remained concentrated in the Middle Ob' producing area (Gustafson, 1989).

In the final analysis, the early development of the West Siberian basin would be completely dominated by a small number of very large oil fields. Among these, the super-giant Samotlor was the most important. Production from this one field alone exceeded the preliminary production quota for the entire Province of Tyumen (Table 5.12). By 1975, the crude oil flows from Samotlor had reached 87 million tonnes per annum – 56 per cent of West Siberian oil production, and well over 17 per cent of the crude oil flows for the USSR. When evaluated solely on the basis of contemporary crude oil flows, Gosplan's rapid development programme had been an astounding success. Despite severe shortages of labour, and rising production costs, according to Riva (1994, pp. 16–17): 'the rapid exploitation of the Volga-

Urals and the West Siberian oil provinces (and specifically of the super-giant *Romashkino* and *Samotlor* fields) propelled the USSR into the position of the world's largest oil-producing country in 1974'.

Inspired by the new 'international' distinction, the central planners began the tenth five-year planning period with considerable enthusiasm for the energy industry. In short, while technical experts had, in fact, paid a good deal more attention to Soviet energy policy, the discussions appear to have been limited to the abundance of resources, and the OPEC oil crisis which had upset the Western world, and which had simultaneously benefited Soviet hard currency reserves, in 1973. In October 1974, Brezhznev addressed the nation with pride, and with a prophecy of limitless accomplishments in the energy industry (Goldman, 1980).

Less than a month later, in November 1974, energy was the main topic at the annual meeting of the USSR Academy of Sciences. Gustafson (1989, pp. 23–24) reports:

> the source of the speakers' heightened awareness of energy appears to have been the oil crisis that had struck the West the year before. Their words carry no sense of a Soviet problem but rather an air of unhurried positioning for the future. If experts in Moscow did not yet perceive a crisis, political leaders and planners saw even less reason to worry. . . . At the Twenty-fifth Party Congress in February 1976, Brezhnev gave little more time to energy than he had in earlier speeches.

The tenth five-year plan called for, and hinged on, the following increments to primary energy supplies: (1) a 30 per cent increase in crude oil flows to 640 tonnes per annum in the year 1980; (2) a 12.7–15.5 per cent increase in coal production; and (3) an 8.2 per cent annual increase in gas supplies to 435 bcm per annum by 1980 (Hewett, 1984). This expansion, however, was to be accomplished with only a minimal increase in energy investments. Soviet investments in energy supplies totalled only 11.4 billion roubles in 1976, and 12.0 billion in 1977 – a constant 28.1 per cent of total Soviet investments in industry (Gustafson, 1989).

Apparently, the second OPEC energy crisis was not anticipated by the planners. By 1977, the sudden and unplanned shortage of primary energy supplies had threatened the fulfilment of critical industrial production quotas, creating bottlenecks and shortfalls in virtually every sector of the Soviet economy (see Table 5.1). The changes in the international market were, by coincidence, matched by declines in domestic coal and oil production. To summarize the complex sequence of events in the coal industry, the Donets Basin – the major European coal basin in the USSR –

reached its peak production level in 1976, only to collapse in a prolonged episode of accelerated decline. Twelve months later, in June 1977, Soviet coal production had risen to only 1.978 million tonnes per day – an annual increase of less than 1.5 per cent, and 30 000 tonnes below the 2.008 million tonne production quota. According to Hewett (1984, p. 85): 'In subsequent years the plan was revised downward, yet output was lower than planned, leading to yet another revision for the next year's plan. Obviously, planners were badly mistaken about the capabilities of this sector'.

The news from West Siberia was equally grim. From 1967 to 1977 – and for the first time ever – Western Siberian geologists had failed to meet the annual quotas for industrial reserve additions. In the super-giant Samotlor oil field a sharp reduction in the flow rates of new wells was reported. Numerous forecasts concerning the poor future prospects of the Siberian oil industry were, in the words of Gustafson (1989, p. 90) validated by 'abrupt increases in the West Siberian depletion ratio (that is, the percentage of new output required to compensate for declines in older fields) and, above all, sometime in 1977 the beginning of a decline in the growth rate of West Siberian output'. Deprived of the usual abundance of new capacity from the West Siberian producing region, Soviet crude oil flows reached only 10.92 MMb/d (543.5 million tonnes per annum) in June 1977, 80 000 b/d below the 11.00 MMb/d production quota (547.5 million tonnes per annum) (Hewett, 1984).

To add insult to injury, in April 1977 the US Central Intelligence Agency (CIA) released a series of reports predicting that the Soviet oil industry would reach a peak flow rate of approximately 12 MMb/d in the early 1980s, only to enter a sharp, and sustained period of decline. The reports cited two major deficiencies in the Soviet oil programme: (1) the excessive use of water flooding in the major Soviet oil fields and Samotlor in particular (Goldman 1980); and (2) the neglect of the exploration programme meaning that 'large new oil fields could not be found and developed rapidly enough to provide sufficient reserves to sustain production' (Riva, 1994, p. 27) In retrospect, the forecast was particularly astute. Although Soviet crude oil flows were developed at a slightly slower pace than envisioned by the CIA, reaching a peak production rate of 12.4 MMb/d in 1983, the super-giant Samotlor did, in fact, enter a period of decline after 1980 (Gustafson, 1989).

The reaction of the Politburo was swift and decisive. At the December 1977 plenum of the Central Committee, Brezhnev delivered an inspirational speech, stressing the vital importance of the Tyumen producing region, and proposing a crash programme to save the West Siberian five-year oil output target (Gustafson, 1989) Discounting some minor resistance from Prime Minister Kosygin, and Nikolai K. Baibakov, the Chairman of Gosplan, the energy industry, and specifically West Siberia, became an absolute priority for the government. In the spring of 1978, Brezhnev paid a personal visit to Siberia, in an attempt to inspire geologists and oil men to put their full effort towards the fulfilment of the annual production quota. In December of the same year, an 'enlarged session' of Gosplan officials was convened to review the practical issues of speeding up energy development in Siberia (Gustafson, 1989).

The 'new' energy policy was supported by a significant injection of capital funds. Soviet investment in the energy sector reached 88.6 billion roubles in the years 1976–80, a 45 per cent increase over the 61.0 billion roubles reported in 1971–75. The energy industry 'suddenly' accounted for over 40 per cent of the Soviet Union's total investments in industry – a 4 per cent increase over the 'immutable' 35–36 per cent that had been earmarked for energy in the eighth and ninth five-year planning periods (see Table 5.2). The spending was accelerated in the oil industry, where capital investment (including the costs of expanding the pipeline transportation network) reached 32.2 billion roubles in the tenth five-year plan, 14.68 per cent of total Soviet investments in industry, and a significant 57 per cent increase over the 20.5 billion roubles reported for 1971–75 (see Table 5.10). The increments to annual investment statistics are noteworthy. According to official estimates, annual investment in the oil industry reached 8.7 billion (pre-1982) roubles in 1982, a 111 per cent increase over the 4.12 billion (pre-1982) roubles reported in 1976 (Gustafson, 1989).

With new resources made available and above all Leonid Brezhnev's personal endorsement, the Oil Ministry wasted no time in the development and implementation of an accelerated crude oil production programme. Development drilling reached 65 million metres in the years 1976–80, a 72 per cent increase over the 37.8 million metres reported in the ninth five-year planning period (see Table 5.13). The bulk of the incremental effort was concentrated in West Siberia, where the pace of development drilling tripled in only three years. According to figures provided by the Oil and Gas Ministries, the level of development drilling in Siberia reached 12.9 million metres in 1980: a 239 per cent increase over the 3.8 million metres completed in 1977. According to Gustafson (1989, p. 90): 'Whereas in 1977 less than 25 percent of the oil industry's drilling effort had gone into West Siberia, by 1982 West Siberia's share was well over half'.

The programme was augmented by a massive transfer of labour, and physical resources to the remote northern territories of Siberia. To cite only one example: the number of permanent oil drilling brigades assigned to Glavtiumenneftegaz, the MNP's agency in the Province of Tyumen, grew by 97 per cent between 1977–81, from the 70 brigades which had been attached to the agency at year-end 1976 to 138 brigades at year-end 1981 (Gustafson, 1989).

Despite these 'permanent' additions to the Siberian labour force, development drilling took precedence over the construction of basic social amenities so that the region continued to suffer from a high rate of labour turnover. In 1977, Gosplan was forced to adopt the expensive *Ekspeditsionnyi Metod*; that is flying in temporary drilling teams from Bashiriia and Tatariia in the Volga-Urals to supplement deficiencies in the West Siberian work-force. Gustafson (1989, p. 92) reports: 'All told, drilling teams on loan from outside West Siberia contributed 10 million meters of hole (mainly development) by the end of 1981, roughly one-fourth of the total in West Siberia'.

To the relief of Gosplan, Brezhnev's emergency cash rescue programme would be sufficient to stabilize production throughout the tenth five-year planning period. Soviet crude oil flows reached 603.2 million tonnes in 1980, a mere 36.8 million tonnes below the aggressive 640 million tonne production target (see Table 5.4). Not surprisingly, the West Siberia producing region was responsible for the bulk of the increment. Crude oil flows from West Siberia reached 322.46 million tonnes in 1980, a 108 per cent increase over the 154.68 million tonnes reported in 1975, and a remarkable 53 per cent of total Soviet production. The region's 'stellar' performance was sufficient to satisfy even the central planners. By year-end 1980, West Siberian production lay safely within the boundaries of the tenth five-year planning quota, slightly in excess of the low end of the target set out in 1976 (Gustafson, 1989).

As anticipated by central planners and particularly given the massive diversion of labour and materials to West Siberia, most other producing regions had registered a decline. To cite only two examples: (1) in the Volga-Urals, where overstretched drilling teams battled with ageing oil fields and jet lag, crude oil flows fell by 17 per cent from the peak flow rate of 224.9 million tonnes reported in 1975 to only 185.9 million tonnes in 1980; and (2) crude oil flows from the Ukraine fell to 6.38 million tonnes in 1980 (a 45 per cent reduction) from the 11.54 million tonnes reported in 1975.

Needless to say, the diversion of scarce financial resources, labour and equipment to Siberia drew sharp criticism, and reports of dismay, from the established production associations. In the words of Hewett (1984, p. 60):

A 1981 article co-authored by the current and previous directors of Tatneft' –
the oil production association with the largest volume of oil production in the
Volga-Urals fields (specifically in the Romashkino field) – indicates that there is
reason for concern. The authors [castigated] central authorities for starving
Tatneft' of materials and labour while keeping on the pressure to maintain
output levels. . . . The older equipment in Tatneft' requires an increasing amount
of repair work, yet with Tatneft's current repair capacity and equipment
deliveries, they are unable to make up even for real depreciation.

The authors correctly predicted that the number of idle wells and shut-ins in
the Volga-Urals would grow steadily throughout the eleventh five-year
planning period.

Detailed reports from the oil fields of Tyumen betrayed a similar sense
of foreboding. The excessive focus on the production quota had, once
again, diverted valuable time and resources from the exploration effort, and
from the construction of basic infrastructure and social amenities. To
further complicate matters, the 'new' energy programme had been
implemented with little regard for costs, so that drilling and production
costs rose precipitously throughout the tenth five-year planning period.

As the MNP had, given the production-oriented incentives of the
planning system, devoted most of its resources to development drilling, the
burden of exploration effort was shifted gradually and inexorably to
Mingeo. Mingeo's share of all exploratory drilling in the USSR rose
steadily throughout the ninth and tenth five-year planning periods
(Gustafson, 1989). Yet by the end of 1980, Mingeo had completed only 0.9
million metres for exploration purposes in West Siberia.

According to estimates provided by Gustafson (1989), the total volume
of exploratory drilling in West Siberia was increased by only 32 per cent
throughout the ninth and tenth five-year planning periods, from 2.65 million
metres in 1971–75 to 3.5 million metres in 1976–80. Gustafson (1989,
pp. 95–96) asserts: 'Even within exploration, the bias was towards near
term results'. Approximately 66 per cent of the total Siberian exploration
effort, including both oil and gas exploration wells, fell under the category
of *razvedochnoe burnie*, that is the proving up of commercial oil reserves
for immediate development.

The *razvedochnoe* drilling programme was, however, further
concentrated in established oil fields so that the development of new
Siberian oil fields fell far behind schedule. The tenth five-year plan required
22 new fields to be brought on line in the Province of Tyumen. Yet, by the

end of the planning period, the production association had successfully tied in only 19 new oil fields. According to Hewett (1984, p. 53):

> Of these new fields only four were accessible by road, and only 7 were connected to central power. Consequently, much of the costly infrastructural investment necessary to precede increments to oil output still lay ahead in 1980, and since then inevitable problems have developed in meeting output plans for western Siberia.

The distortion in the Soviet exploration effort was exaggerated by the natural ageing of the Siberian basin, and increasing scarcity and remoteness of promising new locations. The net effect was a sharp reduction in the rate of growth of industrial reserve additions. With the noticeable absence of any 'super-giant' discoveries, a typical (average) new oil and gas field discovered in the years 1976–80 is estimated to have contained only 12.3 per cent as many industrial reserves ($A+B+C_1$) as the average new field discovered before 1966 (Gustafson, 1989).

A similar tale of superficial prosperity was developing on the international oil market, where the Soviet Union's progress was, to all outward appearances, unambiguously positive. As mentioned above, Soviet oil exports to both communist countries and the West had been rising steadily since the Khrushchev revolution and the USSR's tentative emergence as a participant on world oil markets in 1954. By 1972, the volume of crude oil and petroleum product exports had reached 106.51 million tonnes per annum, an 18 million tonne increase over the 88.5 million tonnes reported in 1968 (see Table 5.17). Approximately half of the total, 53.75 million tonnes, was delivered to the West.[5] Assuming the official Soviet exchange rate, the value of hard currency earnings from Soviet oil exports reached US$567 million in 1971, representing 21.6 per cent of the total hard currency earnings for the entire USSR (see Table 5.18) (Elliot, 1974). Despite the ongoing tensions of the cold war, this figure was destined to double in only one year.

The Soviet alliances in the Yom-Kippur War, and hence the subsequent Arab Oil embargo, were established well in advance of the actual armed conflict. According to Goldman (1980, pp. 88–89):

> Prior to and during the war, the Soviets increased not only the flow of arms but also economic advice. The main thrust of the latter was that if war broke out the Arabs should withhold petroleum from the non-communist Western and Japanese world. Such an embargo, the Soviets argued, would cause a 'major commotion in all the countries of the capitalist world'. The Soviets

kept up their call throughout the summer and fall. Radio Moscow on 25 September 1973, a few days before the invasion, urged the Arabs 'to use oil as a weapon against Israeli aggression', and four days later by extension to 'use oil as a political weapon against imperialism'.

These efforts would be handsomely rewarded. In the months from October 1973 to March 1974, the Arab oil embargo, and its resulting fourfold increase in world oil prices, provided a political and economic windfall for the USSR. The political benefit was self evident: as Goldman (1980, p. 89) observed: 'The longer the embargo continued, the more it would hurt the non-OPEC nations, especially the United States, which along with the Netherlands was subject to the most severe restrictions'.

Table 5.17: Soviet crude oil production and international trade statistics, 1971–82

				(million metric tonnes)	
Year	Production of Crude Oil and Condensate	Total Exports Crude Oil and Petroleum Products	Annual Growth Rate (%)	Soviet Exports of Crude Oil and Products to the West	Soviet Exports of Crude Oil and Products to the CMEA
1971	377.1	105.02	n.a.	53.75	51.27
1972	400.4	106.51	1.42	53.75	52.76
1973	429.0	118.46	11.21	55.75	62.71
1974	458.9	116.47	(1.68)	50.27	66.19
1975	491.0	128.91	10.68	57.74	71.18
1976	519.7	147.83	14.67	70.68	77.15
1977	545.8	155.79	5.39	75.16	80.63
1978	571.5	164.25	5.43	79.64	84.61
1979	585.6	163.25	(0.61)	72.67	90.59
1980	603.2	162.76	(0.30)	70.68	92.08
1981	608.8	162.26	(0.31)	70.18	92.08
1982	612.6	169.23	4.29	83.62	85.61

Sources: Goldman (1980); Stern (1987); Hewett (1984).

On the economic front, Soiuznefteexport, the foreign trade organization in charge of petroleum exports, wasted no time in the initiation of an aggressive marketing campaign expanding the Soviet Union's influence, and profits, in the global oil market. By 1973, Soviet oil exports had reached 118.46 million tonnes, an impressive 11.95 million tonnes higher than the 106.51 million tonnes reported in 1972, and the largest annual increase in the history of the USSR (Goldman, 1980). While the bulk of the increase, some 9.95 million tonnes, was exported to CMEA member nations, special 'emergency' shipments were also supplied to the US and the Netherlands. Soviet oil exports to the West reached 55.75 million tonnes in 1973, an annual increment of 2 million tonnes.

Table 5.18: Hard currency earnings from Soviet oil exports, 1971–85

| | (Nominal US$) | | |
Year	Foreign Exchange Earnings (Millions of US$)	Expressed as a % of Total Hard Currency Earnings	Annual Growth Rate (%)
1971	567	21.6	
1972	556	19.6	(1.64)
1973	1248	26.1	124.46
1974	2548	34.1	104.17
1975	3176	40.5	24.65
1976	4514	46.4	42.13
1977	5275	46.5	16.86
1978	5716	43.4	8.36
1979	8932	45.7	56.26
1980	10803	46.7	20.95
1981	11222	58.3	3.88
1982	13747	61.4	22.50
1983	14085	64.2	2.46
1984	13669	61.6	(2.95)
1985	10867	53.8	(20.50)

Sources: Stern (1987); Ericson and Millar (1979).

The emergency shipments, essentially a permanent increase in the Soviet Union's share of western oil markets, were supplemented by OPEC's ongoing commitment to rising oil prices. According to figures provided by Hewett (1984), the average price paid by free world (non-CMEA) countries for Soviet crude oil and product exports reached US$11.45 per barrel in 1975 – a 373 per cent increase over the US$2.42 per barrel reported in 1972 (see Table 5.19). By year-end 1975, the value of hard currency earnings from Soviet oil exports had reached US$3176 million or 40.5 per cent of the total hard currency earnings for the USSR.

Table 5.19: Average prices paid for Soviet oil exports

		(US$/Bbl)	
Year	Average Price Paid by CMEA Member Countries	Average Price Paid by Free World Countries	Spot Price Middle East Light Crude 34 Degree
1971	2.28	2.42	1.69
1972	2.73	2.42	1.82
1973	2.96	4.58	2.81
1974	3.29	11.23	10.98
1975	6.40	11.45	10.43
1976	6.68	12.31	11.63
1977	8.61	13.88	12.57
1978	11.15	13.41	12.91
1979	13.01	25.52	29.19
1980	14.59	34.76	35.85
1981	17.20	35.81	34.29
1982	n.a.	33.00	31.76

Source: Hewett (1984).

Three years later, in December 1978, the Iranian Revolution would set the stage for yet another significant increase in the level of world oil prices. Cries for the immediate abdication of Mohammed, the Shah of Iran, were accompanied by severe oil workers' strikes and the political execution of Paul Grimm, the expatriate assistant manager of operations for the Iranian

Oil Service Company. By the end of 1978, Iranian petroleum exports had ceased altogether (Yergin, 1991). Less than a year later, without prior warning, a squadron of Iraqi troops attacked a dozen cities and key installations in Iran. As suggested by Yergin (1991, p. 771) Iran's oil infrastructure was a major target of Iraqi air operations:

> On September 23, 1980, the second day of the [Iran/Iraq] war, Iraqi war planes began a sustained assault against the Iranian refinery at Abadan, the world's largest, and proceeded over the next month to damage it severely. They also carried their attack to every Iranian oil port and oil city. The Iranians counterattacked against Iraqi oil facilities, completely choking off Iraqi oil exports through the Gulf.

By year-end 1980, the spot price of $34°$ Middle East light crude had reached US$35.85 per barrel, a 178 per cent increase over the US$12.91 per barrel reported in 1978 (see Table 5.19).

The value of hard currency earnings from Soviet oil exports reached US$13.74 billion in 1982, four times more than the $3.176 billion reported in 1975, and over 60 per cent of the total hard currency earnings for the USSR. Ironically, these gains were realized despite a significant reduction in the level of Soviet oil exports. In short, the domestic consumption of Soviet oil supplies rose rapidly in the years 1977 to 1981, placing severe restrictions on the quantity of crude oil available for export. By the early 1980s, Gosplan was in no position to sponsor an aggressive Soiuznefteexport marketing campaign. Instead, Soviet oil exports were reduced gradually from a peak of 164.25 million tonnes in 1978 to only 162.26 million tonnes in 1981. This was, of course, a forced retrenchment at the worst possible opportunity.

The deficiency has been credited as the main inspiration for yet another significant change in Soviet energy policy. The targets for the eleventh five-year plan were announced at the Twenty-Sixth Party Congress in November 1981. Its guidelines may be summarized as follows: (1) a 45–47 per cent increase in the production of natural gas from 435 billion cubic metres in 1980 to 630–640 billion cubic metres in 1985;[6] (2) a 2.8–6.1 per cent increase in the production of crude oil from 603 million tonnes in 1980 to 620–640 million tonnes in 1985; and (3) a 7.5–11.7 per cent increase in the production of coal from 716 million tonnes in 1980 to 770–800 million tonnes in 1985. The subtle change in the growth rates of the oil and gas targets underscored a distinct preference for gas in the Soviet fuel balance. By 1985, the share of oil in the Soviet fuel balance would decline from 44 to 39 per cent; the decline was to be made up with natural gas (Gustafson, 1989).

The priority given to the energy industry was unmistakable. According to Gosplan directives, the Soviet energy sector was scheduled to absorb an additional 44 billion roubles of investment funds in the years 1981–85, an intimidating 66 per cent of all new Soviet investment in the eleventh five-year plan. In the words of Gustafson (1989, p. 36):

> The share of energy in the planned increment of industrial investments came to a whopping 85.6 percent! One can imagine the long faces in many an industrial ministry in Moscow as the implications of the new energy investment targets for the Eleventh Plan sank home.

In recognition of the difficulties facing the West Siberian production associations, and rising crude oil production costs, the largest share of the Soviet Union's aggressive energy investment, approximately 43 billion (1973) roubles, was earmarked for the oil industry (see Table 5.3).

Unfortunately, the windfall oil profits of the 1970s would never be realized by the domestic oil industry. While Soiuznefteexport had collected a handsome profit from hard currency oil exports, domestic crude oil and product prices, and those charged to the CMEA member nations, were maintained at 1971 levels throughout the Arab embargo and beyond (see Appendix D). To aggravate matters further, the Soviet export market was fully segregated, so that the Ministry of Petroleum was not permitted to reflect (or claim) the hard currency or CMEA revenues accruing from its own production effort (Goldman, 1980). Gross profits fell steadily throughout the decade of the 1970s – from 6.24 roubles per tonne in 1970 to a mere 2.55 roubles per tonne in 1981.

As mentioned previously, at least 90 per cent of all trade with CMEA member nations was conducted with soft currency (TRs), and governed by the strict terms of complex CMEA barter agreements. In short, the prices for all goods to be included in the CMEA agreements were set, at five-year intervals, on the basis of world market prices that prevailed in some previous five-year planning period. Specifically, the oil prices utilized for CMEA trade in the years 1971–75, were based on the world oil prices that had been realized in the years 1966–70. This rigid CMEA pricing policy would prove to be unduly cumbersome – and expensive – in periods of rising oil prices. The rigid pricing policy was put in place to make managing the planning process easier but in the process led, at times, to forgone opportunities or disgruntlement for buyers when prices were higher than those at which similar goods could be obtained in the international market (Henderson and Kerr, 1984–85). There was a fundamental conflict between the planners' need for stability in economic relationships for 'planning' actually to take place and the inherent volatility of market prices that was never solved when the two were forced to interface.

By 1974, the average oil price charged to CMEA member countries had reached only US$3.29 per barrel – a mere 30 per cent of actual world oil prices (approximately US$11.00 per barrel). By the early 1970s, allegations that the Soviet Union was charging a higher price to captive satellite nations than to 'free world' Western European consumers were simply nonsensical. The multi-dimensional puzzle of 'discriminatory' Soviet pricing practices had been resolved to the benefit of the CMEA member nations. According to Hewett (1984, p. 162):

> In effect the Soviet Union is giving Eastern Europe what Michael Marese and Jan Vanous have called an 'implicit subsidy'. Energy and materials are sold to Eastern Europe at below world market prices in exchange for machinery and equipment that is purchased for higher than world market prices . . . To eliminate the subsidy, the Soviets would have to ask that energy be purchased only for dollars, which would be tantamount to dismantling the current procedures governing intra-CMEA trade. . .

At the same time, the magnitude of the 'implicit subsidy' to CMEA member countries was rapidly becoming unacceptable. One year later, in 1975, Gosplan switched to a 'moving average' base price. Intra-CMEA oil prices were to be negotiated annually on the basis of world oil prices prevailing in the preceding five years. As was the case with most of the economic reforms undertaken in the Brezhnev era, the revision would not be sufficient to ease the gross distortion in CMEA subsidies. This was complicated further because the TR exchange rate was also distorted. Intra-CMEA crude oil and product prices lagged sadly behind world oil prices throughout the decade of the 1970s. By 1980, the implicit East European subsidy had reached levels as high as US$17.8 billion – literally absorbing the entire value of hard currency earnings from Soviet oil exports (Hewett, 1984).

The crude oil transportation programme was, by most accounts, remarkably successful. In 1970, over 50 per cent of the total volume of crude oil and petroleum products was still being transported by the archaic channels of river, sea and railways. In recognition of the economic advantages of pipeline transportation, Gosplan had organized a comprehensive pipeline construction programme, so that by 1980 pipelines were to have 'taken over the transportation of oil and oil products almost completely' (Elliot, 1974, p. 110). The programme, which envisioned the construction of over 22 000 km of new trunk line in the ninth five-year planning period, was disrupted by delays (primarily blamed on the shortage of steel supplies), so that only a disappointing 7700 km had been completed at the end of 1972.

Still, Gosplan was determined to reorganize the archaic oil transportation network. A Ministry of Construction was created to accelerate the pace of pipeline construction in Siberia. The pipeline construction programme was enhanced significantly, and from 1972 to 1980, 35 long-distance oil pipelines had been completed, and brought into service. These included: (1) the 1220 mm, 818 km Aleksandrovskoe-Anzhero-Sudzhensk pipeline in 1972; (2) the 720 mm, 1334 km Tuimasy-Omsk II pipeline in 1972; (3) the 1220 mm, 1813 km Ust-Balyk -Kurgan-Ufa-Almetevsk pipeline in 1973; (4) the 1220 mm, 2245 km Nizhnevartovsk-Kurgan-Kuybyshev pipeline in 1976; and (5) the 1220 mm, 1268 km Surgut-Perm pipeline in 1978 (see Appendix C). According to a joint Organisation for Economic Cooperation and Development/International Energy Agency study (OECD/IEA, 1995, p. 125) the pipelines from east Samotlor to Anzehero-Sudzhensk, Samotlor to Almetyevsk, and Nizhnevartovsk to Samara 'increased the transmission capacity between West Siberia and the European portion of the former USSR by about 175 mt per year'.

In summary, an incredible 18 375 km of new long-distance oil pipelines were completed, and made operational, in the ninth and tenth five-year planning periods. By 1980, 91 per cent of all crude oil production was transported by pipeline.

Despite tremendous progress in the sheer volume of crude oil production, and the pipeline transportation sector, Brezhnev's new energy policy would not be sufficient to turn the tide of events for the ill-fated Soviet oil industry. The policy, which was reminiscent, in some respects, of the 1965 and 1973 reforms had, once again, failed to address major deficiencies in the planning mechanism, pricing and incentive structure. These fundamental planning problems, which had, in fact, formed the foundations for the first Soviet energy crisis (1977), would continue to undermine the long-term prospects for the crude oil industry, setting the stage for a second oil crisis in the decade of the 1980s. Brezhnev, however, would not live to see the long-run complications that arose from his 'cosmetic surgery'.

NOTES

1. As reported in Ebel (1970, p. 86).
2. An oil field receives the title (classification) of 'giant' if it contains reserves in the range of 700 million to 3500 million barrels. Similarly a 'large' oil field must contain reserves in the range of 35 million to 699 million barrels (Hewett, 1984).
3. The TR was essentially a 'clearing' currency, designed specifically for the bilateral trade agreements. According to Hewett (1984, pp. 160): 'Oil sold to Czechoslovakia, for example, generates earnings in TR's which are in fact not transferable: they are good only as credits against the purchase of a limited range of Czechoslovakian goods'.
4. A milliard is equal to 1000 millions.
5. The 'West' has been broadly defined to include all USSR trading partners that were not members of the CMEA. It is important to note the fact that Finland was not a CMEA member country, and, as a result, has been included in this category despite the fact that it conducted the majority of its oil trade with the Soviet Union on a barter basis.
6. The gas targets included the construction of six major gas trunk lines between West Siberia and the European USSR.

6. Desperate Measures

6.1 CORRUPTION TO PERESTROIKA: THE LONG ROAD TO REFORM

> On November 10, 1982, Leonid Brezhnev, in bad health for years and increasingly enfeebled, died of a heart attack. In what was considered by many in the Soviet Union as a triumph for the system, it took only fifty-four hours for Yuri Andropov to emerge as Brezhnev's successor. But the smooth succession from Brezhnev to Andropov did nothing to solve a far bigger succession crisis – the transfer of power from one generation to another. The generation that had ruled the Soviet Union since Stalin's death had become a gerontocracy as well as an oligarchy, a development that owed much to Brezhnev's stress on stablilty. . . The problem with the Brezhnev era was that stability had turned into stagnation. Unlike Stalin, who threatened everyone with prison or death, or Khrushchev, who rocked the boat with his egalitarianism, appeals to popular sentiment, and utopian or unworkable schemes, Brezhnev guaranteed the elite's status, privileges, and life style. Consequently, the Soviet Union was rendered impervious to reform (Kort, 1993, p. 275).

Yuri Andropov, a shrewd political administrator, had given careful consideration to his own personal advancement. A ruthless proponent of Brezhnev throughout the post-Khrushchev succession wars, Andropov was well rewarded for his loyalty. He was appointed to the head of the KGB in 1967, and granted elite status as a full member of the Politburo less than six years later, in 1973. Witness to the darkest secrets of the politburo, Andropov's alliances changed gradually, and inexorably in the last years of his tenure, and in Brezhnev's declining years, he seized every possible opportunity to discredit the ruling elite that comprised Brezhnev's inner circle.

Corruption, which had grown rapidly throughout the Brezhnev years, had reached chronic proportions in the early 1980s. While it was extremely costly for the economy, it could be used as a lever against those 'on the take'. Given that corruption was endemic, it meant that few were not vulnerable. Brezhnev's own daughter was implicated in schemes involving diamond smuggling, bribery, and currency speculation (Kort, 1993). According to Moynahan (1994, pp. 221–2):

> To succeed in anything one had to be sly, to lie, to violate rules and laws. What was the right way to live? To answer the question, one at least had to know the true state of affairs. And no one knew what it was like – both those who had been lied to, and those who had lied.

As head of the KGB secret police, Yuri Andropov had access to facts and figures that would have astounded most members of the Politburo. Incredibly, Brezhnev kept his own Politburo in the dark so it could not blame him for the building economic disaster (Coleman, 1997). Embarrassing stories were 'leaked', with increasing frequency, by a KGB intent on stopping corrupt government officials at the highest level of the Soviet hierarchy.

At the same time, powerful allies in the military and critical Soviet interest groups quietly promoted Andropov's selection as General Secretary, a distinction that Brezhnev had, in fact, (privately) reserved for Konstantin Chernenko. At the earliest opportunity, the death of Suslov in 1982, Andropov resigned from the KGB and assumed the powerful position of Party ideologist and kingmaker. Months later, at an extraordinary session of the Central Committee in November 1982, he was elected General Secretary of the Soviet Union.

His mandate – to stem the tide of corruption that was eroding the very foundation of Soviet society – was absolute. As pointed out by Coleman (1997, pp. 138–9) after years of faithful reporting to Brezhnev, Yuri Andropov:

> knew the bitter economic truth. Top-secret KGB analyses convinced him that Brezhnev had left the economy in ruins – and his heirs no choice other than to accept the political risks of change. Equally important, Andropov had what it took to push reform through. According to Soviet sources who worked closely with him in the Party and the KGB, including some who often disagreed with him, Andropov was brave, bright, talented, and ruthless.

No expense was spared in a massive campaign to promote a new era of honesty and efficiency throughout the USSR.

In January 1983, Party activists in red armbands searched cafes, baths, department stores and cinemas for people who should have been at work; a raid on a Moscow bathhouse netted two generals.   The crackdown infuriated women, who had no alternative but to shop in working hours because of the lengthening queues. Trying to win back favour, Andropov slashed the soaring price of vodka. Thousands, however, were found guilty of illegal economic activities, and given harsh sentences, including capital punishment. The latter was a clear sign of desperation as Andropov's previous record as head of the KGB showed a clear move away from the more violent tactics of intimidation (Coleman, 1997).

Far less consideration was given to genuine economic reform, which often appeared to be almost an afterthought. Indeed, while a wide variety of reform packages had been debated at great length throughout the Brezhnev era, stability and the preservation of the status quo prevailed. Only a few, in the end, were deemed worthy of implementation. The two noticeable exceptions to this general rule included cosmetic, Kosygin style, economic reforms, and a new long term energy policy (Kort, 1993). Inspired by the rapid increase in domestic oil consumption, and a forced retrenchment in oil exports between 1977–81, Brezhnev had devoted considerable time and attention to the promotion of energy conservation. Energy savings targets were introduced into the planning process in 1976, and in 1982 Gosplan organized an interagency programme to determine exactly how much energy (fuel) a typical Soviet enterprise should consume. Energy efficiency norms were established for each of the major production processes – and virtually every machine and gadget using energy – and incorporated into the five-year planning process (see Appendix D).

In 1979 A. P. Aleksandrov, the President of the Academy of Sciences, was commissioned to lead a panel of Soviet energy experts in the formation of a new long-term energy programme. The draft proposal, which was completed in 1982, provided an excellent opportunity for instant, pre-fabricated reform. Gustafson (1989, p. 42) reports that: 'Andropov seized upon the draft energy program as one of his first policy initiatives.' In his opening speech as General Secretary, in November 1982, he hinted that he planned to put conservation and fuel switching ahead of energy supply. Less than five months later, in April 1983, the draft energy policy received an official stamp of approval from the Politburo.

In essence, the Aleksandrov Commission envisioned a gradual transition towards 'conservation', and the creation of a balanced, and efficient, Soviet energy policy. The draft programme, authenticated by a multitude of definitive, and highly technical, energy studies, outlined the following, two-phase blueprint for conservation:

1. *Phase I: Preparation (1983–1990)* – In Phase I the Soviet economy was to be thoroughly prepared for the implementation of a comprehensive domestic conservation programme. The new policies included the stabilization of oil production at contemporary levels, significant increases in the production of natural gas and nuclear power, improvements in the measurement of domestic consumption, the creation of energy-saving incentive programmes (including price increases and accountability), and rapid fuel switching with an emphasis on the replacement of oil by gas in industrial power plants (see Appendix D); and

2. *Phase II: Active Implementation (1990–2000)* – In Phase II energy conservation was to be achieved through the modernization and electrification of technical processes and industry. To cite only a few examples, Phase II included policies for accelerating the construction of cogeneration plants, the relocation of energy-intensive heavy industries and a reorganization of the complex energy transportation system (Gustafson, 1989).

The Aleksandrov Commission called for a significant increase in energy investment to approximately 20–22 per cent of total Soviet investment for 1985–90. In Phase I, the bulk of these incremental investment funds were to be devoted to development of energy supplies – specifically oil, natural gas, and nuclear power. In Phase II, the investment funds were to be shifted gradually to the demand side (essentially a crude version of demand side management). According to Aleksandrov directives, the abundance of nuclear power and natural gas supplies developed in Phase I would be sufficient to fuel the Soviet economy until such time as the beneficial results of Phase II, the actual conservation stage, could be fully realized. It is interesting to note that the Commission was sharply divided on the question of oil production. According to Gustafson (1989, p. 243): 'The published supporting papers dwelled on the high cost and uncertainty of oil and appeared to hint that there could be worse choices than allowing its output to fall. Aleksandrov himself, however, called publicly for further increases in oil output. Other senior experts appeared similarly divided.'

The Aleksandrov proposal was submitted to the Party Central Committee in June 1983. Shortly thereafter, the Politburo issued a number of official announcements emphasizing the benefits of conservation and fuel switching, so that the main elements of the new energy policy, that is Phase I, could be implemented immediately. A long-anticipated programme to convert power plants to gas was approved instantaneously, and the

November 1983 plenum of the Central Committee adopted a number of new policies to aid in the conservation of petroleum products throughout the USSR.

Andropov's absolute commitment to the gradual reorientation of Soviet energy policy to conservation and the efficient utilization of alternate energy sources, specifically natural gas and nuclear power, was underscored by a noticeable and planned reduction in the level of oil investment. According to information provided by Gustafson (1989), approximately 9.1 billion roubles were invested in the oil industry in 1983, an average annual increase of only 4.6 per cent over the 8.7 billion roubles invested in 1982 (see Table 6.1). One year later, in 1984, capital investment in the oil industry was reduced to 8.9 billion roubles. These figures represented a dramatic deceleration in the annual growth rates of Soviet oil investment, which averaged over 11 per cent throughout the tenth five-year plan, and a significant 13.2 per cent in the first two years of the eleventh five-year planning period.

Table 6.1: Capital investment in the Soviet oil industry, 1976–85

Five-Year Plan	Soviet Investment in Industry (Billions of pre-1982 Roubles)	Soviet Investment in Energy (Billions of pre-1982 Roubles)	Energy's Share of Soviet Investment in Industry	Soviet Oil Investment (Billions of pre-1982 Roubles)
1976–80	221.4	65.7	29.7	26.4
1981	49.5	16.8	33.9	8.1
1982	50.9	17.7	34.8	8.7
1983	53.7	18.7	34.8	9.1
1984[a]	61.9	22.2	35.9	8.9
1985[a]	65.5	25.3	38.6	10.4
1981–85	n.a.	n.a.		45.2

Note: [a] reported in post-1982 prices.

Source: Gustafson (1989).

As foreshadowed by grim reports from the production associations, Andropov's policy of fiscal austerity for the oil industry had come at the worst possible moment for the energy industry which was, in fact, entering into a second, and considerably more severe, period of crisis. The nuclear power programme, a critical component of Phase I of the Aleksandrov proposal, was besieged by severe shortages of critical construction materials, and cost overruns. Production was falling far behind schedule, and the industry required emergency infusions of capital (Ebel, 1994).

The coal industry suffered from rising production costs and an unplanned acceleration in the rate of decline from ageing coal fields. Production fell far below Gosplan's most conservative expectations, reaching only 726 million tonnes in 1985, 49 million tonnes below the revised 775 million tonne production target. As mentioned in Chapter 5, the shortfall was exacerbated by a dramatic deterioration in the quality of Soviet coal supplies. To cite only two examples: (1) by 1983 the heat content of the 'leading category of Donbas steam coal' had reached levels as low as 4947 gigacalories, a 20 per cent reduction from the 6180 reported in 1965 (Gustafson, 1989); and (2) the 'easily recoverable' reserves of the Kansk-Achinsk field, a promising new discovery in southwestern region of Siberia' had a very low caloric content, only 3300 kilocalories per kilogram (kg), a very high moisture content, and a tendency to self-ignite (Hewett, 1984).

The deterioration in the quality of Soviet coal supplies caused a variety of problems for power plants. For example, in the early 1980s coal was far more expensive to transport than either oil or natural gas. Efforts to reduce the costs of coal transportation were frustrated by the lack of innovation, and problems associated with the incorporation of new technology in the rigid, and cumbersome, central planning process. At the same time, the poor quality of coal supplies damaged boilers, and produced only faulty and sporadic combustion performance, leading to frequent power outages. By the mid-1980s Minenergo was forced to re-open ageing and inefficient oil-burning power plants, simply to maintain a steady stream of electricity to the regions most severely afflicted by brownouts (Gustafson, 1989).

While tremendous progress had been made in the gas industry, efforts to launch Phase I of the Aleksandrov proposal, the switch to gas, were frustrated by lack of critical infrastructure and storage facilities. Large investments (37 billion roubles) were made in the gas industry in the eleventh five-year plan (see Tables 6.2 and 6.3). As mentioned in Chapter 5, the bulk of these funds was devoted to the construction of six major gas pipelines between West Siberia and the European USSR. Despite numerous

Table 6.2: Energy investment in the ninth, tenth, and eleventh five-year plans (billions of pre-1982 roubles), 1971–85

	Ninth Five-Year Plan 1971–75	Tenth Five-Year Plan 1976–80	Eleventh Five-Year Plan 1981–85	
	Actual	Actual	Planned	Actual
Oil	16.0	26.4	43.0	45.2
Coal	7.3	9.8	12.0	11.6
Gas	8.3	10.3	22.0	14.5
Electricity	17.0	19.4	23.0	24.3
Gas Pipelines	6.5	12.1	27.0	22.5
Oil Pipelines	3.0	3.5	5.0	3.0
Exploration	n.a.	3.5	n.a.	4.6
Total	n.a.	85	n.a.	125.7

Sources: Gustafson (1989); Hewett (1984).

Table 6.3: Capital investment in the Soviet energy economy, 1976–90

Five-Year Plan	(billions of roubles 1 Jan. 1984 prices)			
	1976–80	1981–85	1985–90	1991[a]
Total Soviet Investment	644.3	760.8	953.8	n.a.
Total Investment in Industry	251.5	300.7	399.1	n.a.
Energy	99.7	141.0	177.9	48.3
Oil	35.6	56.5	73.3	21.4
Gas	27.1	41.1	51.4	6.7
Oil's Share of Total Soviet Investment (%)	5.5	7.4	7.7	
Oil's Share of Soviet Investment in Industry (%)	14.2	18.8	18.4	
Oil's Share of Soviet Investment in Energy (%)	35.7	40.1	41.2	
Gas' Share of Soviet Investment in Energy (%)	27.2	29.1	28.9	

Note: [a] 1991 prices
Source: Yesterday, Today and Tomorrow of the Russian Oil and Gas Industry (1995).

difficulties, including an American embargo on the delivery of turbine components for pipeline compressor stations in 1981, the Ministry of Gas, and MNGS (the Ministry for Construction of Oil and Gas Enterprises), would exceed the planners' expectations for the priority targets of production and transportation in the eleventh five-year planning period (Ebel, 1994).

By 1985, the production of natural gas supplies had reached 643 billion cubic metres (bcm), a remarkable 13 bcm above the 630 bcm production target, and all six pipelines had been competed ahead of schedule (see Table 6.4) (Gustafson, 1989). Sadly, the excessive concentration of Soviet investment funds on field development and pipelines had left little in reserve for low priority infrastructure, such as housing in West Siberia, and critical storage facilities. In short, while the eleventh five-year plan called for an 82 per cent increase in storage capacity – from 22 bcm in 1980 to 40 bcm by 1985 – the funds for construction simply were not available (Gustafson, 1989). An extreme shortage of gas storage facilities, specifically those located in close proximity to end-use consumers, frustrated Phase I efforts to displace *mazut* fuel oil in power plants, placing an additional 'unplanned' burden on the Soviet oil industry.

Reports from the Tyumen oil fields should have set off the alarm. In 1982, and for the first time ever, the Province of Tyumen failed to meet its annual production quota. The planners adjusted (corrected) the target downward, and the matter was not made public (Gustafson, 1989). Still, discrepancies (shortfalls) grew steadily throughout the eleventh five-year planning period, causing frequent revisions in the annual production targets. To the astonishment of central planners, the Province of Tyumen reached a peak production rate of only 366.2 million tonnes in 1983–84, only to enter a disturbing period of 'temporary' decline. By 1985, Tyumen crude oil flows had fallen to 352.7 million tonnes; 22.3 million tonnes below the lower boundary of the 375–80 million tonne production quota (see Table 6.4).

As projected by the CIA, the bulk of the reduction, approximately 38.8 million tonnes (780 000 b/d), was attributed to reduced flows – an accelerated rate of production decline – at the super-giant Samotlor oil field (Ebel, 1994). Deprived of critical production from West Siberia, crude oil flows from the USSR would reach a peak flow rate of 616.3 million tonnes (approximately 12.38 MMB/d) in 1983, before falling gradually to 595.3 tonnes in 1985 – an unacceptable (in terms of planning margins of error) 34.7 million tonnes below the 630 million tonne production target (see Table 6.5).

Table 6.4: The eleventh and twelfth five-year plans, 1985–90

	1980 actual	1985 plan	1985 actual	1990 plan	1990 actual
National Income (Utilized)	(1976–80)	(1981–85)	(1981–85)	(1986–90)	(1986–90)
Annual Growth Rate (%)	4.4	3.4	3.0	na	1.46
Industrial Production	(1976–80)	(1981–85)	(1981–85)	(1986–90)	(1986–90)
Annual Growth Rate (%)	4.5	4.7	2.9	na	2.1
Gross Value of Agricultural	(1976–80)	(1981–85)	(1981–85)	(1986–90)	(1986–90)
Production Annual Growth Rate (%)	1.3	4.7	2.1	na	1.9
Oil (million tonnes)	603	630	595	635	570
Oil from the Province of Tyumen	303.8	375–380	352.7	410–425	
Gas (billion cubic meters)	435	630	643	850	815
Coal (million tonnes)	716	775	726	795	703
Electricity (billions of kilowatts)	1,294	1,555	1,544	1,860	1,726

Note: The five-year planning targets for coal, oil, gas and electricity are the final versions published in November 1981 for the eleventh five-year plan, and June 1986 for the twelfth five-year plan.

Sources: Hewett (1984); IMF (1992); Gustafson (1989).

Table 6.5: Production, productivity and idle wells, 1975–85

Year	Crude Oil and Condensate Production ('000 mt)	Production Per Active Well (mt)	Number of Active Wells	Active Wells as a Percent of Total Wells (%)	Total Wells	Idle Wells	Coefficient of Utilization of Active Oil Wells
1975	490,799	7,520	65,264	0.96	68,074	2,801	0.954
1980	603,207	7,101	84,949	0.97	87,976	3,027	0.955
1981	608,820	6,683	91,098	0.97	94,007	2,909	n.a.
1982	612,551	6,369	96,172	0.96	100,481	4,309	n.a.
1983	616,343	5,984	102,994	0.96	107,677	4,683	n.a.
1984	612,710	5,579	109,828	0.95	115,112	5,284	n.a.
1985	595,291	5,090	116,949	0.94	123,839	6,890	0.950
1986	614,752	4,822	127,495	0.96	132,817	5,322	n.a.
1987	624,177	4,560	136,887	0.96	142,008	5,121	0.953
1988	624,326	4,265	146,385	0.96	152,181	5,796	0.955
1989	607,254	3,941	154,073	0.95	161,780	7,707	0.952
1990	570,468	3,640	156,722	0.92	169,745	13,023	0.949
1991	515,530	3,253	158,484	0.90	176,402	17,918	0.947

Sources: The Oil and Gas Industry of the USSR (various issues); Stern (1987).

154

Despite numerous warnings from Nizhnevartovskneftegaz[1], Andropov remained committed to the 'new' Soviet energy programme. As mentioned above, Soviet oil investments were reduced gradually to only 8.9 million roubles in 1984. The bulk of these funds, approximately 50 per cent of the total, was earmarked for West Siberia (Gustafson, 1989).

Quite obviously, Gosplan was, once again, planning to offset declines from the ageing producing regions (that is, the Volga-Urals, and the Caucasus) with increased flows from West Siberia. The pressure placed on the Tyumen oil workers was immense. By the early 1980s, the rampant decline in the ageing producing regions had placed the future of Soviet oil industry in jeopardy. Any downward deviation, however slight, from aggressive 1985 West Siberian production targets could result in the unimaginable: an actual reduction in the level of Soviet crude oil flows.

At the same time, problems in the nuclear power, coal, and gas storage industries had underscored the necessity for increased oil supplies. In the final analysis, planners in Moscow had little alternative but to utilize every available method to ensure increased crude oil flows from Siberia. Disputes over the five-year Tyumen production target were settled quietly, at the upper boundary of what were considered 'reasonable' expectations for the oil industry (Gustafson, 1989).

The failure to satisfy even 'revised' annual production targets was greeted with disbelief, and outright hostility from Moscow. In the words of Tchurilov (1996, p. 109): 'As late as the middle of the 1980s, government officials still believed that Tyumen held unlimited oil resources. This point of view was encouraged by the new first Secretary of Tyumen Obkom, Gennadiy Bogomyakov'. According to Tchurilov (1996, p. 186) by the early 1980s, Bogomyakov

> knew that everything was not going smoothly in the [Tyumen] oil fields and encouraged [N.A.] Mal'tsev [the Oil Minister] to sweep through the area every now and then like an avenging angel. Both men believed unblinkingly that all problems resulted from slack discipline. That's why whenever the Minister visited the Ob' region he left behind some sacked oilmen, as though such dismissal were his visiting card. He hit out at the top. Usually his victims were the directors of some producing or drilling unit.

The 'purges' were concentrated in the 'most serious trouble spots', so that, in the years 1982 and 1983, 'all but one or two of the twenty administrative heads in Nizhnevartovsk were fired' (Gustafson, 1989, p. 105).

Unfortunately for Gosplan, the problems in West Siberia would prove to be far more serious than the simple case of 'shirking' or 'laziness' suggested by the Oil Ministry. By the mid-1980s, the Soviet Union was experiencing its second oil crisis (1984–86). The foundations for the unscheduled production disturbances – a growing repertoire of industrial complaints and inadequacies – included, but was by no means limited to the five problems outlined below.

6.1.1 The Excessive Concentration on Production, and A Shortage of Critical Investment Funds

As mentioned frequently throughout this book, the Soviet oil 'strategy' of the late 1970s and early 1980s was concentrated, almost exclusively, on the rapid development of the established West Siberian oil fields. The eleventh five-year plan called for the following improvements in the Soviet drilling programme: (1) an 83 per cent increase in MNP exploration and development drilling to 130 million metres by the year 1985. Approximately 58 per cent of the total, some 75–77 million metres, was to be completed for development purposes in West Siberia (see Table 5.13); and (2) the rapid development of 80 new West Siberian oil fields (Tchurilov, 1996).

The aggressive West Siberian drilling targets, which were maintained despite the abrupt deceleration in oil investment in the years 1983–84, would prove unattainable. In short, the 'easy' exploration and development prospects of the 1970s had long since been exhausted. Problems attributed to the 'natural ageing' of the Soviet oil industry (including by this time the ageing of established West Siberian oil fields) continued to take their toll on drilling costs, and an exhausted Soviet work-force. By 1985, the average depth of development wells had reached 2100 million metres, a 4.8 per cent increase in less than five years (see Table 6.6).

At the same time, the productivity of drilling crews suffered from 'management problems', sporadic shortages of labour, and the inferior quality and scarcity of essential material inputs. According to figures provided by the Oil and Gas Ministries, the average number of metres drilled per crew (rig team) reached 13 153 in 1985 – only 3366 metres higher than the 9787 metres reported in 1980. The 34.4 per cent increase represented only a minor improvement over the productivity gains – approximately 31.5 per cent – that had been achieved in the period 1975–80. Needless to say, the 'actual' increase in the productivity of drilling crews fell far below the targets specified by Gosplan (Hewett, 1994). According to Gustafson (1989, p. 109): 'Pipe, bits, and muds were of poor quality; rigs were open to the cold; road builders failed to prepare

Table 6.6: Average well depth, drilling productivity, and production costs in the USSR, 1960–85

Year	Average Well Depth (Metres)		Drilling Productivity Metres per Crew[a]	Costs of Production (Roubles Per Metric Tonne)	Capital Investment in New Oil Industry Capacity (Rbls/Tonne of New Capacity)	
	Development	Exploration			USSR	Tyumen
1970	1,728	2,492	6,070	4.28	29.3	22.4
1975	1,821	2,718	7,442	5.70	23.8	14.8
1980	2,003	2,788	9,787	7.21	46.0	33.9
1981	2,010	2,842	11,031	8.01	48.6	36.7
1982	2,040	2,768	11,320	9.40	54.7	42.5
1983	2,066	2,807	12,152	10.82	56.0	43.3
1984	2,079	2,745	12,512	12.26	n.a.	n.a.
1985	2,100	2,813	13,153	13.68	62.4	46.1
1986	2,152	2,782	14,846	14.88	72.0	53.9
1987	2,178	2,822	15,998	15.90	81.7	61.8
1988	2,212	2,831	16,943	17.12	91.3	69.6
1989	2,240	2,825	16,645	18.96	109.8	84.9
1990	2,238	2,873	15,693	21.13	120.9	100.2
1991	2,227	2,794	16,461	57.39	346.8	115.5

Note: [a] Drilling productivity statistics, metres per rig team, are given as the total for exploration and development drilling.

Sources: The Oil and Gas Industry of the USSR (various issues); Khartukov (1995); *Soviet Energy: An Insider's Account* (1991).

the way; line crews lagged in laying power lines; drillers sat waiting for key parts to arrive – these stories are familiar from similar complaints throughout the economy'.

Regardless of their actual validity, complaints from the Tyumen Obkum and Oil Ministry fell on deaf ears in Gosplan, and the same old 'temporary' (tried-and-true) solutions were called upon to salvage the plan. Notable among these were: (1) a heavy reliance on rough and ready technologies, such as the turbo-drill, and cluster wells;[2] and (2) the increased utilization of the permanent (*Vakhtovyi*), and temporary (*Ekspeditsionnyi*) *Metods* to expand the West Siberian labour force.

The ability of the Soviet drill teams to rely on antiquated technologies such as the turbo-drill was facilitated by the lower average well depths required in West Siberia. According to Goldman (1980, 127–8):

> Most frustrating for Soviet petroleum exploration has been the lack of high-quality drill bits and better quality pipe. . . . These handicaps limit the drilling range of most Soviet efforts to under 2,000 meters. That was adequate in West Siberia where oil was found at a depth of 1.6 to 2.4 kilometers. But in future fields, such as the deep part of the Caspian Sea, the Black Sea, Timan-Pechora, and the Viliui Basin where the greatest promise seems to be at depths of 2,500 to 5,000 meters, the Soviet turbo-drill, and seismographic system cannot penetrate.

As 'drilling' productivity suffered from substandard technology and sporadic shortages of material inputs, an increasing number of workers were required simply to make up the deficiency. This was a crude substitution of labour for capital at best and a poor use of manpower in a tight labour market. By 1985, the number of drilling brigades that were 'permanently' attached to Glavtiumennefte005gaz had reached 282, a 136 per cent increase over the 119 reported in 1980, and a significant 63 more brigades than the 219 that had been specified by Gosplan in the original version of the eleventh five-year plan. Still, the Tyumen producing region continued to suffer from a severe shortage of skilled labour. The number of 'flown in' workers rose steadily throughout the five-year planning period, reaching levels as high as 40 000–50 000 workers in 1985. By 1985, the 'temporary' jet-lagged work-force was responsible for 40 per cent of Tyumen drilling, and 25 per cent of well restoration and repair work (Gustafson, 1989). The massive increase in the number of workers swamped the sector's meagre resources for housing and services, causing morale to drop and the labour turnover rate to increase. All this hampered efforts to improve productivity.

The implications for drilling costs were significant. According to figures provided by the Oil and Gas Ministries, the average cost of 'development' drilling reached 132.71 roubles per metre in 1985, an 11 per cent increase over the 119.56 roubles per metre reported in 1980 (see Table 5.14). This tendency was even more pronounced in the exploration sector. While the average depth of exploration wells reached only 2813 metres in 1985 – a 25 metre increase over the 2788 metres reported in 1980 – average 'exploration' drilling costs rose significantly. As average well depths had not yet become a problem in West Siberia, the increase in exploratory drilling costs over this period has been attributed to the remote, and inhospitable location of promising, albeit complex, geological formations.

Still, despite these difficulties, the focus on development reigned supreme, and important achievements were produced by the Oil Ministry. MNP completed 143.8 million metres in 1981–85; 13.8 million metres above the 130 million metre drilling target (see Table 5.13). In 1985, 11 984 new oil wells were commissioned in the USSR, a 70 per cent increase over the 7046 completed in 1980 (see Table 6.7).

Table 6.7: New oil well completions and productivity in the USSR, 1980–90

Year	New Oil Well Completions	Average Well Productivity (Tonnes/Month) *
1980	7,046	621
1981	8,641	585
1982	8,950	555
1983	9,698	525
1984	11,219	489
1985	11,984	447
1986	13,789	420
1987	14,979	396
1988	15,859	369
1989	15,019	339
1990	13,016	306
1991	11,091	273

Note: * Average monthly well productivity has been estimated by multiplying daily productivity by 30.
Source: The Oil and Gas Industry of the USSR (1986 and 1992).

As mentioned above, the inflated West Siberian drilling target was simply unattainable. Inauspiciously, the bulk of the shortfall was concentrated in 'new' oil fields. While the initial version of the eleventh five-year plan called for the development of 80 new West Siberian oil fields, the need for an immediate and affordable increase in West Siberian crude oil production was absolute, and the target was subsequently revised downwards to 30. By 1985, only 24 new oil fields had been brought on line, resulting in an implied long-term opportunity cost – an unplanned shortfall in crude oil production – of approximately 10–15 million tonnes per annum (Gustafson, 1989).

As always, an excessive focus on production diverted valuable resources from the Soviet exploration effort. By 1985, the Oil and Gas Ministries and Minego had fallen short of the 34.4 million metre exploration target by a substantial 2.3 million metres. According to figures provided by the Oil and Gas Ministries, a mere 32.1 million metres were completed for exploration purposes over the period 1981–85, less than 20 per cent of the total Soviet drilling effort of approximately 161.5 million metres (see Table 6.8).

Table 6.8: Exploration and development drilling in the USSR for oil and gas

	(million metres)				
	All Ministries				
Five Year Plan	Exploration	Development	Total	Exploration (Plan)	Shortfall Plan minus Actual Exploration Drilling
1971-75	26.1	41.8	67.9		
1976-80	26.7	66.0	92.7	28.0	1.3
1981-85	32.1	132.3	161.5	34.4	2.3
1986-90	38.9	187.3	226.2	44.9	6.0

Sources: Hewett (1984); Gustafson (1989); *Oil and Gas Industry Statistical Handbook* (various issues).

6.1.2 Deficiencies in the Quality and Availability of Critical Materials and Supplies

As was the case with most post-Stalin attempts at economic reform, Yuri Andropov's heroic efforts to revitalize Soviet industry would prove sadly ineffective.   Despite significant investments in modern technology (including a doubling of investments in industrial robots) (Kort, 1993), the growth rate in industrial production would rise only fleetingly, from 3.3 in 1982 to 4.0 in 1983, before collapsing to levels well under the critical 2.0 per cent mark (Hewett, 1984). According to Kort (1993, p. 278):

> It also proved impossible, even for the ubiquitous Soviet police, to end the absenteeism and lax work habits of tens of millions of workers. They soon reverted to their old ways that reflected the motto: 'They pretend to pay, we pretend to work'.

National income, and the majority of the broad economic indices, fell short of their targets in the eleventh five-year planning period.   While Gosplan had envisioned a 3.4 per cent increase in national income in the years 1981–85, the official statistics suggested 3.0 per cent – a 0.4 per cent shortfall. Despite its promising performance in 1983, industrial production grew at an average annual rate of only 2.9 per cent in the eleventh five-year plan; 1.8 per cent below its aggressive 4.7 per cent quota (see Table 6.4).

As might have been anticipated – and more so given a gradual deterioration in Soviet relations with the US – large quantities of critical supplies were diverted to high priority end-users such as the military and agriculture (Gustafson, 1989). Deprived of both capital investment and material inputs, the performance of the civilian machinery sector, and other low priority industrial sectors – specifically those providing industrial support for the oil industry – was woefully inadequate. As always, the deficiencies were most pronounced in the overall quality, rather than the quantity, of production.

The 'support' sector for the Soviet Union's oil and gas industries was born in the city of Baku, Azerbaijan, in the early 1920s. To cite only a few examples: the first machine building plant to support the fledgling crude oil industry, the Sattarkhan Plant, was constructed in Azerbaijan in the year 1922. In its early years, 1923–25, Sattarkhan was renowned for the construction of down hole pumps, rotary drilling rigs, and Christmas trees. By 1925, the factory had been expanded to facilitate the large-scale production of pumping rods. Years later, in 1943, Kishlinski, the largest machine-building enterprise in the USSR, was completed in Azerbaijan SSR.

Despite the cost, and inconvenience, to the West Siberian oil workers, and discounting a small number of factories that were constructed in the Volga-Kama-Basin, the centre of the oil and gas machine-building industry would remain in Azerbaijan throughout the entire history of the USSR. In the late 1980s, 70 per cent of Soviet equipment for producing oil and gas still came from Azerbaijan, principally from the city of Baku (Gustafson, 1989).

At the time of the second Soviet energy crisis (1982–86) the Ministries responsible for the manufacturing of oil and gas service equipment included: (1) Minkhimmash (the Ministry of Chemical and Petroleum Machine-building) which supervised the construction of 'most' machinery facilitating the operation of oil and gas wells; (2) Mintiazhmash (the Ministry of Heavy and Transport Machine-building) which supervised the construction of drilling rigs; and (3) Minchermet (the Ministry of Ferrous Metallurgy) which supervised the construction of pipe.

Glavneftemash, the industrial association responsible for the construction of petroleum machines, fell under the direct jurisdiction of Minkhimmash. By the mid-1980s, Glavneftemash was comprised of 15 enterprises, all located in close proximity to Baku, and employed approximately 20 000 workers (Gustafson, 1989). This one industrial association alone was responsible for 67 per cent of the oil-field equipment in the entire USSR.

According to long-established Gosplan tradition, the allocation of scarce investment funds was dictated to a great extent by the priorities of the Central Committee. While the vast hierarchy of 'government' priorities was complex, and often confusing, a few directives (guidelines) were straightforward and worthy of mention. As noted frequently throughout this book, the military and agriculture were given clear priority over the construction of civilian machinery and consumer goods. In the energy sector, the extractive ministries – such as Mintiazhmash – took precedence over Minkhimmash, a mere oil and gas support association (Gustafson, 1989). Minkhimmash, one of the lowest government priorities in a low priority sector, received only a smattering of residual capital funds. For example, in the years 1975 to 1985 inclusive, the investment funds allocated to Minkhimmash were equal to approximately one-fiftieth of the investment resources allocated to the Ministries it supplied (Gustafson, 1989). A tiny fraction of these funds, approximately 2 billion roubles, was allocated to the construction of oil- and gas-field machinery.

Despite a severe shortage of capital, Glavneftemash would exceed Gosplan production targets throughout the tenth and eleventh five-year planning periods. Battling primitive and outdated production facilities, the industrial association managed to triple its production in less than 10 years – from 1978 (the beginning of Brezhnev's crash energy programme) to 1988 (Gustafson, 1989). The prime directive – to satisfy unreasonable quantitative production targets – was absolute, so that extreme measures and short-cuts were taken whenever possible. Deficiencies in the quality of Glavneftemash equipment grew steadily with the demands on its overworked staff and factories. By the mid-1980s approximately 33–50 per cent of Glavneftemash's well finishing and operation equipment was rated as defective once put into use (Gustafson, 1989). In addition, Mintiazhmash often produced heavy, unreliable drill rigs; and Minchermet continued to supply low quality drill and casing pipe.

Under continuous pressure from the central planners, the oil and gas ministries had little alternative than to divert valuable labour and resources to the construction of their own well-repair and equipment plants. Shipments of defective equipment were delivered directly to these facilities so that minor repairs and upgrades could be completed immediately upon arrival. By 1986, the Tyumen oil industry had 8000 people employed in these activities and their numbers were expanding rapidly (Gustafson, 1989). For example, the Iugansk oil production association constructed its own well-repair factory in the Province of Tyumen in the early 1980s and SoyuztermNeft – the science and research organization in charge of all enhanced recovery projects for the Soviet oil industry – constructed the Neftetermash manufacturing plant near Krasnodar to produce steam-gas generators and accessories, steam generators, insulated tubing and production pipe, packers, downhole completion units, separators, and surface well completion units.

According to Yurko and Reitman (1990, p.43):

A significant accomplishment for SoyuztermNeft is the manufacturing plant (Neftetermash) at Chernomorsk about 50 km from Krasnodar. The plant was developed because equipment from Soviet factories was very hard to get or was of very poor quality. They now design and manufacture, to their own specifications, a variety of equipment used for thermal stimulation recovery. The main benefits of manufacturing their own equipment were the reduction of initial investment, access to modification and cost and quality control.

As always, a portion of the deficiencies from domestic suppliers was supplemented by an increase in technical imports. The value of oil-related technical imports rose steadily throughout the eleventh five-year planning period from 194.6 million roubles in 1981 to 354.0 million roubles in 1985. In light of tensions arising from the Cold War, the bulk of these expenditures, approximately 253 million roubles or 71.5 per cent of the total, were devoted to oil-equipment imports from Romania (Gustafson, 1989).

6.1.3 An Excessive Use of Water flooding, and Reluctance – Inability – to Adopt Alternative Enhanced Recovery Technologies

The Soviet Union's belief in, and heavy reliance on, the use of water injection as a 'superior' method of enhanced oil recovery has been well documented. Indeed, the method proved highly successful in the Volga-Urals region where water was injected into wells in the earliest stages of production. As early as 1961, members of an American delegation to Russia noted the 'glittering successes in a few select oilfields' (Ebel, 1961, p. 32). The 'select oilfields' included wells that had been injected with large quantities of water in the early development stage.

By the late 1970s, the Soviet Union's unique (premature) use of water injection was celebrated as having significantly 'enhanced' ultimate recovery in the Volga-Urals region. While water flooding was an established technique in many other oil producing countries around the world, it was generally employed to sustain production in the final stages of an oil field's life cycle. The Soviet Union use of the method in the initial (or early) stages of production was, to some extent, unique, and the source of considerable controversy among oil men and petroleum analysts. According to Goldman (1980, p. 120):

> The use of water injection extended production considerably in the Volga-Ural region. Even though ultimate recovery from wells was only a fraction of the anticipated 80–90 per cent hoped for, the yields would have been much worse if it were not for the use of water and the [submersible] pumps.

However, as more and more water is injected into a well, the volume of water in the reservoir is increased and water can rise above the oil in the reservoir and flow into the producing wells. At this point, the production association had the choice of two options: (1) drilling additional in-fill wells into the remaining pockets of oil; or (2) installing high-capacity submersible pumps to lift the large quantities of oil and water that would be necessary to sustain crude oil production levels (Riva, 1994, p. 54).

Encouraged by the apparent achievements in the established producing regions, the Oil Ministry introduced waterflooding to the West Siberian oil fields at the earliest possible opportunity. However, Gosplan's repeated upward revisions to the Tyumen' production quota led to excessive water flooding and the extraction of oil from West Siberian reservoirs at rates far greater than the maximum efficiency rate of recovery. While it had taken approximately 18 years for the water-cut (the percentage of water in the total quantity of fluid lifted from an oil field) to reach 10 per cent at Romashkino, the water-cut at Samotlor would reach 10 per cent in less than three years. To the amazement of central planners, the policy, which had worked reasonably well in the Volga-Urals, worked poorly in Siberia. As early as 1974, the Academy of Sciences had recognized, and was lamenting the fact that the uncontrolled and excessive use of water injection at Tiumen' had permanently lowered the area's ultimate output (Goldman, 1980).

In contrast, by the mid to late 1970s Western oil analysts were drawing global attention to the unique Soviet method of water flooding and its potential devastating effects on the ultimate recovery of reserves. As mentioned in Chapter 5, in April 1977, the CIA issued a series of reports warning that excessive water flooding at giant Soviet fields would lead to accelerated decline rates, and a sharp, and sustained reduction in Soviet oil production. In short: while the majority of Soviet oil men claimed that water flooding had the potential to enhance the ultimate recovery of an oil field, Western analysts remained sceptical, suggesting that excessive water flooding could, in fact, shorten the long-term productive life of an oil field, reducing total recovery, and artificially elevating crude oil production costs (Gustafson, 1989).

The repeated warnings, from both Western analysts and respected Soviet scientists, would be met with incredulity, and even suspicion, by the Oil Ministry. Indeed, the 'early' evidence suggested otherwise. In the late 1970s and early 1980s, unprecedented oil recovery rates had been reported in the Volga-Urals despite the fact that the water cut had been driven to levels as high as 50 per cent. Respected Soviet publications boasted that the Western analysts were mistaken, and that water flooding would produce higher final recovery rates in the USSR than could be achieved by any of the other (competing) oil producing countries around the world. Indeed, in the early 1980s Soviet oil officials were projecting a total recovery rate for the Soviet Union of 43 per cent; nearly10 per cent higher than the 33.3 per cent expected in the United States (Gustafson, 1989). In the final analysis, the Soviet Union's belief in, and commitment to, its unique water recovery programme was unshakable and would remain so throughout the decade of the 1980s.

It is important to note that the Oil Ministry was still projecting world class final recovery rates in the early 1990s. Table 6.9, which was presented to an Alberta Oil Sands and Research Technology USSR Exchange Tour, illustrates the Oil Ministry's perception of Soviet recovery rates in January 1990.

Table 6.9: The USSR's perception of ultimate oil recovery around the world

Region, Country	Final Recovery
The United States and Canada	0.33–0.37
Latin America	0.26
Europe	0.37
Africa	0.27
Far East	0.24
Iran	0.16
Saudi Arabia	0.35
The Oil Industry of the USSR (projected)	0.45

Source: Yurko and Reitman (1990).

Tables 6.10 and 6.11 provide a brief illustration of the rapid development of the Soviet crude oil industry, and its heavy reliance on water flooding in the years 1950–87 inclusive. In 1950, a time when the oil industry in the USSR was dominated by crude oil flows from Baku, the Soviet water flooding programme was in the initial stages of its development. According to estimates provided by the Oil Ministry, only one out of every 41.5 producing oil wells had been injected by water, and the volume of water injected per one tonne of oil produced was estimated at well under 0.25 m/t.

As the centre of the Soviet oil industry shifted gradually to the Volga-Urals, and eventually to West Siberia, the Soviet water flooding programme was accelerated and new oil wells were injected in the earliest stages of

Table 6.10: Water flooding in the USSR, 1950–87

Year	Production of Crude Oil and Gas Condensates (million tonnes)	Total Wells	Number of Production Wells per one Injection Well	Injected Water Per One Tonne of Oil (m³/t)	Water-Oil Ratio	Average Well Productivity (Tonnes/Month)[a]
1950	37.8	20,186	41.5	0.24	1.92	186
1955	70.7	29,807	15.4	1.20	1.69	243
1960	147.7	37,014	11.1	1.30	1.27	372
1965	242.8	45,924	8.5	1.40	0.63	510
1970	353.0	56,514	8.7	1.60	0.76	588
1975	490.8	71,875	7.1	2.00	0.90	651
1980	581.7	90,651	5.6	2.60	1.34	645
1981	587.6	97,505	5.6	2.80	1.49	606
1982	592.6	104,465	5.5	3.10	1.64	576
1983	596.3	112,401	5.3	3.40	1.85	543
1984	592.4	120,891	5.2	3.60	2.06	501
1985	570.7	128,988	5.1	3.90	2.31	456
1986	586.6	139,662	5.0	4.10	2.50	432
1987	593.4	153,376	4.9	4.40	2.70	399

Note: [a] Average monthly well productivity has been estimated by multiplying daily productivity by 30.

Source: Yurko and Reitman (1990).

Table 6.11: Water flooding in West Siberia, 1965–87

Year	Production of Crude Oil and Gas Condensates (million tonnes)	Total Wells	Number of Production Wells per one Injection Well	Injected Water Per One Tonne of Oil (m³/t)	Water-Oil Ratio	Average Well Productivity (Tonnes/Month)[a]
1965	0.95	47	10.7	0.02	0.01	696
1970	31.40	1,211	5.0	1.43	0.05	2,670
1975	148.00	4,362	4.4	1.75	0.17	3,561
1980	312.60	13,737	4.3	2.08	0.35	2,406
1981	334.40	17,089	4.1	2.23	0.46	2,088
1982	352.90	19,172	3.6	2.46	0.62	2,013
1983	370.10	25,043	3.5	2.86	0.80	1,623
1984	377.90	28,387	3.6	3.04	1.04	1,458
1985	366.00	40,934	3.5	3.30	1.29	987
1986	384.60	44,132	3.5	3.69	1.52	963
1987	403.00	51,909	3.9	3.55	1.96	837

Note: [a] Average monthly well productivity has been estimated by multiplying daily productivity by 30.

Source: Yurko and Reitman (1990).

168

their development. By 1975, one of every 7.1 producing oil wells had been injected by water, and the volume of water injected per one tonne of oil produced had increased to 2.0 m/t. Nevertheless, the average production from a typical Soviet oil well had risen steadily from 186 tonnes per month in 1950 to 651 tonnes per month in 1975 – a 250 per cent increase in only 25 years – and the unique Soviet water-injection programme was hailed as an 'astonishing success'.

By 1980 the average water to oil ratio for the Soviet Union had reached 1.34, implying an aggregate water-cut of 57 per cent. By this time, however, the difficulties suggested by the CIA were apparent, even in the aggregate productivity statistics. The average well productivity in the USSR fell to 645 tonnes per month in 1980, a one per cent reduction from the 651 tonnes per month reported in 1975. The statistics were even more discouraging in West Siberia, where the average water-cut had risen dramatically to 26 per cent in 1980 representing a 20 per cent increase in only 10 years. By 1980, the productivity of a typical West Siberian oil well had fallen by 32 per cent from the peak productivity rate of 3561 tonnes per month reported in 1975 to only 2406 tonnes per month (see Table 6.11).

Undeterred by 'productivity' problems at Samotlor – and the dire predictions of the Western oil analysts – the Oil Ministry maintained an aggressive water-injection schedule throughout the decade of the 1980s. By 1985, one of every 5.1 producing oil wells had been injected by water, and the volume of water injected per one tonne of oil produced had increased to 3.9 m/t. The aggregate Soviet water-cut reached an unprecedented 70 per cent in 1985; a 13 per cent increase in only 5 years. In West Siberia, where one out of every 3.5 oil wells had by this time been injected with significant quantities of water, approximately 3.30 m/t, the average water-cut reached 56 per cent, and average well productivity had plummeted to only 987 tonnes per month.

The Soviet Union's excessive reliance on water flooding has been attributed to some extent to a resistance to adopt experimental and 'expensive' enhanced oil recovery technologies and inability (reluctance) of the Soviet industrial ministries to produce the necessary technical equipment. Indeed, the USSR had been experimenting with alternative methods of enhanced oil recovery (EOR) since the early 1920s. In 1924, the success of lab research on the *in situ* gasification of coal inspired two young Soviet scientists to apply a similar procedure to heavy oil deposits. Less than ten years later, in 1933, the first *in situ* thermal method (fireflood) was attempted on the Neftegorsk oil field.

The project was so successful that Stalin was persuaded to visit the well site in the years prior to World War II. Disrupted by war, and assigned a low priority by the Khrushchev administration, the project was discontinued until the early 1960s (Yurko and Reitman, 1990). In 1964, SoyuztermNeft implemented the first steam soak (cyclic steam stimulation method) at the Zybza oil field. Promising results at the Neftegorsk and Zybza oil fields led to further investigation, and the utilization of a number of enhanced oil recovery methods throughout the USSR.

In the years immediately following World War II, and leading up to the first oil crisis (1945–75), the Petroleum Ministry of the USSR was primarily concerned with primary (waterflood) production and exploration for light and medium crude (Yurko and Reitman, 1990). As fields in Azerbaijan, Tartaria, the Ukraine, and eventually West Siberia, began to enter into an extended 'unplanned' episode of accelerated decline, the Ministry was forced to consider more expensive production alternatives including the development of smaller oil fields, heavy oil and bitumen, and enhanced recovery methods. According to Gustafson, (1989, p. 130), by the mid-1970s, tertiary or enhanced recovery methods had

> suddenly became fashionable in Moscow. Enthusiasm for enhanced recovery was especially high among technologists at MNP headquarters, at the State Committee for Science and Technology (whose responsibility included publicizing successful Western innovations), and above all in Gosplan, where it was seen as a way to take advantage of the established infrastructure in the older oil regions.

In 1976, the Petroleum Ministry inaugurated a 'special' enhanced recovery programme, to inspire the use of sophisticated tertiary recovery methods throughout the USSR. The methods under consideration involved the treatment of oil-bearing rocks with detergents and carbon dioxide, and the injection of reservoirs with steam and natural gas, instead of water. As the industrial ministries lacked the expertise to produce efficient state-of-the-art machinery, a significant portion of the technical equipment and chemicals were to be imported from the West. The Soviet industrial ministries, however, were slow to produce critical supporting machinery, that is, compressor stations, and natural gas feed lines, and the programme failed dismally.

By the end of the tenth five-year planning period in 1980, the volume of incremental crude oil production attributed to EOR had reached only 2.7 million tonnes per annum, 1.2 million tonnes more than the 1.5 million tonnes reported in 1975 (see Table 6.12). Deputy Minister E. Khalimov, the minister in charge of the 'newly created' EOR Department, would, as reported by Goldman (1980, p. 124) blame the Ministry of Chemical Machinery Construction 'for producing only 2 per cent of the equipment needed'. His 'excuse' fell on unsympathetic ears in the Gosplan, and he was fired less than one year later, 1981 (Gustafson, 1989).

The eleventh five-year plan included an EOR production target of 8 million tonnes by the year 1985 and a significant 196 per cent increase over the 2.7 million tonnes reported in 1980. A total of 500 million roubles, approximately 1 per cent of total fixed capital investments in the oil industry, were devoted to EOR projects. Three new institutes were created to promote EOR research and development throughout the USSR (Gustafson, 1989).

Table 6.12: Incremental oil produced by EOR methods in the USSR, 1975–90

(millions of tonnes)		
Year	Planned	Actual
1975	n.a.	1.5
1980		2.7
1981		2.6
1982		3.0-3.4
1983		4.0
1984		n.a.
1985	8.0	5.0
1986		5.0
1987		5.7
1988		6.7
1989		n.a.
1990	15.0	14.4

Note: The figures do not include EOR production from Hydrodynamic methods: approximately 26.9 million tonnes in 1986; 30.0 million tonnes in 1987; 42.5 million tonnes in 1988; and 39.0 million tonnes in 1990.

Sources: For 1975–85, Gustafson (1989); for 1986–90, Yurko and Reitman (1990).

As in the tenth five-year plan, the programme was frustrated by the failure of industrial ministries to produce the necessary equipment. Fearful of the repercussions from an expensive EOR equipment failure, the oil and gas production associations resisted the adoption of the 'newfangled' technical procedures at every possible opportunity. By the end of the planning period in 1985, enhanced oil recovery methods were responsible for a mere 5 million tonnes of incremental crude oil production. This represented a dismal 3 million tonnes below the official quota. In the final analysis, and despite considerable pressure from Gosplan, the tried and true method of water flooding would persevere as the only reliable method of 'enhanced oil recovery' throughout the 1980s.

6.1.4 The Growing Burden of Well Maintenance and Repair Work, and Rising Crude Oil Production Costs

As mentioned above, the absolute priority attributed to development drilling, that is increased production, resulted in an explosion of new well completions throughout the USSR. According to figures provided by the Oil and Gas Ministries, a total of 50 492 new oil wells were completed in the Soviet Union in the years 1981–85, nearly as many as the entire number of oil wells that existed in the USSR in 1970 (approximately 55 539) (see Tables 5.16 and 6.7).

Not surprisingly, the requests for well maintenance and repair jobs grew steadily with the number of new well completions. Neither the Oil Ministry, nor the ill-equipped maintenance and repair crews were prepared for the onslaught. According to Gustafson (1989, p. 112) they were: 'Long neglected by the industry, repair crews in the early 1980s complained that they received lower pay and fewer prerequisites than drillers or operators, that they were allocated less equipment and support'. The complaints, duly registered with the growing repertoire of similar 'minor' irritations throughout the Soviet industrial sector, drew little in the way of sympathy, or support, from the Oil Ministry. Production associations were forced to shut-in promising oil wells for the want of minor repairs, and the number of idle wells grew steadily throughout the eleventh five-year planning period. By 1985, the number of idle oil wells had reached 6890, an increase of nearly 4000 over the 2909 that had been reported by the Ministry in 1981 (see Table 6.5).

Remarkably, for a system which was renowned for the stability of prices, crude oil production costs continued to rise steadily throughout the five-year planning period. Discounting an official adjustment to wholesale enterprise (producer) prices in 1982 (Hewett, 1994), the increase reflected the following considerations: (1) expenditures associated with the natural ageing of the Soviet oil industry; (2) the migration of the centre of Soviet oil production to the harsh territories of West Siberia; (3) the increasing difficulty of operating conditions; (4) expenditures associated with excessive water flooding (electric submersible pumps); (5) an increase in technical imports (gas lift equipment); (6) reductions in the productivity of both drilling crews and crude oil wells; and (7) increased expenditures associated with labour shortages (fly-ins). According to figures provided by the Oil and Gas Ministries, the average production cost for crude oil in the USSR reached 13.68 roubles per metric tonne in 1985; a 90 per cent increase over the 7.21 roubles per metric tonne reported in 1980 (see Table 6.6).

Once again, it is necessary to point out that these figures were maintained primarily for accounting purposes, and as such, do not reflect the true costs of production. To cite only a few of the more obvious examples: in the early 1980s, rent charges, royalties (for the raw mineral deposits), charges for the outright purchases of land, the costs of acquiring raw materials, fixed interest charges on capital, and finders' fees (geological exploration costs) had yet to be adequately represented as basic or essential costs of production. In short, despite the beneficial influence of a half-hearted price reform in 1982 (see Appendix D), Soviet energy prices failed to reflect either the true costs of production, or the domestic and international opportunity costs of energy. The discrepancy led to the gross misallocation of scarce investment resources throughout the lifetime of the USSR, and beyond. The longer the planning system remained in place, the more obvious the difficulties prophesied by von Mises in the 1920s became. With no way to determine an efficient allocation of resources, the planners' task became increasingly ad hoc.

To cite only one hypothetical example: if crude oil prices had been permitted to reflect the true costs of production – or at least an approximation for the international opportunity costs such as the world oil price – Brezhnev and Andropov might have been saved considerable time and expense in the drafting and implementation of the new long-term energy (and domestic conservation) programme. At the same time, inspired by the fiscal benefits arising from higher domestic oil prices, the Oil Ministry might have been persuaded to increase the amount of investment funds allocated to EOR, and the development of new but 'seemingly' unprofitable West Siberian oil fields.

6.1.5 Poor Reservoir Management, and/or the Inability of the Central Planning Process to Accommodate the Unique and Idiosyncratic Properties of Individual Reservoirs.

Gosplan's obsessive concentration on rapid production, the growing scarcity of investment funds and critical material inputs led the Oil Ministry and production associations to take decisions that were detrimental to the efficient management of unique, and idiosyncratic, reservoirs. Consider the following examples:

CASE I – Samotlor: By the mid-1980s, the exhaustive inventory of technical problems in the super-giant Samotlor oil field had been exacerbated by poor reservoir management, and difficult, often hostile, drilling and operating conditions. For example, to minimize the difficulties and expense associated with drilling in the swamplands of Siberia, Nizhnevartovskneftegas was forced to drill development wells 'directionally', in clusters of approximately 12 to 18 wells, from individual islands. The conditions below the surface were equally formidable. According to Riva (1994, p. 164): 'the reservoir sandstones contained large amounts of feldspar that has been weathered to clay minerals that become suspended in the oil. The clay, along with silt and sand in suspension and the deposition of salts, hampers oil production'.

Needless to say, the initial high rates of production demanded by Gosplan were achieved and sustained, however temporarily, by the excessive utilization of water flooding. The massive volume of water bypassed productive reservoirs in the early stages of development, so that, as mentioned above, the water-cut exceeded 10 per cent within three years. To further aggravate matters, the electric submersible pumps required to maintain the large volume of crude oil flows at Samotlor were of poor quality and easily damaged by silt and sand. As a result, they had to be replaced frequently: every two months.

While the pumps at Romashkino were good for approximately a year (Goldman, 1980), the brief two month life-span of electric submersibles at Samotlor rendered them overly expensive, and unduly burdensome to the Oil Ministry. Despite these difficulties, however, the number of Tyumen oil wells with electric centrifuge pumps increased significantly in the eleventh five-year planning period from 18 per cent of all Tyumen oil wells in 1980 to an impressive 45 per cent by 1985 (Gustafson, 1989).

A promising 'new' alternative to water flooding – the gas lift technique – involved the injection of associated gas back into wells to improve the gas pressure in ageing reservoirs. Indeed, the Soviet Union had been experimenting with imported gas lift equipment since 1972. The results, particularly in West Siberian oil fields, were promising and the eleventh five-year plan called for the installation of 4500 gas lift units throughout the USSR. The bulk of these – some 3000 to 4000 units – were earmarked for Samotlor (Gustafson, 1989). While the lift units themselves were to be imported from Western suppliers, domestic suppliers (the notoriously unreliable oil and gas service industry) were to produce the necessary gas feeder lines, and ancillary compressor stations.

Not surprisingly, the oil and gas machine building sector would turn out to be a weak link in an otherwise promising reservoir development strategy. Gustafson (1989, p. 111) suggests that: 'the designated domestic subcontractor did not produce the compressor stations, and the crews that were to install the gas feeder lines were drawn off to work on the big Siberian gas trunk lines (a side effect, incidentally, of the American embargo of 1981–82)'. By 1985, the number of gas lift units installed and in operation at Samotlor had reached only 1400, less than one half of the 3000 units originally specified by Gosplan.

Frustrated by the deficiencies in both electric submersible pumps and gas lift equipment, the Oil Ministry had no choice but to seek alternative, albeit considerably less efficient, methods of sustaining the volumes of crude oil flows. Water injection was utilized at every possible opportunity, and the growing number of oil wells that could no longer sustain production, were simply abandoned. Brand new in-fill wells were drilled into the few remaining pockets of oil. According to Riva (1994, p. 164) in the 'frantic effort to sustain production . . .well workover activity was virtually abandoned, leaving thousands of wells idle in the field . . .[and] Samotlor's best reservoirs were overdeveloped, while others were bypassed'. By 1985, production from the super-giant Samotlor oil field had fallen to 114.5 million tonnes per annum, a 38.8 million tonne reduction from the peak flow rate of 153.3 million tonnes reported in 1980 (Ebel, 1994).

CASE II – Uzen'skoye: The giant Uzen'skoye oil field was discovered in 1961, on the Mangyshlak Peninsula in Kazakhstan. With estimated reserves of approximately 7.5 billion barrels of 'in-place oil' it was a promising favourite of the Oil Ministry. Total recovery was planned at an ambitious 45 per cent, and, according to the established Soviet procedure, the field was injected with water in the early stages of its development. Despite the best intentions of the production association, the reservoir contained complex, non-homogeneous producing horizons, and was severely damaged by poor reservoir management, the inappropriate and excessive use of water injection, and overproduction of the easier, more permeable production horizons during the first 14 years of its development. According to Riva (1994, p. 55):

> The [Uzen'skoye] field was waterflooded from the beginning of production because of satisfactory waterflooding results in marine sandstones in other areas. Ignorance of the reservoir bed framework caused the most permeable channel sands commonly to be penetrated by development wells or injection wells only. In some cases, overproduction from development wells led to formation water breakthrough and, in others, the injection of large amounts of water caused significant displacement of oil beyond the oil–water contact. The declining reservoir pressure and the cooling of producing beds resulted in the degassing of the oil and considerable precipitation of paraffin. Such deteriorating reservoir properties led to the complete plugging of the less permeable horizons. The development of the Uzen'skoye field can be described as the poor realization of a poor plan. Large-scale heating of the injection water did not begin until 14 years after the field was put on stream. By then it had become impossible to achieve the planned recovery efficiency of 45 per cent. The main output was derived from the more-permeable horizons, while the beds with lesser permeabilities became almost unproductive. The ultimate recovery from Uzen'skoye now [1994] is estimated at 1.875 billion bbl, only 25 per cent of the original oil in place. Thus as much as 1.5 billion bbl of oil may have been lost.

To further aggravate matters, Andropov's ability to deal successfully with the problems and complexities underlying the second Soviet oil crisis was undermined by the delicate condition of his health. Diagnosed with kidney failure in the summer of 1983, Andropov was undergoing dialysis treatments, and literally disappeared from the public arena. In the winter of 1983, the little that was left of Andropov's time and energy was diverted to a series of escalating and demanding international crises.

To cite only a few examples: (1) the war with Afghanistan dragged on interminably, claiming Soviet soldiers, and valuable financial resources; (2) the Soviet Union backed Syrian forces in Lebanon, frustrating US attempts to end civil war; (3) in October 1983 the United States invaded Grenada, overthrowing a radical Marxist regime. The intervention, which was intended to reduce Soviet influence in the Caribbean, was successful; (4) US President Ronald Reagan initiated the development of SDI (Strategic Defence Initiative) – an American defence system designed to intercept and destroy Inter Continental Ballistic Missiles in the air (before they could reach their targets). The Soviet Union had to respond, requiring even more scarce resources to be allocated for military purposes; and (5) in 1983, the US responded to the Soviet deployment of SS20 intermediate range missiles in Eastern Europe, with a deployment of newer, and more accurate Pershing II and cruise missiles in Western Europe (Kort, 1993).

In a few short years, relations between US President Ronald Reagan and the nation he had nicknamed the 'Evil Empire', had deteriorated perceptibly. Incensed by the SDI initiative, and the US deployment of advanced missiles in Western Europe, the USSR walked out of three sets of arms control negotiations. For the first time in two decades the United States and the Soviet Union were not even discussing arms reduction. The Reagan government was simply willing to force the Soviet Union into unsustainable competitive spending on arms until it collapsed.

Ironically, neither Andropov, nor his new long-term energy programme, would survive long enough to witness the full implications of the second Soviet 'oil crisis'. By December 1983, Andropov was too ill to attend critical meetings of the Central Committee and the Supreme Soviet. He died less than two months later, on 4 February 1984. It was:

> the year of the heaviest losses in Afghanistan, with 2,343 officers and men returning home as 'military cargo 200' – in coffins. Andropov's successor, Konstantin Chernenko, breathless and wheezing with advanced emphysema, was too weak to raise his hand to his predecessor's coffin as it was carried into Red Square. He was dead by March 1985 (Moynahan, 1994, p. 222).

Too ill to argue with an ageing Politburo, Andropov had been cheated of the opportunity to implement any meaningful economic reforms. Nevertheless, braver, and much more confident than any of his predecessors, he had the foresight and ability to groom a worthy successor. According to Coleman (1997, p. 139): 'Like Andropov, [Mikhail Sergeevich] Gorbachev understood that the Brezhnev legacy meant reform simply had to be risked again. Brezhnev's caution was now more dangerous than Khrushchev's boldness. Still, Brezhnev's ageing cronies in the Politburo didn't see it that way – yet'. Facing severe opposition from dutiful, and by this time apprehensive, proponents of the Brezhneva era, Gorbachev was unable to rally the votes necessary to win the General Secretaryship.

Locked in tense party negotiations for nearly four days, the exhausted Politburo 'selectorate' elected Konstantin Chernenko, a Brezhnev appointee who had been denied the position less than two years earlier in November 1982. The decision, which had been delayed by a deep and widening division at the upper echelons of the Soviet hierarchy, was facilitated by a unique and disturbing compromise. While Chernenko had been named

General Secretary, enough power had been placed in Gorbachev's hands, according to Kort (1993, p. 280) 'that *Pravda* at one point referred to him as the second secretary'.

The opposition continued throughout Konstantin Chernenko's brief thirteen-month tenure. The reform process slowed and in some cases stopped. Policies and international relations, swayed to and fro by the ongoing debate at the Politburo, were inconsistent and often contradictory. In the spring of 1984, the Soviet Union shocked the world by boycotting the US summer Olympics. A fresh and determined effort was made to limit 'unsupervised' contact between Soviet citizens and the few foreigners that were 'permitted' to visit the USSR. In the meantime, Gorbachev, and his policy of *novoe myshlenie* (new thinking), had gained both political influence and recognition. By December, the abrasive foreign policy had been almost completely reversed, and the USSR had agreed to resume disarmament negotiations with the Reagan government.

Gorbachev's views on Soviet energy policy, and specifically the relative 'priority' of the oil industry, had been established well in advance of his promotion to second secretary. In 1983, on the conspicuous occasion of Lenin's birthday, Gorbachev delivered a speech stressing the vital importance of the machinery sector placing it well ahead of the energy industry in a revolutionary new ranking of critical government priorities (Gustafson, 1989). The speech, partially inspired by Andropov's commitment to conservation, and (undoubtedly) the disturbing spectre of rising energy costs, triggered a fierce battle over 'investment' priorities that would rage throughout 1984 and the first half of 1985, frustrating planners' attempts to draft a comprehensive version of the twelfth five-year plan.

The new energy policy set out in the Aleksandrov Commission's two-phase energy programme was announced to the public one month after Andropov's death, in March 1984. On this round, at least, the dedicated Brezhenv (supply-side) conservatives had clearly won the day. The programme, which had been submitted to the Central Committee in June 1983, had been worn down by years of debate, controversy and resistance. As related by Gustafson (1989, p. 45):

> In particular, the published program called for oil and condensate output to keep growing to the year 2000 (some of the Aleksandrov Commission's experts, on the contrary, had stressed oil's high cost and uncertainty, hinting that output should be allowed to peak), and it was vague about the conservation measures to be undertaken in the second half of the 1980s. In short, the program showed several signs of retreat to an energy policy still heavily based on oil,

undoubtedly reflecting a combination of weakened leadership at the top and the worrisome events of 1983 and 1984.

In November 1984, N.K. Baibakov, the chairman of Gosplan, announced the guidelines for the 1985 (annual) plan to the Supreme Soviet. Once again, energy supplies had been given clear priority over the machine sector. By 1985, Soviet investments in the oil industry had risen to 10.4 billion roubles, a significant 17 per cent increase over the 8.9 billion roubles reported only one year earlier in 1984 (see Table 6.1). The age-old conflict between short- and long-term planning goals had, yet again, been resolved in favour of the immediate short-term. The Gorbachev alternative, conservation, and the modernization and revitalization of the Soviet machine sector, would be delayed until the eleventh hour and his inevitable succession to General Secretary. In the meantime, energy, 'the lifeblood of the Soviet empire', had become a financial burden – another item in the long list of claims (urgent requests for capital) on the scarce financial resources of the USSR.

6.2 THE BEST LAID SCHEMES OF MIKHAIL SERGEEVICH GORBACHEV

Gorbachev had not been properly informed about the task ahead. Georgi Shakhnazarov, his closest political advisor, told me that Gorbachev had to wait until he took over the top job as general secretary of the Soviet Communist Party before he got his first look at the full KGB assessment of the impending economic disaster. Earlier access had been denied him, first as a Politburo member, then as second secretary and the clear heir apparent. Both Andropov and Chernenko, like Brezhnev before them, had kept their Politburo colleagues in the dark. Even Mikhail Gorbachev, one step from the pinnacle of Kremlin power, was ill prepared for the enormous economic challenge he would have to confront (Coleman, 1997, p. 140).

6.2.1 Fast Out of the Gate

In January 1985, Konstantin Chernenko was admitted to hospital with an advanced case of emphysema. Mikhail Gorbachev who was enjoying a tour of Great Britain with his wife Raisa, wasted no time arranging for the flight back to Moscow. His presence was essential to chair the weekly (Thursday) meetings of the Politburo. Fully prepared, he would not have to wait long for his next chance at the General Secretaryship. At 7:20 p.m. on the evening of 10 March 1985, Chernenko died of heart failure. Barely three hours later, at 10:30 p.m., Gorbachev summoned the Politburo to the Kremlin for an emergency resolution of the 'succession issue'. His quick action meant that it was impossible for three old guard opponents to attend on time and he was elected General Secretary by a slim majority of four to three (Coleman, 1997).

The ceremonial 'nomination' session of the Central Committee was, by all accounts, unusually harmonious. The outmanoeuvred old Brezhnevites had no alternative but to make the best of it. They agreed to present a façade of unity to the world. On 11 March 1985, Gorbachev was unanimously elected General Secretary of the Soviet Union. His first words to the Politburo were carefully chosen to inspire a perpetuation of the leaders' newfound solidarity and stressed the fragile foundations on which the new regime had been built.

The struggle for 'democratic unity' among the Soviet elite was undermined by the old guard majority of six to four in the Politburo and would frustrate the reform process throughout Gorbachev's entire six-year tenure (Coleman, 1997). Still, he was determined to succeed. Only 54 years old, and healthy, he was the first General Secretary since Lenin to have acquired a law degree. He was also the first to be well groomed in the art of international diplomacy.

In a manner strangely reminiscent of his patron Andropov, Gorbachev had given greater consideration to ideology, and succession, than to an actual concrete reform programme. Guided by little more than ideals, and a vision, he set out on the perilous journey of *glasnost*, *perestroika*, and *demokratizatsia*.[3] The need for radical economic reform was absolute. According to Moynahan (1994, p. 227):

> The old work ethic 'building socialism', was long since buried. Workers had had too many demands, too many lies, too much terror. . . . 'Our rockets can find Halley's comet and reach Venus', said Gorbachev. 'But our fridges don't work.'

In April 1985, General Secretary Gorbachev delivered his first policy speech to the plenum of the Central Committee. While the USSR could boast of tremendous accomplishments in virtually every aspect of the Soviet lifestyle, for example, all citizens had the benefits of a developed economy, employment, and social security, 'further changes were needed in order to achieve a 'qualitatively new state of society' (White, 1991, p. 269).

The list of mandatory 'changes' was exhaustive, and, to the consternation of Gosplan, overly ambitious and vague. It included: (1) the acceleration of economic growth (Coleman, 1997); (2) the efficient utilization of scarce natural resources including labour; (3) the decentralization of the command economy – including the devolution of the responsibility of 'cost accounting' to the level of the Soviet enterprise, and the establishment of a closer connection (link) between wages and the amount of 'work' that was actually being accomplished; (4) the modernization of the economy – specifically the modernization and reconstruction of existing enterprises (as opposed to the construction of new factories); (5) the redirection of capital and resources towards conservation of natural resources including oil; and (6) the development of socialist democracy and popular self government (Gustafson, 1989; White, 1991).

As foreshadowed by Gorbachev's first 'economic' policy speech (delivered on the 113th anniversary of Lenin's birthday 22 April 1983), a revolutionary new 'ranking' of investment priorities placed the machinery sector first, agriculture second, consumer goods third, and social welfare fourth. The status of the energy sector, noticeable only through its omission, was not mentioned at all (Gustafson, 1989).

With no history to guide the process – indeed, no Stalinist command style economy had ever been transformed to a workable, modern socialist democracy – the details of this vague policy prescription were to be forged, painstakingly and painfully, in the months and years to follow (Coleman, 1997). The challenge – to effect a radical political transformation without civil war, and/or to achieve democratic unity in a Politburo deeply divided between old guard conservatives and radical economic reformers – required the utmost in diplomacy, and accommodation. Tragically, the key elements in the reform programme – economic acceleration, the modernization of the machine sector, the efficient allocation of natural resources, and investment priorities – would be undermined by the very elements that had made them possible (the excessive accommodation of political resistance), and by harsh economic reality.

The minor portions of Gorbachev's new reform programme that had been prefabricated – and indeed were ready for implementation – concerned the improvement of what was perceived as 'a lack of discipline' in the workplace. Incompetent managers were to be fired without hesitation at the earliest possible opportunity and the leading culprit, 'the green snake' (alcoholism) was attacked in a comprehensive, albeit highly unpopular, campaign against alcohol. The state production of alcoholic beverages was reduced. Sixty-seven per cent of liquor stores were closed and the fine for public drunkenness was increased significantly (Kort, 1993).

The firing purges for incompetence and alcohol abuse were supplemented by Gorbachev's attempts to diminish the ranks of 'old-guard' conservatives, and to broaden and consolidate his power base. The 'cold purge' affected thousands of government and party officials, at all levels of the vast Soviet hierarchy, including: (1) prime minister Nikolai Tikhonov who was replaced by economic specialist Nikolai Ryzhkov; (2) Foreign Minister Andrei Gromyko who was replaced by Eduard Shevardnadze; and (3) Moscow party chief Victor Grishin who was replaced by Boris Yeltsin (Kort, 1993).

Serious, non-administrative reforms would take considerably longer. In short: as Gustafson (1989, p. 48) reports, the Politburo 'rejected draft after draft of the Twelfth Five-Year Plan as Gosplan and the ministries resisted shifting resources on the scale Gorbachev demanded'. As might have been anticipated in the midst of the second Soviet energy crisis, debates over the oil industry, and specifically the rapid development of West Siberia, were particularly severe. A torrent of opinions, complaints, and technical papers were submitted to Gosplan, and even the Politburo, under the protective umbrella of *glasnost*.

Siberian reformers, ever looking out for the interests of their region, lobbied for a balanced regional development programme. This included: (1) a reduction in the share of investment devoted to the energy industry; (2) a relocation of the Soviet Union's most 'energy intensive' industries such as mineral fertilizers, plastics, synthetic fibres, chemistry, petrochemicals, and heavy machinery to Siberia; and (3) increased investment in Siberian infrastructure, housing, and social amenities (Gustafson, 1989). Gosplan, seeking to balance the plan, and undoubtedly to benefit from the established infrastructure in the older producing regions, argued for the 'rational' development of West Siberia, and greater utilization of enhanced oil recovery in the Volga-Urals.

At the same time, familiar problems associated with corruption, misinformation and lies had infested the oil industry, severely complicating an already intricate planning dilemma. The incentives, fully specified and enforced by the rigid central planning process, were undeniable. Mingeo, responsible for the fulfilment of overly ambitious reserves targets, tended to overstate reserves. The Oil Ministry (MNP), hoping to attain and achieve reasonable production targets, tended to understate the potential of oil fields. According to official estimates reported by Gustafson (1989, p. 126), in 1982 'the forecasts of Siberian oil prospects by various agencies differed from one another by 50 per cent and more'. To cite only one conspicuous example: in the early 1980s, a renowned Soviet geologist, Farman Salmanov, sent

> a flurry of urgent telegrams. . . to Moscow, heralding the discovery of a field as large as Samotlor at Salym. His extravagant claim was based on the results of tests made at a mere five exploration holes. Salmanov claimed that the oilmen were ignoring the new giant. Mal'tsev immediately jumped into action. Roman Kuzovatkin, by then head of Yuganskneftegas, was made personally responsible for commissioning these five wells. He had to call the Minister each evening and report on the day's progress. . . . within a year, all five wells were flooded with water. You would expect Salmanov to have been embarrassed, but nothing like that happened. In fact, he blamed the oilmen for the disappointing well performance. Mysteriously, 860-mmt of additional reserves appears on Tyumen's books. Salmanov, discoverer of Siberia's oil wealth, had resorted to finding fictional oil just to save face (Tchurilov, 1996, p. 110).

The super-giant Bazhenov oil field (commonly referred to as Salym) was, according to Riva (1994, pp. 30–31), part of 'a vast and complex bituminous shale formation that occurs over a large part of Siberia. Although some oil was produced from fractures in the shale on the Salym uplift, the Soviets were unable to economically extract oil from the shale itself'.

The discovery, which had been enthusiastically reported throughout the USSR, sent oil shares plummeting on the New York Stock Exchange (Riva, 1994). By June 1985, it had been exposed as an error in evaluation. Aside from being a considerable source of embarrassment to the USSR, the ruse had created difficulties in the planning process, and a considerable misallocation of critical scarce resources.

The tendency towards statistical misrepresentation (lies) by the oil workers was exacerbated by unreasonable production targets, the firing purges of 1984–85, and political pressure from the Oil Ministry (Gustafson, 1989). In 1985 alone, the Nizhnevartovsk production association received official visits, that is, warnings, from 300 commissions from Moscow. According to Gustafson (1989, pp. 106–107):

> They extracted promises from the local managers to pump harder, then flew off again. The ministry made constant nervous changes in targets. But the one thing that did not change was the constant pressure to produce. It was felt to such a point, indeed, that in some new fields desperate managers were producing oil from exploratory wells. Local officials reacted by systematically misreporting outputs, and phoney data flowed up the ranks in response to the threats flowing down. As the central press poured on criticism of the disarray in the oil fields,

the local Party apparatus in Nizhnevartovsk (the site of the worst trouble) attempted to squelch it, an early case of local resistance to glasnost.

In September 1985, Gorbachev flew to Siberia to facilitate the tedious and highly controversial planning proceedings. In the words of Lev Tchurilov (1996, pp. 191–192):

> Everyone at the Oil Ministry was bubbling with excitement when it was announced that Mikhail Sergeyevich Gorbachev planned an official visit to Tyumen. Perhaps the government might really do something concrete to help the industry. [The Oil Minister V.A.] Dinkov decided to mount an exhibition in Surgut and Nizhnevartovsk, demonstrating to Gorbachev that the oil field equipment produced in the USSR. left much to be desired. If you visited this show, you could not avoid the conclusion that without better support from the machine builders, the oil industry would not survive, let alone flourish.

Gorbachev, himself, viewed the visit as an opportunity to promote the new state policies of *glasnost* and *demokratizatsia*, and to negotiate a compromise. No feelings were spared in a scathing criticism of corruption and 'gross' negligence.

As reported by Gustafson (1989, p. 107) on the issue of errors in reserves and production statistics he accused local officials of lying, and he wagged his finger at them, exclaiming, 'That won't do, comrades!'

On the issue of Siberian living conditions:

> He urged the Party authorities 'not to give in to any economic planners if they forget' about the priority of providing 'normal living conditions' for the workers; and he warned that unless Siberian oil and gas construction projects also provided for people's comfort, 'the consequences and the damage are unavoidable'. We shall have production facilities, capacities, and jobs standing idle, producing nothing (Gustafson, 1989, p. 178).

> It is embarrassing ...when a drilling foreman addresses us and says that the greatest incentive in Nizhnevartovsk is to be given a ticket to see a movie (Gustafson, 1989, p. 120).

On the issue of declining reserves additions, and the excessive development of the super-giant oil fields, Riva, (1994, p. 31) reports Gorbachev as saying:

> It has now become clear that the time of golden gushers, of easy oil is coming to an end. It is necessary to switch to forced extraction of oil, to move to more difficult areas with fields providing lower yields and to develop more complex

deposits. Central departments, the Communist party, and the government and economic management organizations in Tyumen Province displayed sluggishness and took the road of least resistance in overcoming problems. They decided to compensate for their own shortcomings by increasing the producing burden on giant fields.

On the issue of defective equipment: Gorbachev promised 'to take all the necessary steps to revitalise . . . [the] equipment suppliers' (Tchurilov, 1996, p. 192). What exactly that would entail was never specified.

In the final analysis – despite Gorbachev's initial determination to give 'top' priority to the machinery sector, conservation, and the living conditions in Siberia – the requirement for increased oil supplies was absolute. Given the desperate condition of the coal and nuclear power industries, more oil was needed to fuel the economy, and balance the five-year plan. It was a necessary compromise, which under the new rules of *glasnost* had to be impressed on Tyumen oil workers.

'Tiumen's misfortunes give the economy a fever,' he said, [endorsing] 'higher targets for oil output' and . . . confirming the fact that the Politburo had decided on a significant 64 per cent increase in West Siberian energy investment, from around 50 billion roubles in the Eleventh Five-Year Plan to 82 billion in the Twelfth (Gustafson, 1989, p. 50).

The first draft of the twelfth five-year plan was presented to the public in the autumn of 1985. It reflected the full extent of the 'compromises' that would be necessary to achieve *demokratizatsia* and a willing participation in the process of reform.

First and foremost – in anticipation of the beneficial influence of reform – the plan called for a significant (positive) increase in gross national income which was scheduled to accelerate over a fifteen-year time period, literally doubling by 2000 (Gustafson, 1989; IMF, 1992). The targets for the energy sector were equally ambitious, and included: (1) a 5.9–7.6 per cent increase in the production of crude oil from 595 million tonnes in 1985 to 630–640 million tonnes by 1990; (2) a 29.9–32.3 per cent increase in the production of natural gas from 643 billion cubic metres in 1985 to 835–850 bcm by the year 1990; (3) a 7.4–10.2 per cent increase in the production of coal from 726 million tonnes in 1985 to 780–800 million tonnes in 1990; (4) a 19.2–21.8 per cent increase in electricity from 1544 Bkw-hr in 1985 to 1840-1880 Bkw-hr by 1990; and (5) a 133.5 per cent increase in nuclear power from 167 Bkw-hr in 1985 to 390 Bkw-hr in 1990 (see Table 6.13). The aggregate Soviet energy target envisioned a 16 per cent increase in the production of fuels throughout the USSR. The final version of the plan was published almost a year later in June 1986, and specified production targets in the middle to high end of the ranges proposed by the draft (Gustafson, 1989).

Table 6.13: Energy targets for the twelfth five-year plan, 1985–90

	1985 Actual	1990 (Draft Targets)	1990 (Final Targets)
Oil (million tonnes)	595	630–640	635
Gas (billion cubic metres)	643	835–850	850
Coal (million tonnes)	726	780–800	795
Electricity (Bkw-hr)	1,544	1,840–1,880	1,860
Hydropower (Bkw-hr)	215	n.a.	245
Nuclear (Bkw-hr)	167	390	390

Source: Gustafson (1989).

The targets for the oil industry were supported by a detailed, and ambitious, exploration and development programme. The twelfth five-year plan called for the following improvements to the Soviet oil industry: (1) a 39 per cent increase in MNP exploration and development drilling from the 143.8 million metres completed in 1981–85 to 200 million metres; (2) a 39 per cent increase in 'oil and gas' exploration drilling from all three ministries (the oil and gas ministries, and Mingeo) from the 32.4 million metres reported in 1981–85 to 44.9 million metres in the years 1986–90 (see Table 6.8). The bulk of the exploration effort was to be concentrated in West Siberia, where Mingeo was to increase its exploration effort by 90 per cent; (3) the rapid development of 77 new West Siberian oil fields (Gustafson, 1989); (4) a significant expansion of the offshore drilling programme; and (5) an increase in enhanced oil recovery in the ageing producing regions. Excluding production from hydrodynamic methods, the target for incremental EOR production was set at 15 million tonnes in 1990 representing a 200 per cent increase over the 5 million tonnes reported in 1985 (see Table 6.12).

In the final analysis, the absolute necessity for increased production – and the continued rapid development of the large Siberian oil fields – was to be offset (compensated) by an accelerated exploration and small field development programme, enhanced oil recovery, and offshore development – all factors which could be expected to improve the long-term prospects for the Soviet crude oil industry.

The five-year investment targets were surprisingly realistic. In recognition of rising production costs throughout the USSR, the energy sector was to receive a significant 35 per cent increase in capital investment. Discounting questions concerning the accuracy of Soviet financial statistics, 180 billion roubles were to be invested in the energy sector in the years 1986–90; 47 billion roubles over the 133 billion roubles invested in the eleventh five-year plan (Gustafson, 1989). A significant portion of these funds, approximately 82 billion roubles or 46 per cent of the total, was earmarked for West Siberia. Given the extreme degree of uncertainty concerning oil and gas production costs, the exact level of investment planned for the individual oil and gas sectors was never fully specified (Gustafson, 1989). However, the bits and pieces of information that are available suggest that the oil industry was once again scheduled to receive the largest share of the Soviet Union's total energy investments (Stern, 1987).

Like Gorbachev's first official proposal for economic reform (presented to the Central Committee in April 1985), the targets for the conservation of Soviet energy supplies were ambitious and loosely defined. By the mid-1980s, the Soviet Union had fallen far behind the Western nations in energy conservation and the efficient utilization of scarce natural resources. Brezhnev's complex list of energy-saving targets (introduced in 1976), and Andropov's intricate demand side management programme – essentially Phase I of the Aleksandrov proposal (introduced in 1983) – had, so far, failed to produce the desired results (see Appendix D). By 1985 Gorbachev, and even the most conservative members of the Politburo, recognized that drastic measures would be required to make up the difference and attain energy efficiency parity with the Western nations.

According to Gosplan directives, the energy intensity of the Soviet economy was scheduled to fall by 8.5 per cent in the twelfth five-year plan; a significant acceleration over the 5.4 per cent reduction in energy intensity reported in the years 1980–85 (Gustafson, 1989). It is important to notice the fact that estimates of energy intensity in the Soviet Union are calculated on the basis of the increase in energy consumption per rouble of national income. This practice represents a noticeable departure from established

The Russian Oil Economy

Western procedure in which energy intensity has traditionally been measured in terms of gross national product (GNP) or gross domestic product (GDP). The discrepancy in basic accounting practices, and poor quality of Soviet data, severely complicated attempts to calculate energy/GNP ratios for the Soviet Union (Stern, 1987). The implied energy-national income (E/NI) elasticity of approximately 0.56 represented a noticeable improvement over the E/NI elasticities reported in the late 1970s (0.83), and early 1980s (approximately 0.76) (Hewett, 1994). In short: while Gosplan targets specified a 22.1 per cent increase in national income in the years 1986–90, domestic energy consumption was to increase by only 12.4 per cent over an identical time frame (see Table 6.14). The aggregate conservation target for the year 2000 of an E/NI elasticity of only 0.43 was considerably more ambitious.

Table 6.14: Energy conservation in the Soviet Union

The Five and Fifteen Year Targets for the Energy Intensity of National Income, 1986–2000

	Twelfth Five-Year Plan (1986–90)	Fifteen-Year Plan (1986–2000)
Growth Rate of National Income (%)	22.1	100
Reduction in Energy Intensity (%)	8.5	40
Implied Rate of Increase in Energy Consumption (%)	12.4	43

Source: *Gustafson (1989).*

The exact means by which the conservation targets were to be accomplished may be adequately described by two themes that were, by this time, familiar elements in Gorbachev's revolutionary new reform proposal:

1. *The Modernization of the Economy*:   As mentioned frequently throughout this book, the failure of the central planning system to accommodate technical progress and innovation led to an excessive

reliance on antiquated, energy guzzling, technology throughout the USSR Indeed, while the Western world had responded to the oil price shocks of 1973 and 1978-80 with the adoption, and utilization, of fuel efficient technologies, there was little incentive to do so in the Soviet Union where domestic energy prices were maintained at artificially low levels – primarily for accounting purposes – and failed to reflect either the true costs of production, or the opportunity costs inherent in world energy prices. Gobachev's proposal to modernise the Soviet economy – and revitalise the machine building sector would provide a partial solution to the 'planning' problems associated with both; (1) lagging industrial productivity, and (2) the conservation of scarce natural resources; that is energy.

As mentioned above, the 'modernization' proposal had been submitted to the Central Committee in April 1985. Gustafson (1989, p. 244) reports that: '"Revolutionary advances" were needed, [Gorbachev] told his audiences. The country must replace its ageing capital stock, and every sector of the economy must be re-equipped with up-to-date machinery'. By June 1986, a few of the 'implementation' details had been finalized, and Gorbachev's economic reform policy was beginning take shape. The twelfth five-year plan called for a number of concrete and quantifiable modernization and conservation initiatives. To cite only two examples:

- antiquated electric power stations, a significant number of which were still not equipped to burn any other fuel than mazut fuel oil (Stern, 1987) were to be converted (modernized) to accommodate natural gas and coal (Gustafson, 1989);
- the science and technology section of the plan was to be restricted to the consideration of energy efficient technologies; that is 'Machines that [used] 8–10 percent less energy than their predecessors' (Gustafson, 1989, pp. 146). Investment funds were to be reallocated 'from [the] extractive industries (and also from defence, although reformers were discrete on this point until 1988) to civilian machine building and manufacturing, particularly to advanced sectors such as computers, electronics, robotics, and modern chemistry' (Gustafson, 1989, p. 145).

Given the significant increase in energy investment scheduled for the twelfth five-year plan, the sole indication that Soviet investment resources were to be 'reallocated' (shifted gradually from the extractive energy sector to machine building), may be found in the aggressive five-year investment targets for civilian machinery. According to estimates provided by Hewett et al., 1987), investment in the civilian-machine building complex was scheduled to increase by 68 per cent in the twelfth five-year planning period.

2. *The Decentralization of the Command Economy*:   Gorbachev's proposal for 'radical' economic reform was announced to the Twenty-Seventh Party Congress on 25 February 1986. While Gosplan would retain the responsibility for the general 'guidance' of the economy, factories and farms would be given much more freedom to determine their own priorities. In the future, 'retail and wholesale prices would have to reflect the costs of production much more closely so that enterprises could be guided by "economic", rather than "administrative" regulators and so that the massive subsidies that held down the cost of basic foodstuffs could be reduced' (White, 1991, p. 271). This gradual decentralization of the 'planning' process was expected to aid in the efficient allocation (and conservation) of scarce natural resources throughout the USSR. It was not, however, an attempt to do away with either central planning or communism (Gustafson, 1989; White, 1991).

By the mid-1980s, it was generally accepted, even in the Soviet Union, that the major impediment to conservation and the efficient allocation of energy resources was the perception that energy supplies were abundant, and relatively inexpensive (Stern, 1987, p. 13) and that domestic fuel prices would be maintained at levels well below the international oil price for the foreseeable future. Attempts to change these perceptions in order to raise the domestic energy price to a level which might reflect the true (and rising) costs of production were severely complicated by the complexities and idiosyncrasies of the central planning process (see Appendix D).

Hence, while some progress had been made on the issues of decentralization and price reform, the sheer enormity of the problem – the reorganization of the entire Soviet planning matrix – had made it impossible to complete a comprehensive policy prescription by June 1986; the final deadline for the twelfth five-year plan.   Instead, the details were to be announced sporadically throughout the five-year planning period (see Table 6.14). In the meantime, 'pre-planned' reforms such as well repair, quantitative conservation initiatives, *glasnost*, the campaign against corruption and alcohol, and the modernization of the civilian machine sector, would receive top priority.

The directives from Gosplan were clear. As in the first oil crisis of 1977, no expense was to be spared in a massive campaign to revive the Soviet oil industry. A major (1985) initiative to revive the growing number of idle oil wells, primarily by flying an even larger number of well-repair and drilling crews to West Siberia, was accelerated (Gustafson, 1989). Political pressure was applied to ensure that the domestic oil workers would receive an adequate, and operational, inventory of machines and equipment. In the words of Tchurilov (1996, p. 192): 'Shortly after Gorbachev came back from Tyumen, a special decree was issued ordering the technical refurbishment of the oil industry'. Months later, in January 1986, the Minister of Minkhimmash announced that the oil and gas sector (the Glavneftemash production association) had become a top priority. According to Gustafson, 1989, p. 191): 'There followed a familiar flurry of bureaucratic activity; half the association's plant directors were fired; a blizzard of investigative commissions descended on Glavneftemash and other suppliers'.

In the same year, the planners were instructed to reallocate 500 million rubles of MNGS's Twelfth Plan construction budget to housing, schools, and other social facilities countrywide (Gustafson, 1989). Minenergo (the power ministry) was ordered to increase the portion of its investment funds to be allocated to housing in Tyumen by 50 per cent.

The intense pressure from Moscow, and additional investment funds, resulted in a significant quantitative improvement in oil field equipment and machines, electrical supplies and electricity, and even housing. As reported by Gustafson, 1989, p. 120):

> By the end of 1986 . . . pressure from the leadership had forced up the pace of housing construction dramatically . . . Gaslift equipment and electric submersible pumps began to arrive in large numbers. The supply of electricity improved when the gas-fired Surgut-2 power plant began operation.

In the first year of the programme, 1986, the gross value of oil field equipment produced in the Soviet Union reached 248 million roubles, an 8.3 per cent increase over the 229 million roubles reported only one year earlier.

Despite MNP's heroic attempt to establish a new quality-control office in Azerbaijan, Glavneftemash continued to turn out significant quantities of low quality, defective equipment. According to Tchurilov (1996, p. 192):

> Among the many promises Gorbachev made was a pledge to take all necessary steps to revitalize our equipment supplies. But the defects in equipment coming

from the big Azerbaijani plants actually increased. Some special inspectors were sent to Baku to complain and advise, but they might as well have been whistling in the wind. The machine builders did not have the faintest interest in how their equipment performed once it left the factory gates.

This was a fundamental weakness of the Soviet system. Unlike market economies where poor quality supplies would have resulted in a switching of suppliers, Soviet firms were 'tied' to their suppliers. They were not allowed to switch and as a result, individually or collectively, they had no way to sanction shoddy workmanship. This lack of market discipline meant that poor quality was an endemic feature of Socialist economies (Hobbs et al., 1997).

As always, quality, innovation and technical progress suffered from the race to fulfil unreasonable quantitative production targets. To cite only one example: in 1987 Minkhimmash (the Ministry of Chemical and Petroleum Machine-building) produced only 39 prototypes for new technical equipment in the 'fuels and energy sector', a 30 per cent reduction from the 56 prototypes that had been introduced in 1985 (Gustafson, 1989).

In recognition of these blatant deficiencies, the supply of domestic machinery was supplemented by a significant 63.8 per cent increase in technical imports. Excluding pipe and refinery equipment, the value of oil-related technical imports reached 579.9 million roubles in 1987, 225.9 million roubles over the 354 million roubles reported in 1985 (Gustafson, 1989).

Further, even as early as 1986, Gorbachev appears to have introduced a subtle change in import policy. In light of a considerable easing of tensions related to the Cold War, the 'Soviet Union's excessive reliance on imports from Eastern Europe satellite nations was tentatively reduced (Kort, 1993). According to the official trade statistics, Romania's share of oil-related technical imports was reduced gradually from 71.5 per cent of the total in 1985 to 66.3 per cent in 1986, and 59.7 per cent in 1987. The bulk of the reduction was 'made-up' (displaced) by an increase in technical imports from France.

Encouraged by a substantial increase in investment funds – and undoubtedly the endless procession of emissaries (commissioners) from Moscow – the Oil Ministry launched an aggressive production and development programme. MNP exploration and development drilling reached 46.1 million metres in 1988, a significant 40 per cent increase over the 29.7 million metres reported in 1985. The aggregate (oil and gas) drilling statistics from all three ministries registered a slightly smaller 38 per cent increase, from 35.574 million metres in 1985 to 49.078 million metres in 1988. As might have been anticipated, particularly given the aggressive quantitative production targets, the share of oil and gas drilling devoted to development increased from 80.7 per cent of the total in 1985 to 82.7 per cent in 1988 (see Table 6.15).

As specified by the twelfth five-year planning targets, the bulk of the effort was concentrated in West Siberia. According to estimates provided by the Centre for Foreign Investment and Privatization (Moscow), the oil production associations completed 29.235 million metres for exploration and development purposes in West Siberia in the year 1988 alone (see Table 6.16). Approximately 95.8 per cent of the total, some 28 million metres, were drilled in the Province of Tyumen. In summary, the drilling activities of the Tyumen production associations accounted for over 60 per cent of the 1988 MNP drilling programme for the entire USSR

As in the previous five-year planning period, the massive Tyumen drilling effort was further concentrated on the rapid development of previously discovered oil fields. The Tyumen oil production associations completed 27.602 million metres for development purposes in the year 1988, 98.6 per cent of their entire 28 million metre drilling programme.

Table 6.15: Exploration and development drilling in the USSR

	(million metres)		
	All Ministries		
Year	Exploration	Development	Total
1985	6.875	28.700	35.574
1986	7.302	34.115	41.418
1987	7.902	37.587	45.489
1988	8.475	40.603	49.078
1989	8.000	39.612	47.613
1990	7.182	35.378	42.560
1986–90	38.86	187.30	226.16

Source: Oil and Gas Industry Statistical Handbook, (various issues).

While the Soviet exploration and development programme would appear to be a simple repetition of previous 'errors in judgement', that is, the excessive focus on the rapid development of existing Tyumen oil fields, it should be noted that some attempt was made to address a few of the fundamental problems undermining the oil industry, in particular, to repair the growing number of idle oil wells and stem the decline in the ageing producing regions. To cite only a few examples, the twelfth five-year plan called for the following quantitative improvements to Soviet oil production: (1) an increase in production from Azerbaijan (Baku), from 13 million tonnes in 1985 to 14.5 million tonnes by 1990; (2) an increase in production from Kazakhstan, from 22.8 million tonnes in 1985 to 39.1 million tonnes by the year 1990; and (3) a reduction in decline rates from the Volga-Urals (the second Baku) (see Table 6.17). The decline rate in Volga-Urals production had been rising steadily, from 15 per cent in the years 1975–80 to 26 per cent in the years 1980–85. The twelfth five-year plan, however, called for a minor 43 million tonne reduction in crude oil flows from the Volga-Urals producing region, from 143 million tonnes in 1985 to 100 million tonnes in the year 1990. The implied production decline rate (only 30 per cent over five years or approximately 6.9 per cent per annum) represented a significant deceleration in rampant decline rates for the ageing Volga-Urals producing region.

To accommodate these minor (secondary) requests for improvement, the MNP exploration and development effort was cautiously accelerated in the following producing regions: (1) the North Caucasus, where exploration drilling reached 180.4 thousand metres in 1988 and a significant 34 per cent of the total 536 thousand metre drilling effort; (2) the Volga-Urals, where exploration drilling reached 1.171million metres in 1988, approximately 14 per cent of the total 8.627 million metre drilling programme; and (3) the Far East, where exploratory drilling reached 83.8 thousand metres in 1988, approximately 40 per cent of the total 210 million metre drilling effort.

The Gas Ministry, which had been granted an exclusive jurisdiction over the offshore oil and gas development programme from 1978 to January 1988, accelerated its drilling programme in Azerbaijan, Kazakhstan, and Komi, where tentative efforts were made to accelerate the exploration and development of promising offshore production regions.

Despite these minor improvements, however, the focus on development reigned supreme. According to official statistics, development drilling from all three ministries (the oil and gas ministries and Mingeo) reached 40.6 million metres in 1988, a 41.5 per cent increase over the 28.7 million metres reported in 1985. Over an identical time frame, exploration drilling had risen by only 23 per cent, from the 6.875 million metres reported in 1985 to only 8.475 million metres in 1988. The share of exploration, however, had fallen to only 17.3 per cent of the total Soviet oil and gas drilling effort in 1988, a 29 per cent reduction from the 46.3 per cent that had been reported in the ninth five-year plan, 1971–75.

As in the eighth, ninth, tenth, and eleventh five-year planning periods, a significant portion of the meagre Soviet exploration programme fell under the category of *razvedochnoe burenie* – exploratory or outlining wells drilled for the purpose of 'proving up' and developing previously discovered oil reserves. As noted above, Mingeo had been assigned the responsibility for exploration in West Siberian, and a significant portion of its drilling programme was still of the deep-drilling (pre-development) variety. This was primarily designed to fulfil the ambitious five-year targets for industrial reserve additions $(A+B+C_1)$. While the oil production associations had accelerated their West Siberian exploration effort in 1987–88, from 366.9 thousand metres in 1987 to 433.7 thousand metres in 1988, it may be safely assumed that these efforts were also directed to the immediate 'preparation' of previously discovered oil fields for rapid development.

In the words of Lev Tchurilov (1996, p. 192), who was employed at the Department of Oil Production at the time: 'After [Mal'tsev was dismissed in early 1985], we were under enormous pressure to bring new fields on line and any discussion about production methods was just swept under the carpet'.

As in the first Soviet energy crisis, the combination of increased capital investment and political pressure including Gorbachev's personal appearance in Tyumen, had a considerable degree of success. According to information provided by the oil and gas ministries, 15 859 new oil wells were commissioned in the USSR in 1988; a 32 per cent increase over the 11 984 commissioned in 1985. A significant portion of these was drilled in new oil fields. According to estimates provided by Tchurilov (1996), approximately 70 new deposits were proceeded with annually, and as many as 20 of these lay in West Siberia. Of course not all of these would prove economic to develop and others would prove negative.

As in the past, quality and productivity were sacrificed in the haste to develop new oil fields. A growing number of increasingly marginal oil wells were commissioned prematurely – before the reservoirs had been thoroughly delineated – and equipped with poor quality machinery, which would frequently be subject to breakdowns (Tchurilov, 1996). According to official (and of course biased) statistics, the productivity of new oil wells fell by 17 per cent in the first three years of the planning period; from 447 tonnes per month in 1985 to only 369 tonnes per month in 1988 (see Table 6.7).

Table 6.16: Exploration and development drilling activity for oil and gas by the oil production associations for select regions in the Russian Federation

(thousand metres)				
Region	1987	1988	1989	1990
Baltic				
Exploration	2.5	4.6	4.6	9.8
Development	52.0	46.2	49.2	45.1
North Caucasus				
Exploration	189.8	180.4	194.9	185.9
Development	352.4	355.2	374.5	326.8
Volga-Urals				
Exploration	1,109.4	1,171.2	1,132.7	1,011.0
Development	7,288.6	7,456.2	7,077.7	6,469.5
Timan-Pechora				
Exploration	50.2	67.7	48.4	38.4
Development	803.5	737.6	607.1	473.3
West Siberia				
Exploration	366.9	433.7	361.7	294.1
Development	26,066.0	28,801.5	27,497.1	25,207.8
Tyumen*				
Exploration	324.3	402.2	331.3	268.2
Development	24,891.3	27,602.4	26,334.7	24,177.3
Far East				
Exploration	91.6	83.8	74.8	58.0
Development	120.7	126.6	120.7	108.1

Note: * The Tyumen producing region includes all West Siberian oil production associations except Tomskneftegaz.

Source: Barton (1998).

As in previous planning periods, the average productivity of a typical Soviet oil well was undermined by the natural ageing of oil fields, the poor quality of machines and equipment, and the excessive use of water flooding throughout the USSR. According to Yurko and Reitman (1990), the average water-cut reached 73 per cent in the Soviet Union in 1987, a 3 per cent increase over the 70 per cent water-cut that had been reported in 1985. By 1987 the average productivity of a typical Soviet oil well had fallen to only 399 tonnes per month; a 12.5 per cent reduction in two years (see Table 6.10).

Table 6.17: Crude oil and condensates production in the USSR

(millions of tonnes per annum)						
Region	1985	1986	1987	1988	1990	1990 (Plan)
Russian Federation	542.0	561.0	569.6	569.0	516.0	565
European Russia	105.0	99.5	92.5	88.3	79.7	72
Kalingrad Oblast	1.5	1.5	1.4	1.3	1.2	
Komi ASSR	18.0	18.0	17.3	15.6	14.6	17
North Caucasus	10.5	10.0	9.8	9.4	8.4	5
Arkhangelsk Oblast	0.0	0.0	0.0	0.0	1.2	
Volga	75.0	70.0	64.0	62.0	54.3	50
Urals	68.0	66.0	62.0	61.4	59.5	50
Siberia	369.0	397.0	415.1	422.2	373.4	443
West Siberia	365.8	394.0	412.1	419.9	371.1	440
Tyumen	352.7	376.0	398.0	404.0	353.1	425
Tomsk	13.1	13.7	14.1	14.64	14.7	15
Sakhalin	3.0	3.0	3.0	2.3	2.3	2.7
Non-RSFSR	53.8	53.2	54.9	54.6	53.0	70
Ukraine	6.0	5.6	5.6	5.4	5.2	6
Byelorussia	3.0	2.0	2.0	1.9	2.0	2
Georgia	2.0	2.0	2.0	0.2	0.1	2
Kzakh	22.8	23.7	24.5	25.5	25.2	39.1
Turkmen	6.0	5.9	5.8	5.7	5.6	5
Other Central Asian	1.0	1.0	2.0	2.3	2.6	1
Azerbaijan	13.0	13.0	13.0	13.6	12.3	14.5
Not elsewhere stated & statistical difference	(-0.5)	(-0.9)	(-0.3)	(-2.2)	1.46	
Total USSR.	595.3	614.8	624.2	624.3	570.5	635

Sources: Gustafson (1989); OECD/IEA (1995); *The Oil and Gas Industry of the USSR: Statistical Handbook* (various issues); Stern (1992); Barton (1998).

Despite problems associated with the firing purges, alcohol, and jet-lag, Soviet drilling crews benefited from the increased availability of equipment

and machinery (however defective). The productivity of drilling crews
increased by 28.8 per cent; from 13 153 metres per rig-team in 1985 to
16 943 metres in 1988, a noticeable improvement over the 16.2 per cent
increase that had been recorded in 1982–85 (see Table 6.6). At the same
time, considerable progress was made in the well maintenance and repair
sector, and the number of idle oil wells was reduced to only 5796 in 1988, a
15.9 per cent reduction from the 6890 reported in 1985 (see Table 6.5).

From the perspective of crude oil flows alone, this was a 'golden era' in
the history of Soviet oil industry. Despite the poor productivity of oil wells,
crude oil production in the USSR reached 624.326 million tonnes in the
year 1988; a significant 29 million tonne increase over the 595.291 reported
in 1985 (see Table 6.5). Once again, the prolific Tyumen production region
was responsible for the bulk of the increment. Annual production from the
Province of Tyumen reached 404 million tonnes in 1988, 65 per cent of
total Soviet oil flows (see Table 6.17).

To the credit of the Oil Ministry (MNP), considerable progress was
made in the ageing producing regions. Crude oil production from the North
Caucaus and Volga-Urals producing regions reached 132.8 million tonnes
in 1988 representing only a minor 13.5 per cent reduction from the 153.5
million tonnes reported in 1985. The implied annual decline rate, 4.7 per
cent per annum, represented a significant deceleration from the implied
annual decline rate of approximately 5.6 per cent recorded in the eleventh
five-year planning period. The trend was most apparent in the Urals where
the annual decline rate fell significantly from 2.94 per cent in 1985–86, to
less than one per cent in 1987–88, and production was estimated at an
encouraging 61.4 million tonnes in 1988. In Azerbaijan (Baku), crude oil
flows reached 13.6 million tonnes in 1988, a 600 000 tonne increase over
the 13 million tonnes reported in 1985.

6.2.2 The Real Cost of Recovery

The recovery of 1986, and subsequent peak in Soviet oil production, has
generally been attributed to a quality which Mikhail Gorbachev would have
referred to as the 'human factor' – an active, enthusiastic participation in
the process of construction (White, 1991). Indeed, given the extent of the
fundamental problems and challenges facing the Soviet oil industry, the
achievement of a recovery, of any sort, is quite remarkable. However, the
key elements of Gorbachev's reform programme – technical progress
(modernization), and the efficient allocation of scarce natural resources
(exploration) – were abandoned in the obsession with ever increasing

construction. In other words, according to Gustafson (1989, p. 120): 'Of the West Siberian gains in 1986, more than half came from either conventional or one-time expedients (infill drilling – that is, additional drilling of developed fields – and well repair), and less than half from new fields and well mechanisation'.

In the final analysis, recovery had, once again, been achieved at tremendous expense and with little regard for the future prospects of the oil industry. The average production costs for crude oil in the USSR reached an estimated 17.12 roubles per tonne in 1988, a significant 25 per cent increase over the 13.68 roubles per tonne reported in 1985. The marginal capacity costs, roubles per tonne (rbls/t) of new oil industry capacity, registered a formidable 46 per cent increase, from 62.4 (rbls/t) in 1985 to 91.3 (rbls/t) in 1988 (see Table 6.6).

Inexplicably, no attempt was made to reform crude oil producer prices, which had been maintained at approximately the same level since 1982. The wholesale enterprise price for crude oil registered only a slight 1.68 per cent increase from 24.86 rbls/t in 1985 to 25.28 rbls/t in 1988. By 1988, the 'gross profits' of the oil industry had fallen to 8.16 rbls per tonne, 32 per cent of the entire proceeds from domestic oil sales, and a significant 4.5 per cent below the normative after-tax rate of 36.5 per cent that had been established as a basic accounting procedure (landmark) for the industry in 1982.

At the same time, investments in the energy sector had been rising precipitously. According to estimates provided by Goskomstat, the level of gross fixed investment in the oil sector grew by 10.6 per cent in 1986, 14.4 per cent in 1987, and 3.5 per cent in 1988, implying a significant 31 per cent increase in oil investment in the first three years of the twelfth five-year planning period. The level of gross fixed investment in the coal industry was increased at approximately the same rate (31 per cent) over an identical time period (1986–88). The statistics for the natural gas industry, which suggest a 37 per cent increase in gross fixed investment in the years 1986–88, were even more alarming (see Table 6.18).

By 1988, the investments required for the production of 'primary' energy supplies in the Soviet Union had exceeded Gosplan expectations, placing an unsustainable, and growing, requisition on over-stretched government revenues. In the words of G.P. Bogomiakov (the First Party Secretary for Tyumen) 'the recovery of 1986…was a classic example of the much-denounced policy of "The plan at any price!" Extensive methods of development remain the order of the day' (Gustafson, 1989, p. 199).

By the mid-1980s, Gosplan had effectively planned its way into a quagmire, a vicious cycle of rising targets and expenditures, from which there was no easy escape. First and foremost was the absolute requirement for increased energy supplies. The alternative, constant or falling domestic fuel supplies, would have resulted in the unthinkable – a domestic energy shortage (lower exports and/or increased imports of energy) – which in turn would have threatened government revenues and the fulfilment of annual targets in industries throughout the USSR. At the same time, rising marginal production costs and unplanned increments to energy investment had effectively eliminated the possibility of a diversion of investment funds from the extractive industries to the civilian machine-building industry, further frustrating the future prospects of the energy industry.

6.2.3 The Collapse of International Oil Prices

The precipitous Soviet financial situation was aggravated by a sudden change in OPEC production policy, and the collapse of world oil prices in the mid-1980s. To summarise the complex developments on international oil markets: the second oil price shock of the late-1970s resulted in a steady increase in 'new' crude oil flows from the non-OPEC producing nations (including a significant increase in net exports from the Soviet Union), and conservation and fuel switching in the developed nations. According to Reinsch et al. (1988, p. 6): 'Between 1979 and 1985, world demand for crude oil [outside the Centrally Planned Economies (CPEs)] fell by over 5 MMB/d, while non-OPEC supply (including net exports from the CPEs) grew by some 6 MMB/d.'

In the meantime, OPEC had assumed full responsibility for the defence of high oil prices and had been cutting back its production steadily since 1979. By the summer of 1985, OPEC production had fallen to only 14 MMb/d; less than 45 per cent of the peak flow rate (31.8 MMb/d) that had been reported in the third quarter of 1979.

The official OPEC production policy was amended shortly thereafter, and Saudi Arabia launched a campaign to reduce world oil prices, and recapture market share by driving the marginal non-OPEC producing nations out of the international oil market (Reinsch et al., 1988). By 1986, the spot price of Arabian light crude oil had fallen to US$13.64 per barrel, less than 50 per cent of the US$28.08 per barrel that had been reported only two years earlier in 1984.

It was a state of affairs (calamity) that had profound implications for petroleum activities around the globe. With the oil surpluses and the

Table 6.18: Growth of gross fixed investment by sector, 1976–90

(Average Annual Growth Rates in Comparable Prices)

Sector	Tenth Five-Year Plan 1976–80	Eleventh Five-Year Plan 1981–85	Twelfth Five-Year Plan 1986–90	1986	1987	1988	1989	1990
Total	3.3	3.5	6.2	8.4	5.6	6.2	4.7	0.6
Industry	3.4	4.2	7.0	8.4	5.7	5.9	7.8	-7.9
Electricity	3.9	5.1	0.7	0.1	2.1	3.9	-3.3	-14.8
Coal	3.9	3.6	6.9	6.8	10.7	10.8	-0.2	-6.9
Oil	12.2	8.8	8.2	10.6	14.4	3.5	4.5	-13.7
Gas	2.9	12.5	18.8	14.4	8.7	10.0	45.6	-29.4
Machinery	4.0	4.0	5.0	12.4	0.9	9.2	-1.8	1.8
Construction Materials	-0.2	1.8	9.9	-2.8	21.6	12.3	10.0	5.6
Agriculture	2.7	1.1	5.1	6.5	2.4	6.3	5.2	6.8
Transport	4.9	3.8	-1.5	3.5	4.4	4.6	-16.6	13.1
Housing	1.9	5.9	7.6	9.9	8.7	6.2	5.8	3.7

Source: IMF (1992).

Table 6.19: Soviet crude oil and condensates production and international trade statistics

Year	Crude Oil and Condensate Production (million tonnes)	Total Exports Crude Oil and Petroleum Products	Crude Oil and Petroleum Products Exports for Hard Currency *	Total Imports Crude Oil and Petroleum Products	Domestic Oil Consumption	Net Exports
1985	595	167	60	14	442	153
1986	615	186	69	17	446	169
1987	624	196	76	16	444	180
1988	624	205	90	22	441	183
1989	607	184	62	15	438	169
1990	570	159	68	11	422	148

Note: * Oil exports to Finland are not included in the category of hard currency exports. In short, Finland conducted a significant portion of its trade with the Soviet Union on a barter basis, so that its inclusion with the 'hard currency' nations would be misleading.
Domestic Consumption has been derived in the following manner: Consumption equals total production minus total exports plus total imports. Net exports are equal to total exports plus total imports.

Source: Smith (1993).

collapse of oil prices in the mid-1980s, foreign investment in the energy sector waned. Petroleum-producing countries were faced with increasing competition for the investment funds needed to sustain production levels. Even countries that had state-owned petroleum monopolies found that development previously funded by oil revenues now required outside funding. A new era emerged in which countries began to pursue liberalized investment policies that in some instances included privatization of state-owned companies.

The implications for Soviet hard currency receipts were grim. While Soviet oil exports had been rising steadily throughout the eleventh and twelfth five-year planning periods, from 162.26 million tonnes in 1981 to 205 million tonnes in 1988, the financial gains, measured in terms of hard currency receipts, were undermined by the dramatic reduction in world oil prices (see Tables 5.17 and 6.19).

The unanticipated 'shortfall' in gross oil revenues was aggravated by the following considerations:

1. *The reservation of a significant portion of oil exports for CMEA barter agreements, and/or soft currency exports to political satellites.* It is interesting to observe that Soviet oil exports to Eastern Europe remained relatively constant over the period 1980–88, despite Soviet efforts to, according to Stone (1996, p. 87), 'scale back its commitments in Eastern Europe and to compel the East Europeans to take more responsibility for their own supplies of raw materials'. According to estimates provided by the Vienna Institute for Comparative Economic Studies (VICES, 1990), the volume of crude oil exports to Eastern Europe was reduced only slightly, to 77.41 million tonnes in 1988, a minor 4.3 per cent reduction from the 80.89 million tonnes reported in 1980 (see Table 6.20).

   The exact value of these exports, and the magnitude of the implicit 'oil subsidy' to Eastern Europe, is considerably more difficult to determine. As mentioned above, all trade with CMEA member nations was conducted with soft currency (TRs) and governed by complex five-year barter agreements. The oil prices utilized in CMEA agreements were determined according to the Bucharest formula, a five-year moving average of world oil prices. A strict adherence to this formula in periods of falling oil prices such as 1985–86 would have forced CMEA member nations to pay a higher nominal (TR) value than the prevailing world oil price.

The Soviet Union made no attempt to introduce a formal revision of the rigid five-year oil pricing formula. Instead, 'favourable terms of trade' were bestowed upon East Europeans according to the dictates of increasingly complex bilateral trade agreements (Gustafson, 1989). According to Stone (1996, p. 86), on the odd occasion when disputes arose over the proposed terms of agreement, 'the Soviets simply announced the new terms, and the East Europeans were compelled to accept them. When it chose to do so, the Soviet leadership was still in a position to dictate terms to its satellites'.

Needless to say, recent attempts to estimate the magnitude of the implicit 'oil' subsidies have been hopelessly complicated by the vast array of terms and conditions in the detailed bilateral trade agreements. Still, years of debate have led to an overwhelming international consensus. The implicit CMEA energy subsidy was real, and extremely expensive.

Table 6.20: Soviet oil exports to the industrialized West and East Europe

	(millions of tonnes)			
Year	Soviet Oil Exports to Eastern Europe	Soviet Oil Exports to the Industrialized West (OECD)	World Oil Price Arabian Light Crude Average Spot Market Rrice ($US)	Value of Soviet Oil Exports to the OECD Nations Millions of $ US
1980	80.89	57.0	35.69	14,157
1981	80.13	53.5	34.30	14,066
1982	74.79	69.0	31.76	16,592
1983	72.93	77.9	28.80	17,522
1984	73.70	81.4	28.08	16,596
1985	75.04	67.5	25.02	12,692
1986	81.10	77.9	13.64	7,888
1987	79.90	83.8	17.28	11,214
1988	77.41	95.3	13.43	10,537
1989	56.39*	78.4	16.35	10,613
1990	43.36*	n.a.	21.54	n.a.

Sources: Gustafson (1989); VICES (1990); Smith (1993).

In the years 1985–87, however, the costs of Soviet 'oil' subsidies to the soft currency trading partners were elevated by a significant deterioration in the terms of trade in bilateral agreements with Finland, India and Yugoslavia. According to estimates provided by Smith (1993, p. 139): 'The net effect on the terms of trade was equivalent to an additional indirect cost of approximately $150 million for each dollar per barrel fall in the oil price'.

2. *The Soviet Union's heavy reliance on the spot market – rather than long-term contracts – as the primary vehicle for hard currency oil exports.* The Soviet Union's excessive reliance on the spot market for international oil sales was due to the planners' use of oil exports as a means of maintaining some semblance of order in the highly volatile balance of payments accounts. Soviet energy exports were sporadically manipulated to achieve an equilibrium in the balance of payments accounts in the years 1970 to 1985. To be specific, deficits arising from increased imports, primarily of grain, were often corrected by sudden increases in the level of hard currency oil exports. According to Gustafson (1989, p. 269): 'When world energy prices rose and the trade balance did not require adjustment, Soviet hard-currency oil exports either stagnated or dropped'. As always, the interface between the planner's need for stability and the unpredictability of international markets proved difficult.

   As a general rule of thumb, prices obtained on the spot market are more volatile than posted prices – and certainly those specified by the terms of long-term contracts – and tend to overreact to short-term market pressures. To cite only one example: the posted price of Saudi Arabia Light averaged US$16.15 in 1986, US$2.50 per barrel higher than the average spot price of US$13.64 reported for Arabian Light crude. The implied nominal discrepancy was significant. According to Smith (1993, p. 139): 'At the 1986 level of net exports for hard-currency (54 million tonnes) each dollar fall in the price of a barrel of Soviet crude cost the Soviet Union $400 million directly in lost hard currency revenue'.

   Given the unusual Soviet export policy, and the need for a specific level of gross oil revenues to balance the hard currency payments account, this approximation is likely to be a considerable underestimate of forgone oil revenues. At lower oil prices, the Soviet Union would have to export a larger quantity of oil supplies to maintain a specific value of revenues, further depressing the West European prices for Urals' crude.

6.2.4 Rising Debt and Deficits

The failure of the Soviet Union to reduce its commitments to Eastern Europe, falling international oil prices, and the excessive reliance on international spot markets, had severe implications for the balance of payments account. Hard currency oil exports were increased significantly in the first three years of the twelfth five-year planning period, from 60 million tonnes in 1985 to a peak flow rate of 90 million tonnes per annum in 1988 and literally absorbed the entire increment to Soviet crude oil flows. By 1986, the height of the oil price crash, the hard currency receipts from Soviet oil exports had fallen to US$6851 million – less than 50 per cent of the US$14 720 million reported in 1984 (see Table 6.21). The situation improved slightly when the world oil price began a faltering recovery in the year 1987, and the hard currency receipts from oil exports reached US$9925 and US$9528 million in 1987 and 1988 respectively.

The once lucrative crude oil industry, which had accounted for over 50 per cent of the total hard currency receipts in the Soviet Union in the year 1984, could no longer be counted on as a reliable 'swing' producer which could be called upon at will to 'balance' highly volatile balance of payments accounts. While oil was still the primary source of hard currency receipts in the USSR, its share had fallen significantly from 52 per cent in 1984 to only 34 per cent in 1988. The opening up of the planned economy to more international trade in an attempt to raise living standards through imports and to acquire more technology, thus created a host of new problems and further discredited formally planning the economy.

According to estimates provided by Smith (1993), the Soviet Union incurred a deficit in its hard currency accounts in the years 1985, 1986 and 1988 (see Table 6.22). To make up the deficiency, the Soviet government was forced to enter international capital markets with requests for new borrowing to cover estimated, and growing, balance of payments deficits. By 1988 the level of Soviet 'hard currency' indebtedness to Western financial institutions and governments had reached US$30 465 million, a significant 199 per cent increase over the $10 200 million reported at year-end 1984. In short, by the late 1980s the Soviet balance of payments deficit had become a problem and a growing source of irritation to the over-stretched Soviet economy. Western financial institutions became increasingly worried about lending to the USSR. In part this was because there was no way to secure the loans against tangible assets.

The size of the Soviet government's deficit was even more alarming. According to official estimates, the Soviet budget deficit reached 90.1

Table 6.21: Hard currency earnings from exports in the Soviet Union

(millions of US$)

	1984	1985	1986	1987	1988	1989	1990
Total	28,361	23,694	21,802	25,756	27,826	29,412	33,500
Energy	18,895	15,401	10,898	13,268	12,745	12,690	16,370
Oil *	14,710	11,240	6,851	9,925	9,528	9,329	11,600
Gas	3,736	3,797	3,694	2,725	2,538	2,648	3,602
Arms and Military	3,868	2,937	4,384	4,626	4,798	4,707	3,799
Diamonds and Precious Metals	1,730	1,715	2,228	2,829	4,162	4,866	5,935

Expressed as a Percentage of Total Hard Currency Earnings

	1984	1985	1986	1987	1988	1989	1990
Energy	67	65	50	52	46	43	49
Oil	52	47	31	39	34	32	35
Gas	13	16	17	11	9	9	11

Note: * Oil exports to Finland are not included as a source of hard currency.

Source: Smith (1993).

207

Table 6.22: Hard currency balance of payments and debt in the Soviet Union

(millions of US$)

	1985	1986	1987	1988	1989	1990
Exports	23,694	21,802	25,756	27,826	29,412	33,500
Imports	25,662	23,056	22,882	28,362	34,562	35,100
Trade Balance	(1,968)	(1,254)	2,874	(536)	(5,150)	(1,600)
Services Balance	(1,841)	(1,827)	(1,680)	(3,307)	(3,839)	(6,200)
Current Account Balance	(3,809)	(3,081)	1,194	(3,843)	(8,989)	(7,800)
Gross Debt a)	27,979	33,061	36,653	40,856	51,820	61,152
b)				45,753		
Net Debt to Western Financial Institutions and Governments	14,917	18,292	22,519	30,465	37,320	45,355
Soviet Hard Currency Claims	14,650	18,400	22,800	25,100	28,500	n.a.

Source: Smith (1993).

billion roubles in 1988 – a 400 per cent increase over the 18.0 billion nominal roubles reported in 1985. According to White (1991, p. 276):

> The main immediate causes of this trend are two fold – the impact of the anti-alcohol campaign and that of the collapsed in world oil prices on budget revenue. According to the official budget figures, which may well understate, net annual budget revenue from foreign trade operations fell by around 20 billion rubles in 1985–88.

To these can be added the following components: (1) Gorbachev's enthusiastic commitment to Abel Aganbegyan's programme of 'economic acceleration' which necessitated a significant increase in investment throughout the USSR; and (2) an alarming increase in consumer subsidies, which rose by 55 per cent or 31.8 billion nominal roubles; from 58.0 billion nominal roubles in 1985 to 89.8 billion roubles in 1988.

As outlined above, Gorbachev's prescription for economic reform (acceleration) had been announced to the Central Committee in April 1985, barely one month after his succession to General Secretary. He had not been properly informed about the true status of the economy, and indeed had been denied access to the Soviet budget, and all statistics concerning military expenditures by his patron (Andropov), and Chernenko. While Gorbachev was undoubtedly given full access to all of these statistics in March, it is conceivable that in the first year of his tenure in the Kremlin, he would have had only a limited understanding of the KGB's assessment of impending economic disaster (Coleman, 1997). It is likely that the new politicians with their predominantly military–industrial background did not understand, or even care about something as opaque as a budget deficit, even if they were told.

At any rate, despite heated protest from the Ministry of Finance and the central planners, the final draft of the twelfth five-year plan called for a significant accelerated increase in investments, and expenditures, throughout the USSR. The stolid Soviet approach to finance, which had kept the budget deficit constant at approximately 2–3 per cent of GNP in the years 1965–85, inclusive, had suddenly been relaxed. State expenditures reached 417.7 billion nominal roubles in 1986, an 8 per cent increase over the 386.5 reported in 1985 (see Table 6.23). By 1986, the central deficit had reached 47.9 billion roubles, a significant 166 per cent increase over the 18 billion roubles reported in 1985, and an intimidating 6.0 per cent of GNP (Aslund, 1991).

The impending financial crisis – dire projections of an explosive increase in the budget deficit – was drawn to the attention of the Politburo and immediate measures were taken to stem the dramatic increase in government expenditures. Investment was capped at 80 billion nominal roubles in 1987, and reduced to 69.2 million roubles less than one year later in 1988. It is important to notice that a significant portion of the 'expenditure' reductions were concentrated in the oil industry. According to estimates provided by Goskomstat, the annual growth rate of gross fixed investment in the oil sector was reduced to 3.5 per cent in 1988, a significant 10.9 per cent reduction from the 14.4 per cent reported in 1987 (see Table 6.18).

6.2.5 The Failure of Economic Reforms

As mentioned above, Gorbachev's proposal for 'radical' economic reform was announced to the Twenty-Seventh Party Congress on 25 February 1986. Briefly, the planning process was to be gradually decentralised. Enterprises were to be given more freedom to determine their own priorities, and guided by 'economic' rather than 'quantitative' or administrative considerations. Wholesale trade was to be developed and encouraged. While the state would continue to 'fix' prices, a more flexible pricing system would be developed to promote quality, efficiency, and market balance (equilibrium) (Aslund, 1991), and to reduce 'food' subsidies (losses) (White, 1991).

It is impossible to overstate the extent of the difficulties and challenges facing Gorbachev and his 'teams' of economic reformers. Of particular importance were: (1) the enormity of the 'planning' problem, which included the reorganization and re-estimation of the entire Soviet input–output matrix; and (2) a deep-seated, and uncompromising political resistance to reform. In regard to the former, by 1984, Gosplan alone was in charge of 4000 'aggregate' material balances, while the Ministries were in charge of a staggering 40 000–50 000. In the increasingly complex Soviet economy, the information needs of the planning process far outstripped the resources available for its collection (Hobbs et al., 1997).

By the summer of 1987, a comprehensive economic reform programme had, however, somehow been completed. The critical legislation – the 'Law on State Enterprises' and the 'Basic Provisions for Fundamental Perestroika of Economic Management' – were adopted by the Supreme Soviet and Central Committee in June 1987. One month later, ten supporting 'decrees' were approved on the major components of the Soviet

Table 6.23: State budget expenditures in the USSR.

(billions of nominal roubles)

	1985	1986	1987	1988	1989	1990
Gross National Product	777.0	798.5	825.0	875.4	924.1	990*
Total State Expenditure	386.5	417.1	430.9	459.5	482.6	510*
Investment	70.0	80.0	80.0	69.2	64.2	40.5
Consumer Subsidies	58.0	65.6	69.8	89.8	100.7	110.5
Food Subsidies	56.0	58.0	64.9	66.0	87.7	95.7
Social Insurance and Health Care	83.6	89.3	94.5	102.5	105.5	117.2
Total State Revenues	372.6	371.6	378.4	378.9	401.9	452.0*
State Budget Deficit	18.0	47.9	57.1	90.1	91.8	80.0*
(Percent of GNP)	2.3	6.0	6.9	10.3	9.9	5.8*

Note: * Estimate

Source: Aslund (1991).

economy including banking, prices, technical progress, material inputs, planning, and the organization and activities of branch ministries.

While a detailed evaluation of the conception and evolution of Soviet economic reforms is beyond the scope of this book, it is necessary to mention that the main (operative) components of Gorbachev's proposal had been undermined by excessive debate and compromise. Resistance to the 'decentralization' of the central planning procedure – the development of free competitive markets and a system of pricing which reflected the basis actual costs of production – was particularly severe. As might have been expected, the hard line conservatives argued that free markets would undoubtedly result in chronic unemployment and 'social stratification'. Others were more concerned about the potential loss of power and influence in the vast Soviet hierarchy. In the words of Stepan A. Sitaryan, the First Deputy Chairman of Gosplan:

> Theoretical attempts to reduce the idea of a centralized foundation in the management of the socialist economy and to weaken the role of USSR Gosplan as its leading element are capable of causing irreparable damage. . . . [The] effective functioning . . . of the national ownership cannot be assured without a centralized base in the management of the socialist economy (as reported in Aslund, 1991, p. 125).

In the final analysis, the 1987 reform was limited to subtle changes in the nature of central planning (the transition to 'financial' rather than quantitative indicators), and a reduction of the number of republican ministries, departments, Oblasts, and 'aggregate material balances'.

The Law on State Enterprises outlined the Soviet proposal for the gradual devolution of power and responsibility to the level of enterprises. Managers were given the opportunity and incentives to: (1) participate in the planning process; and (2) conduct their operations according to the principles of self-accounting, self-financing and self-management. As of 1 January 1988, Soviet enterprises were to be made responsible for 'financial' rather than physical (quantitative) indicators. Firms were now expected to cover variable costs out of the 'gross revenues' accruing from the sale of production. According to Smith (1993, p. 106): investment 'would no longer be provided as a "free gift" from the state, but would be financed by [profit and] repayable bank loans on which interest would have to be paid'. Firms and production associations who failed to 'pay their way' under these circumstances could be liquidated (White, 1991).

A new system of state orders was implemented to ease the transition to financial indicators (facilitate the profit steering of enterprises), and to guarantee the interests of the State. In short, the traditional physical production targets (quotas) were replaced by state orders whereby the state would requisition a portion of a firm's production at fixed 'agreed' prices (Smith, 1993). State orders were to be given top priority, accompanied by 'favourable economic conditions' facilitated by the guaranteed availability of good quality material inputs distributed to Soviet enterprises on the basis of competition. In theory, the portion of Soviet production claimed by state order was to be reduced gradually by 1990. However, the reform was frustrated by the low profitability of Soviet firms, political resistance, and/or failure of the Ministries and Gosplan to reduce the volume of mandatory state orders distributed to enterprises (Aslund, 1991). The new system was never given any teeth because a bankruptcy law was never put in place. In any case, as long as prices were set at the centre no significant improvement could take place due to the 'calculation problem' identified by von Mises in the 1920s.

It is important to note that the Law on State Enterprises was implemented on 1 January 1988, the middle of the twelfth five-year plan, and was not accompanied by a 'complementary' reduction in the aggregate five-year planning quotas. As a result, the reform contradicted previous 'government' directives. To further aggravate matters, the twelfth five-year plan had been based on an accelerated production and investment programme, so that the full productive capacity of Soviet enterprises had already been committed. According to Aslund (1991, p. 127): 'for the majority of enterprises and organisations [the production programme for 1988 was] . . . virtually composed of state orders, which do not differ in any way from the ordinary plans, where were previously confirmed from above'.

The minor portion of a firm's output/production that was not claimed by state order could be traded at contract prices negotiated between the enterprise and its customers. In the initial phase of the reform process (1988–91), 'wholesale trade' was limited to new goods and high quality production, the prices of which had yet to be determined (fixed) by Goskomotsen. However, even these 'new' prices were subject to central controls and limits on the permitted levels of mark-ups on the price of goods. As might have been anticipated, producers seized every available opportunity to increase the production of new goods and high quality products, which according to Smith (1993, p. 109): 'frequently embodied very minor alterations to existing products simply to achieve higher prices and revenues'.

The reform had dire (unfortunate) implications for the delicate balance and equilibrium of the overstretched Soviet economy. While the number of material balances and indicators that fell under the jurisdiction of Gosplan was reduced significantly over the period 1986–90, the vast majority of these were simply 'passed over' to Gossnab so that the basic 'centralized' system of planning remained intact.

At the same time, retail prices had not yet been reformed, and indeed remained constant, so that the typical Soviet enterprise could not realize sufficient profits to be self-financing. Instead, the dissolution of the Gosbank monopoly – the loss of Soviet control over the creation of credit (loans) and the money supply – resulted in numerous 'unjustified loans' to Soviet enterprises. The combination of: (1) 'unusable' profits from the sale of new goods; (2) the availability of easy 'unjustified' credit; and (3) the freedom to increase wages under the Law on State Enterprise, resulted in a massive transfer of money or 'excess liquidity' from investment (the enterprise sector) to the consumer sector.

To summarize the complex sequence of events and imbalances: the July 1987 decree on Banking meant that credits to loss-making enterprises would continue almost unfettered (Aslund, 1991), so that only a few firms were declared bankrupt and liquidated. Instead, subsidies to Soviet industry continued in the form of loans from newly formed co-operative banks, and excessive money creation. The few firms that were able to realize a substantial profit from the sale of new goods and/or efficient management, found that they were unable to spend the windfall on investment. As mentioned above, Gossnab maintained an absolute monopoly on the distribution of material inputs, which were reserved for state order production; that is, the fulfilment of unreasonable twelfth five-year planning targets (Aslund, 1991).

Instead, the excess liquidity was distributed directly to consumers in the form of higher wages, resulting in an escalation of inflationary pressures – cost push inflation – throughout the USSR. In short: the Law on State Enterprise effectively granted managers a large number of 'liberties' and no corresponding obligations. According to the new 'principle' of self-management, the directors of enterprises were elected by the workers, and granted the freedom to raise wages with a portion of profits that were no longer simply confiscated by central authorities as tax revenue. As competition in the Soviet labour market intensified, the 'directors' of unprofitable enterprises were forced to raise wages simply to attract and maintain a qualified labour force. The rapid escalation in Soviet wages did not reflect improvements in the productivity of workers, but rather an overflow of excessive liquidity from enterprise accounts to the consumer sector.

At the same time, valuable investment funds were absorbed by the 'inflated' aggregate wage bill, so that the production of consumer goods failed to rise at the same pace as demand. While retail prices were held constant in the years 1987–90, there were few goods available for purchase. Shortages of commodities developed in literally every sector of the Soviet economy, necessitating the strict 'rationing' of consumer purchases in many regions of the USSR (Aslund, 1991). The net effect was a dramatic increase in personal savings rates, the escalation of 'free' and 'black market' prices, and a growing perception that the rouble was worthless. Soviet citizens soon began to lose all interest in working for wages, demanding instead hard currency and commodities as compensation for effort.

The growing shortage of the funds available for investment in Soviet industry was further aggravated by a reduction in government revenues. In the years prior to the reform, the bulk of government revenues had been collected in the form of enterprise taxes, turnover taxes and foreign trade tax. Under the new Law on State Enterprise and a related decree on the 'finance mechanism', the confiscation of enterprise profits was prohibited, and the enterprise tax was reduced to a fixed percentage of a firm's profits. Soviet enterprises were permitted to retain anywhere from 10–100 per cent of 'after tax profits' that had subsequently been confiscated by the state at the end of each year. By 1989, the state revenue from enterprise taxes had fallen to 115.5 billion nominal roubles, a 14.3 billion rouble reduction from the 129.8 billion roubles collected in 1986. The shortfall was exacerbated by the partial liberalization of wholesale and foreign trade, and the difficulties associated with the re-organization of the tax system and collection of tax revenues from recalcitrant Soviet enterprises.

At the same time, the Gorbachev government had been unable (or unwilling) to contain the level of 'politically sensitive' expenditures. Short of a continued debate on the merits of a comprehensive retail price reform, which was not even mentioned in the July 1987 decree on the 'System of Pricing', little was done about consumer (food) subsidies that continued to rise steadily throughout the twelfth five-year planning period. Food subsidies reached 87.7 billion roubles in 1989, a 35 per cent increase over the 64.9 billion roubles reported in 1987. Military expenditures were maintained at 75.2 billion roubles in 1989, an unsustainable 33 per cent of government expenditures and over 8 per cent of GNP (IMF, 1992).

In an effort to contain a rising and unmanageable budget deficit, the Soviet government was, once again, forced to reduce its expenditures (investment) in civilian, that is, non-military, industries. Soviet investment was cut back to 64.2 billion roubles in 1989, a 19.8 per cent reduction from the 80 billion roubles invested in 1987. However, even these cutbacks would be insufficient to contain the large and growing fiscal imbalance (deficit).

While state revenues rose by a mere 6.2 per cent in 1987–89, expenditures soared to 482.6 billion roubles in 1989; a 12 per cent increase over the 430.9 billion roubles reported in 1987. By 1989 the budget deficit had risen to 91.8 billion roubles, a worrying 9.9 per cent of GNP. One year later, Soviet investment was reduced to 40.5 billion roubles; only 7.9 per cent of total Soviet expenditures, and 4.1 per cent of GNP (see Table 6.23). The financial crisis and sustained cutbacks in investment led to (and aggravated) severe production shortages in virtually every sector of the Soviet economy, precipitating the disintegration of the entire centrally planned distribution system.

6.2.6 The End of the Centrally Planned Oil Industry

> It appeared that Gorbachev's glasnost, or 'the sweat air of freedom,' as we called it, had a bad influence on our producers. Worker discipline had deteriorated, and equipment supplies were worse than ever. Nikolai Ryzhkov's government adopted a strangely inconsistent economic policy. While most Soviet industries were given the freedom to sell at least part of their products on the domestic market, the government continued to grab 100% of oil output. Meanwhile, centralized subsidies to the oil industry were drastically cut back. Yet machine builders, already profiting from independent sales of their own goods, continue to get cash handouts from Moscow. Our financial position deteriorated in 1989, and the outlook for 1990 was frightening. It was becoming more and more difficult to fund the import of special equipment. What little hard currency was allocated to the industry was entirely absorbed by imports of meat, bread, medicine, and cigarettes. By the end of 1990 equipment was in such short supply that drilling levels started to fall (Tchurilov, 1996, p. 203).

Perestroika and the Law on State Enterprises had disastrous implications for the ill-fated Soviet oil industry. As crude oil and petroleum products naturally fell under the category of critical material inputs, all facets of the industry remained centralized. Crude oil and fuel supplies were produced, rationed and distributed according to the strict directives of Gosplan (The Oil and Gas Ministries and Gossnab), and the traditional system of material balances. In short, the entire volume of crude oil produced in the Soviet

Union was reserved for state orders and the fulfilment of unreasonable twelfth five-year planning targets.

At the same time, Goskomotsen had retained its monopoly over the pricing of basic fuels and raw materials. The wholesale enterprise (contract) price of crude oil was not reformed, and indeed was maintained at levels that had been determined by the central authorities in 1982. According to official statistics, the average wholesale enterprise price of crude oil crept up only hesitantly from 24.79 roubles per tonne in 1982 to 25.70 roubles per tonne in 1990. The implied annual increase in nominal oil prices, approximately 0.45 per cent per annum, fell far below even the annual rate of inflation that was admitted by the official statistics. The net effect was a real reduction in the level of crude oil prices. When measured in constant 1970 roubles, the domestic wholesale enterprise price for crude oil fell to 20.28 roubles per tonne in 1990, a 12.3 per cent reduction from the 23.13 roubles per tonne reported in 1982.

In the meantime, international oil markets had once again been disrupted by tensions in the Middle East, and world oil prices had risen significantly from the lows recorded in 1986–88. On 2 August 1990, Iraqi troops marched into Kuwait initiating what has come to be known as the fourth oil price shock in less than 17 years (Reinsch and Considine, 1991). The price of Saudi Light 34° crude rose precipitously to US$24.00 per barrel in 1990, a US$10.85 per barrel increase over the US$13.15 reported in 1988. Assuming official exchange rates, Soviet producer prices for crude oil averaged only US$5.94 per barrel in 1990, a meagre 24.6 per cent of Saudi Light 34° crude.

Once again it is necessary to draw attention to the confusion generated by the existence of a diverse and colourful variety of official Soviet statistics. The first steps towards the liberalization of the Soviet foreign exchange system were taken in 1989, when Vneshekonombank (VEB), the state monopoly over foreign exchange transactions, was abolished. Months later, in November 1989, Gosplan established a system of hard currency auctions permitting a limited number of pre-authorized sellers and buyers to trade in hard currency. The reform permitted the utilization of an official 'approximation' to a free market exchange rate in calculations concerning the dollar value of Soviet oil prices. Utilizing the official 'Auction Market' exchange rate of 19.34 roubles per US dollar in 1990, the average Soviet crude oil producer price (a nominal contract price) was valued at 1.329 roubles per tonne or approximately 18 cents (US) per barrel.

While Goskomotsen had jealously guarded the prices of some material inputs –specifically crude oil and fuels, others were increased significantly

(Smith, 1993). As mentioned above, a wholesale market for 'new goods' and 'high quality' production had been introduced by the Law on State Enterprises along with a complementary reform of wholesale prices in 1988. Political resistance to the reform was severe, and effective, so that the majority of the industrial associations responsible for the manufacture of oil and gas service equipment were subordinate to the whims of Gosplan and the Ministries. As a result, they were required to reserve 100 per cent of their production for state orders. However, a few 'fought back' requesting the right to exercise their new-found freedoms, and participate in the newly formed wholesale market.

Notable among these was the Uralmash Machine Building Association, which was a 'personal' favourite of the government and produced high quality rigs for cluster drilling (Gustafson, 1989). According to Aslund (1991, p. 128):

> On the whole, enterprise directors obediently accepted the absence of change, but the gigantic Urals Machine Building Association, Uralmash, in Sverdlovsk insisted on its newly-won legal right to adopt its own plan. After intense strife, Uralmash won its case against its Ministry [1988], but Uralmash's courage, and fortune were boosted by the fact that Prime Minister Nikolai Ryzhkov had been its director general.

The Uralmash victory was supported by Gorbachev in a speech delivered to the 19th Party Conference, and provided the impetus and inspiration for a number of other enterprises to take up the case (Aslund, 1991). The decision on Uralmash, and indeed, all decisions rendered in favour of the industrial associations, contributed to a diversion of material supplies from the fulfilment of state order production, and an increase in wholesale industry prices throughout the USSR (IMF, 1992).

The average crude oil production costs for the Soviet Union had been rising steadily since the 1970s. The trend, which reflected the natural ageing of the Soviet oil industry and more recently the excessive concentration of new development drilling in remote territories of West Siberia, was accelerated by the Uralmash decision, rising labour costs (wage increases), and the general deterioration of economic conditions throughout the USSR. The difficulties included suppressed inflation, the rationing of critical commodities and the gradual disintegration of the state distribution system.

By 1990 the average production costs for crude oil in the USSR had risen to 21.13 roubles per tonne, 4 roubles per tonne higher than the 17.12 million roubles reported in 1988 and an implied annual increase of well over 11 per cent per annum. The marginal capacity cost registered a 32.4 per cent increase, from 91.3 roubles per tonne (Rbls/t) of new industry capacity in 1988 to 120.9 (Rbls/t) in 1990. As in the first few years of the twelfth five-year planning period, conditions in the Province of Tyumen were considerably more severe. The marginal capacity costs for the Tyumen production associations reached 100.2 (Rbls/t) in 1990, a 44 per cent increase over the 69.6 (Rbls/t) reported in 1988.

By 1990, the gross profits from domestic oil sales had fallen to 4.57 roubles per tonne – less than 18 per cent of the entire proceeds from domestic oil sales, and 18.5 per cent below the normative after-tax rate of 36.5 per cent that had been established for the industry in 1982. At the same time the oil industry was still centralized and subject to state profit-related taxes of nearly 100 per cent of gross profits, so that the formal after-tax profitability of the oil industry fell to 0.1 per cent in 1990. According to Khartukov (1995, p. 2) by 1990, 'the national oil-producing industry was being subsidized by the state for one-third of its capital requirements'.

As pointed out above the Soviet government had been cutting back its investments in civilian (non-military) industries in an attempt to contain a growing, and unmanageable budget deficit. By 1990, Soviet investment had been reduced to 40.5 billion roubles, a mere 49 per cent of the 80 billion roubles reported three years earlier in 1987. The cutbacks had unfortunate implications for the Soviet oil industry. The annual growth rate in gross fixed investment in the oil industry was reduced from 14.4 per cent in 1987, to 3.5 per cent in 1988, and 4.5 per cent in 1989. One year later, in 1990, the level of gross fixed investment in the oil industry was reduced by 13.7 per cent.

As in 1983–84, the sudden deceleration in the growth rate of gross fixed investment in the oil industry combined with rising costs of production resulted in an 'immediate' reduction in crude oil flows. According to Ebel (1994, pp. 13–14):

> By late 1988, it was evident that the industry was in serious trouble. This time, however, there was to be no personal intervention by Gorbachev, no additional funds, no stepped-up equipment deliveries or crews flown in from other regions. The country was exhausted physically and emotionally, and it could do nothing more to help.

By 1989, Soviet crude oil flows had fallen to 607.254 million tonnes, a 2.7 per cent reduction from the peak level of 624.326 million tonnes reported in 1988. The bulk of the production was required simply to sustain domestic fuel consumption, so that the USSR was unable to sustain its aggressive crude oil export policy. Total exports of crude oil and petroleum products fell to 184 million tonnes in 1989, a 10 per cent reduction from the peak levels of 205 million tonnes that had been achieved only one year earlier in 1988.

Despite the absolute necessity for an increased volume of hard currency receipts, the level of hard currency oil exports was reduced to 62 million tonnes in 1989. This represented a 31 per cent reduction from the 90 million tonnes reported in 1988. The reduction offset the 'potential' gains accruing from the recent recovery in world oil prices, and hard currency earnings from oil exports fell by US$199 million from US$9528 million in 1988 to US$9329 million in 1989. According to Smith (1993), the deficit in the Soviet current account reached US$8.989 million in 1989, more than double the US$3.843 million deficit that had been reported in 1988 (see Table 6.22). By August 1989, the government had been forced to utilize its foreign exchange reserves to finance the balance of payments account (IMF, 1992).

For the first time in the history of the CMEA, the Soviet Union was unable to meet its contractual energy obligations to the East European satellite nations. The failure to provide 'subsidized' oil exports drew heated protests from Hungary, Poland and Czechoslovakia, and aggravated a noticeable deterioration in intra-CMEA relations. As related by Smith (1993, p. 172):

> The Soviet authorities attributed the failure to meet East European contracts to domestic supply problems, not deliberate policy, a claim which appears to be justified in view of the falling domestic production which necessitated cuts in hard currency exports in1989, while (according to Soviet trade statistics) cuts in oil exports to Eastern Europe for the year as a whole were negligible.

At the same time, the Soviet Union had made little progress in its efforts to extend the principles of *Perestroika* to the realm of international (intra-CMEA) economic relations. The 'Basis Provisions on Economic Reform' which had been adopted by the Central Committee in the Soviet Union in June 1987 had called for a gradual move towards the convertibility of the rouble, primarily within the framework of the CMEA. A more detailed list of Soviet proposals (ideals) included: (1) the creation of a 'Unified Socialist Market' which would be patterned after the 'Single Market' of the European Community to permit the free uninhibited movement of capital, goods, labour and services and capable of generating profits for all participants; (2) the gradual convergence of CMEA prices with world prices; and (3) the full convertibility of CMEA currencies.

Given the existence of considerable subsidies to the CMEA member nations, according to Stone (1996, p. 210): 'the Soviet proposals for perestroika met more open and determined opposition that had any previous Soviet initiative in the CMEA'. By 1988, it was clear that only a moderate reform would be feasible in the thirteenth five-year planning period. The negotiations for the 1991–95 plan were to be guided – directed – by the following recommendations, and proposals: (1) the maintenance of contemporary CMEA pricing arrangements for all goods traded in quotas (that is, contractual trade prices based on a moving average of the world price); (2) the gradual introduction of trade in goods that were not subject to quotas; (3) the 'free' determination of prices for non-quota goods by negotiation between the enterprises making the exchange; and (4) the increased use of national currencies, specifically the rouble, which was to be encouraged 'in all forms of economic co-operation by interested countries' (Stone, 1996, p. 210).

The first of the reformed bilateral trade agreements was successfully negotiated with Czechoslovakia in the spring of 1988 (Aslund, 1991). It represented only a minor victory for reform. According to Stone (1996, pp. 233–34):

> The 1991–1995 plan bound only 40 per cent of the projected USSR–CSSR trade turnover in obligatory quotas. However, all that this meant in practice was that the five-year plan was less comprehensive than it had been in the past, not that the other 60 per cent of trade would be liberalised. . . .The scramble for unilateral advantage that took place during the first phase of negotiations [left] little reason to believe that the central planning agencies would have given their enterprises freer rein in future rounds than they had in the past.

Ironically, any further attempts at gradual CMEA reform would have been pointless. By the end of 1989, communist governments in all of the Eastern European satellite nations had either been desposed or had voted to abolish themselves (Kort, 1993). On 18 October 1989, Erich Honecker, the faithful head of the German Democratic Republic (GDR) Communist Party, was forced to resign. Barely one month later, on 9 November 1989, the GDR announced the removal of all travel restrictions to the West, effectively dismantling the Berlin Wall. Calls for multi-party democratic elections, and radical economic reform, resounded throughout Eastern Europe.

The situation became unmanageable in 1990, when the decline in Soviet crude oil flows accelerated, and production reached only 570.5 million tonnes, a 6.1 per cent reduction from the 607.3 million tonnes reported in 1989. Crude oil exports were diverted from the CMEA to hard currency nations in a desperate attempt to increase hard currency receipts and settle the commercial payments arrears that had emerged in late 1989 (IMF, 1992). The volume of Soviet oil exports to Eastern Europe (excluding the GDR) fell to 43.36 million tonnes in 1990, a 23 per cent reduction from the 56.39 million tonnes reported in 1989 (see Table 6.20). In June 1990, the Soviet Union announced its intention to 'stop' all subsidies to CMEA member nations. As of 1 January 1991, only hard (convertible) currency would be accepted for all exports to the CMEA that were to be delivered at world prices. The complete disintegration of the CMEA trade system followed shortly thereafter (Stone, 1996). The CMEA held its last formal meeting in Budapest, Hungary, on 28 June 1991.

6.2.7   Western Capitalists Riding in on White Horses – Maybe

> Rumour had it that there were more people waiting to get in to the Pushkin Square McDonald's on opening day [January 31, 1990] than there were in Red Square, waiting to see Lenin's Tomb (Cohon, 1997, p. 216).

While the '*Perestroika*' of intra-CMEA relations may have progressed at a snail's pace, all of the CMEA member nations, and particularly the USSR, had been taking decisive measures to attract capital flows from the West. As early as August 1986, Western enterprises were encouraged to submit proposals for joint ventures with Soviet enterprises (Aslund, 1991). On 13 January 1987, the Soviet Union adopted two decrees outlining the conditions and procedures governing the creation of joint ventures with production associations on Soviet territory.

As might have been anticipated given the extreme isolationist history of the USSR, the terms offered to foreign partners were hopelessly complex and, in the perspective of Western investors, poorly conceived (Aslund, 1991). A brief summary of the main conditions, and concessions, included:

1. a Soviet share in the joint venture of at least 51 per cent;
2. the requirement that the chairman of the board, and enterprise director of the joint venture were citizens of the Soviet Union;

3. joint ventures were to be 'subject to' all Soviet decrees, and restrictions 'implying an obligation to obey thousands of unpublished acts, which the foreign partner had no right to see' (Aslund, 1991, p. 144);
4. all joint venture purchases from, and sales to, Soviet enterprises were to be conducted through the appropriate Foreign Trade Organization (FTO) at 'agreed prices which took world market conditions into account' (Smith, 1993, p. 129);
5. the joint ventures were not to be subject to the strict provisions of central planning and production quotas, a double-edged sword because of the difficulties associated with obtaining inputs and disposing of output through the still planned supply chains;
6. a 20 per cent tax on the transfer of profits abroad;
7. a profit tax of 30 per cent;
8. full confidentiality of all joint venture information (Riva, 1994);
9. the free transfer of shares to an approved third party; and finally
10. the terms and conditions that had not yet been 'specified' (regulated) were to be determined by mutual negotiation. The latter included but were by no means limited to labour relations, currency restrictions, taxes, and capitalization.

Perhaps the most important condition for a joint venture in the oil industry was the absolute prerequisite that a licence for the right to exploit mineral resources (oil and gas) be 'transferred' to any joint venture that had been properly registered under the Joint Venture Law of 1987. In the Soviet Union, mineral resources, and indeed all natural resources, were regarded as the natural heritage of the people living in the territory or republic where the resources [were] located. At the same time, the authority over these resources had been 'willingly' delegated to the state, so that the government had exclusive jurisdiction over the rights of exploitation. Licences for the right to exploit natural resources were distributed (granted) to the production associations by the appropriate state mining supervisory body. As a general rule, the Ministry of Oil and Gas (Minneftegaz)[4] and Gazprom were given the first right of refusal on development prospects. Deposits that were rejected, for whatever reason, would subsequently be presented to Mingeo. The production association would not receive title to the oil and gas until it was produced.

While foreign partners were not permitted to obtain an equity interest in a licence, 'exploitation rights' would be transferred to the joint venture after the successful completion of a complex approval process. The arduous 'approval' procedure may be briefly summarized as follows:

1. the foreign partner and Soviet production association were requested to draw up a draft joint venture proposal with an unspecified level of mandatory assistance from the appropriate Ministry – Minneftegas and/or Mingeo;
2. the proposal was to be submitted to the council of ministers of the appropriate Soviet republic;
3. the proposal was to be submitted to the appropriate state mining supervisory body, and the Council of Ministers USSR;
4. a government decree was to be written to sanction the proposed joint venture;
5. the joint venture was to be registered by the appropriate republican Ministry of Finance;
6. the joint venture was required to make all necessary provisions to obtain state approval for exports, and gain access to centralized transmission facilities, that is, pipelines.

The complex maze of negotiations, bureaucratic procedures and state agencies, was complicated by frequent turnovers (firings) of agency personnel, and fundamental differences in the perspectives, and unstated requirements of Soviet and Western participants (Stern, 1992). Prominent among these was the absence of a comprehensive and reliable legislative and fiscal system. For centuries, the Soviet Union (and before it, Russia) had been governed by autocrats (dictators), and administrative fiat (rule by decree). This system had no need of rules that facilitated transactions and provided security for investment.

In the late 1980s, the Gorbachev economic reform programme was still in its infancy, and Russia was barely beginning the complex process of establishing the foundations for the comprehensive legal and fiscal regulatory system that would govern the activities of foreign participation in the oil industry (Stern, 1992). The 'subject' of domestic legal reform, and the creation of a law-based state, had yet to be addressed seriously by the Soviet government (Coleman, 1997).

Western investors, on the other hand, regarded the existence, stability, and predictability of taxes, commercial laws, and legal mechanisms, as an essential pre-requisite to profitable investment (Hobbs et al, 1997). According to Stern (1992, p. 41): 'A common position of potential investors is that nothing can be achieved until a stable legal and fiscal framework has been established, and has been working for a period of time.'

Despite genuine interest by foreign participants, the list of terms, procedures and differences was overwhelming. Applications were deterred by the complexity of the procedures, and the extreme level of political and economic risk. In September 1987 the complex bureaucratic procedure was modified slightly, permitting all-union and republican ministers to grant joint venture approval and joint ventures were permitted to purchase domestic supplies through Gossnab (not simply the FTOs).

By year-end 1987, only 23 joint ventures (JVs), representing an aggregate capital investment of approximately 159 million roubles, had been established in the USSR. In December 1988, the Soviet Union adopted a decree offering a number of incentives to hesitant and bewildered foreign investors. These were:

1. foreigners were permitted to serve as chairman of the board and enterprise director;
2. the requirement that the Soviet share of a JV 'must' exceed 51 per cent was abolished;
3. the JV was given greater powers to hire, and fire, workers;
4. limited tax breaks were offered to foreign partners; and
5. the foreign employees of JVs were given permission to pay for housing and services in Soviet roubles (Smith, 1993).

The limited concessions, the first in what would turn out to be an exhaustive repertoire of JV-related decrees, revisions, tax breaks and corrections, were not sufficient to attract substantial quantities of foreign capital. By June 1990, an aggregate total of only 1754 JVs had been registered in the Soviet Union (Smith, 1993). The combined JV effort, including production and services, amounted to less than 0.5 per cent of GNP in 1990. While there were some glaring success stories – MacDonalds had its first grand opening in Pushkin Square on 31 January 1990 – the victory was too often overshadowed by the arduous negotiation procedure. MacDonalds, for example, was no stranger to the idiosyncrasies of the USSR, and had in fact been negotiating with the Soviet Union (Brezhnev) since 1976 (Cohon, 1997).

The difficulties in the oil industry, which included the obtuse and complex licensing procedures, and the centralized domestic pricing system, were frustrated by the low government priority assigned to JVs in the energy sector as well as unscrupulous entrepreneurs protected by corrupt bureaucrats. According to Stern (1992, p. 29):

At the end of 1989, out of more than 1,000 registered JVs, only two involved the energy sector. . . . Recalling that the dramatic decline in oil production only

began in 1989, we can see clearly that the Soviet authorities did not at that time
believe that energy was a priority sector for JV activity.

If opening up the oil industry to foreign investment was to provide the
means to stave off the impending economic collapse, its timidity meant it
was a complete failure. It also put in place complex systems that would be
hard to dismantle once the planning system was gone. Complex regulations
are a mainstay of corruption (Kerr and MacKay, 1997).

6.2.8 The Collapse of the USSR

> We, the Republics of Belarus, the Russian Federation, and Ukraine, as state-
> founders of the USSR – signatories of the 1922 treaty . . . find that the USSR
> ceases to exist as a subject of international law and geo-political reality
> (Preamble to the agreement forming the Commonwealth of Independent States)

The fortunes of the Soviet oil industry were undermined, to a great extent,
by the sporadic episodes of labour unrest and ethnic violence that
accompanied Gorbachev's reform programme. By the late 1980s, the
policies of *Glasnost* and *Perestroika* had led to the election of independent
and nationalistic minded parliaments throughout the Soviet republics
(Smith, 1993). For the first time since 1922, citizens were free to express
their views on Soviet domination, and to criticize central economic policy.
At first, this took the form of demands for greater republican autonomy but
was subsequently extended to demands for outright independence from the
USSR.

In short, the evolution of democratic principles and 'movements' in the
Soviet Union had exposed, and aggravated, a delicate nationality problem
that had been repressed and/or effectively contained since the Bolshevik
revolution (Kort, 1993). Long-simmering and unresolved grievances
erupted into riots. By the end of 1987, disturbances had emerged in
republics from the Baltic coast to Central Asia, frustrating political stability
and production throughout the USSR.

The thin veneer of stability was shattered by the failure of the state
distribution system, and a growing shortage of critical supplies – including
basic food staples and soap. The deterioration of living conditions
throughout the USSR led to heated protest, and in the summer of 1989 a
series of labour strikes erupted throughout the country. The insurrections,
which included a comprehensive coal miners' strike in Siberia and a two-
week (50 000 worker) Ukrainian coal miners' strike, had serious
implications for the overstretched Soviet energy industry, placing even

more pressure on disillusioned crude oil production associations (Riva, 1994).

Drastic measures were taken to avert a full-scale national disaster. According to Kort (1993, p. 318):

> Desperate to end a potentially crippling crisis, the government promised a package of improvements including pay increases and increased availability of food, medical supplies, and other consumer goods estimated to cost between five and nine billion dollars. Gorbachev announced that local elections would be moved up from the spring of 1991 to the fall of 1990. While these concessions were enough to get the miners back to work, many mines maintained their strike committees to make sure the government delivered on its promises.

Months later, in February and March 1990, local elections were held throughout the USSR. Incredibly, Communist party candidates lost their majorities in three key municipalities – the city councils of Moscow, Leningrad and Kiev. At the same time, Boris Yeltsin was elected to the new parliament of the Russian Republic. Less than two months later, in May 1990, Yeltsin overcame Gorbachev's backing of a rival candidate and was elected by parliament as the President of the Russian Republic (Kort, 1993). His platform included the following 'radical' proposals: (1) accelerated economic reform; and (2) the transfer of considerable powers and liberties to the republics, particularly Russia.

In June 1990 the Russian Parliament declared the supremacy of Russian laws over Soviet laws and Russian control and sovereignty over Russian minerals and natural resources (Smith, 1993). Highly ranked in the exhaustive repertoire of new Russian laws and proposals was the assertion that taxes collected in Russia were the property of the Russian government, and not the Soviet Union. According to Coleman (1997, p. 328):

> The net effect was a constitutional crisis', a 'war of words' between Yeltsin's Russian government and Gorbachev's Soviet regime. Suddenly Yeltsin himself, as the elected president of the Russian parliament, had as much legitimacy as Gorbachev, elected president of the USSR by the Soviet parliament. . . . From then on the momentum would favour Yeltsin.

By the end of 1990, all fifteen Soviet Republics, including Russia, had declared either; (1) political independence, or (2) the total 'sovereignty' of their legal jurisdiction over that of the USSR. It is important to remember that the USSR had originally been established as a voluntary union of 15 sovereign union republics. Under the 1977 Constitution, all union republics were, in fact, sovereign Soviet socialist states with a considerable range of formal powers, including the right to secede and to establish diplomatic relations with foreign powers (White, 1991).

Resistance turned to violence on January 1991, when Soviet troops stormed a television tower in Vilnius, Lithuania, to silence the cries for independence. Thirteen unarmed citizens were killed and 120 more were injured. The incident incited enraged citizens to riot, and a wave of minor inter-ethnic altercations erupted throughout the USSR. In March 1991, Gorbachev held a public referendum on the status of the USSR. The question of whether the Soviet Union should remain intact, was carefully designed to rally support for a 'union' that was rapidly disintegrating. Remarkably, 75 per cent of the participants voted for the continued existence of the USSR as a unified political entity.

Yeltsin seized the initiative for the Russian Republic, placing a second 'referendum' question on the same ballot. Citizens were asked to give their support for the direct 'public election' of the President of the Russian Republic. Seventy per cent of the participants voted for direct Presidential elections in Russia. At the same time, Gorbachev's efforts to discourage (ban) labour strikes throughout the USSR had failed to produce the desired results. Coal miners refused to return to work until wages and living conditions had been improved. Fuel shortages threatened the lives and safety of citizens in the harsh Russian winter. According to Coleman (1997, p. 334): 'In the end, Yeltsin, not Gorbachev, got the miners back to work'. The first democratic elections were held in the Russian Republic on 10 June 1991. Yeltsin was elected with a clear majority, approximately 57 per cent of the vote. On 10 July 1991 he was inaugurated as the President of the Russian Federation.

As was the case with the 1987 economic reform programme, Gorbachev's question on the maintenance of the USSR as a political 'union' had been poorly worded, and the term 'union' had yet to be fully defined. Sensing a critical loophole, the Russian republic, the Ukraine and Kazakhstan rallied the support of their population (approximately 89 per cent of the population of the entire USSR), and set out to define, and achieve, their own political agenda. In April 1991, Gorbachev was forced to negotiate a new draft Union Treaty, granting considerable powers to the Republics, and recognizing the right of six republics – Latvia, Lithuania, Estonia, Moldova, Armenia, and Georgia – to secede (Smith, 1993).

After months of intense negotiation, the new Union Treaty had been completed, and was scheduled for ratification by the nine participating republics on 20 August 1991. The treaty would never receive full signature. On 19 August, a group of hard line conservatives led by Gennadi

Yanayev staged a desperate attempt to maintain Soviet control and domination. On Monday morning, Russia awoke to the sounds of a Tass bulletin declaring that Vice President Yanayev had taken control over the Soviet Union under the authority granted by Article 126, Clause 7 of the Soviet Constitution. Gorbachev, it appeared, had been removed from his position 'for health reasons'. He had in fact been placed under house arrest in the Crimea.

At 10:00 a.m., President Boris Yeltsin arrived at his headquarters, 'Russian White House', to organize televised resistance. In the meantime, tanks had rolled slowly onto the streets of Moscow, encircling the White House. The event, which would come to be known as the second Russian revolution, reached a crescendo at 12:00 p.m., when Yeltsin climbed onto a T-52 tank to rally the support of the people and the military. He asked the people to rebuff the *putsch* and return to the constitution. The pleas were successful, and opposition to the coup spread throughout the Russian Republic (Coleman, 1997). The *coup d'état* disintegrated, and its perpetrators were placed in detention on 21 August 1991.

On 22 August, Gorbachev returned to Moscow to resume his duties as President of the Soviet Union. The struggle for domination was over, and Yeltsin had emerged victorious. Two days later, on 24 August, Gorbachev resigned from the position of General Secretary, and dissolved the Communist Party of the Soviet Union. On the same day, the Ukraine declared its legal independence from the USSR. The secession of the largest non-Russian republic inspired similar declarations throughout the USSR, and the Union Treaty was submitted for re-negotiation.

On 8 December 1991, the governments of the Russian Republic, the Ukraine and Belarus ratified a comprehensive agreement to form a new Commonwealth of Independent States (CIS). The governments called for comprehensive efforts to:

1. preserve and protect a 'common economic space' (IMF, 1992);
2. synchronize economic reforms;
3. facilitate the use of the rouble in inter-republican trade;
4. establish a multi-republican banking union;
5. harmonize taxes; and
6. negotiate a general agreement on the legal and jurisdictional framework for inter-republican economic relations (Reinsch et al., 1992).

On 21 December 1991, the republics of Azerbaijan, Armenia, Kazakhstan, Kyrghyzstan, Moldovia, Turkmenistan and Uzbekistan gathered in Alma-Ata to pledge allegiance to the newly created Commonwealth of Independent States. On Christmas day, Gorbachev resigned from the position of President of the Soviet Union, and the RSFSR adopted the title of the Russian Federation. The Soviet Union had come to an end.

The challenge faced by Yeltsin and his ambitious crew of Russian reformers was temporarily overshadowed, forgotten in the immediate euphoria of victory. Weakened by years of stagnation and neglect, and crippled by the failure of the 1987 economic reform programme, the economy had been laid to ruin by the sudden dissolution of the USSR. A study conducted by the IMF (1992, p. 1) at the request of the former USSR concluded that: 'the economy had entered an acute crisis. The output performance of the country deteriorated sharply (in part on account of a serious strike by coal miners), financial imbalances accelerated, and the external creditworthiness of the former USSR came to be seen as suspect'.

The Former Soviet Union was bankrupt, and its successors, now independent members of the CIS, were refusing to accept any responsibility whatsoever for its outstanding debt obligations. According to estimates provided by the IMF (1992), the government deficit (the sum of the union and republican deficits) had reached 482 billion roubles in 1991, 26 per cent of GDP. Soviet government expenditures had been elevated by the subsidies required to ease labour tensions throughout the USSR, and a failure to reduce military expenditures – which accounted for approximately 43 per cent of total Soviet expenditures in the months of January to October 1991. The IMF (1992, p. 13) comments: 'A deterioration of this magnitude reflects a virtually unprecedented loss of fiscal control, the counterpart of which was the strengthening of inflationary pressures and further increases in undesired money imbalances'.

As before, the cutbacks were restricted to investment in the civilian (non-military) sectors. The former Soviet Union invested a mere 48.3 billion (1991) roubles in the entire Soviet energy sector in 1991. Approximately 44 per cent of the total, 21.4 billion roubles, was distributed to the oil industry (see Table 6.3). When measured in real terms, the level of gross fixed investment in the oil industry fell by 30 per cent in the year 1991, to less than 50 per cent of the planned level of expenditures (IMF, 1992).

As suggested above, the severe financial shortfall was exacerbated by a series of strikes and ethnic altercations in the crude oil producing regions.

According to the IMF (1992, p. 5): 'In recent years, many energy producing regions began to resist further development of the oil industry, demanding increased social services as a pre-condition'.

Deprived of critical Soviet investment funds, and unable to ratify (complete) contracts for Western direct investment (JVs), the oil industry suffered from a severe and growing shortage of critical material inputs. At the same time, the oil and gas support sector was concentrated in 'wayward' republics, particularly Azerbaijan, so that the industry was exposed, and highly vulnerable to 'political' supply disruptions. According to estimates by the IMF (1992), the Soviet oil industry received only 60–70 per cent of the 'domestic' supplies that had been planned by Gosplan in 1991.

Drilling and repair efforts were frustrated by severe shortages of critical equipment and capital, and a growing number of projects were never completed. In 1989, drilling declined because 46 drill crews had to be disbanded (Riva, 1994). According to estimates provided by Minneftegas, the number of new oil well completions fell to 11 091 in year 1991, a 30 per cent reduction from the 15 859 reported in 1988 (see Table 6.7). The number of idle oil wells soared to 17 918 in 1991, more than twice as many as the 7707 reported only two years earlier in 1989 (see Table 6.5). These figures may underestimate the problem (IMF, 1992).

The situation was aggravated in 1991 by a partial reform of producer and agricultural prices throughout the USSR. Prices were amended to improve production incentives throughout the USSR, and to prevent any further deterioration in the Soviet industrial complex. The volume of contract prices to be negotiated between production associations and consumers was increased to 40 per cent in 'light industry', 50 per cent in the machine building industry, and 25 per cent in the raw materials, minerals and energy sectors. While the wholesale enterprise contract price for crude oil was increased to 65 roubles per tonne, a 153 per cent increase over the 25.7 roubles per tonne reported in 1990, the costs of production rose even faster, reaching 57.39 roubles per tonne in 1991. This represented a 171.6 per cent increase over the 21.13 reported in 1990. By 1991, domestic production costs accounted for 98.2 per cent of the average ex-field gate wholesale enterprise price in the Russian Federation.

The rate of decline accelerated in ageing oil fields and Soviet crude oil flows fell to 515.53 million tonnes in 1991, a 15 per cent reduction from the 607.25 reported in 1989. Little progress was made in the critical, albeit expensive, realms of exploration and transportation. According to Ebel (1994, p. 13):

New discoveries were wholly inadequate to offset declines at the older producing fields. The volume of reserves in each newly discovered deposit in the Soviet Union fell to an average of 123 million barrels during 1981–85 and even further to just 58 million barrels during 1986–90. The drop in West Siberia was even more dramatic – to 137 million barrels per discovery, a fraction of past successes.

In the case of transportation, the length of long distance oil pipelines completed in the USSR fell from 6384 km in 1981–85 to 2127 km in 1986–1988. No new long-distance oil pipelines were reported as having been completed in 1989–1990 (see Appendix C). The pipeline maintenance and repair programme suffered from a shortage of inputs and capital, so that a number of the existing pipelines were no longer operational.

Shortly after Gorbachev's official resignation as General Secretary of the Communist Party of the USSR, at which time the impending dissolution of the Soviet Union had become a foregone conclusion, the centralized system of Ministerial control over the oil industry was effectively abolished. On 26 August 1991 Gorbachev dismissed the entire Soviet government (Tchurilov, 1996). Bureaucrats were put in charge until the Ministries could be dissolved.

Work on the creation of a 'new' company to manage the oil industry was started immediately thereafter, and Rosneft was established as a voluntary union of oil enterprises in the Russian Federation. The first meeting was scheduled for 23 September 1991. The founding members gathered for an inaugural dinner and celebration. As reported by Tchurilov (1996, pp. 224–225):

> After dinner we gathered in the conference hall of the Ministry of Fuel and Energy for the formal approval of Rosneftegaz's founding charter. All 53 of the corporation's members signed except Arkticmorneftegazrazvedka and Megionneftegaz. Next on the agenda was the election of a president. From the podium someone announced three possible candidates, Alekperov, Nikitin, and Tchurilov. Each of us made speeches. I said that we had decided to form Rozneftegaz a long time ago. State entities were now collapsing all around us. The creation of new corporations was the way forward. When the matter was put to the vote more than two-thirds of the founders raised their hands in my favour. . . . So it was that Rosneftegaz rose from the ashes of the once mighty Soviet Oil Ministry. Seven decades of central planning were over.

The real question was what was to replace central planning. The simple answer was markets. The creation of a market economy, however, proved to be far more difficult than anyone anticipated (Hobbs et al, 1997).

NOTES

1. Nizhnevartovskneftegaz, which contained the super-giant Samotlor oil field, was the largest, and most successful of all the West Siberian (oil and gas) production associations.
2. In the cluster well drilling technique a number of wells are drilled from one single location, thereby maximizing production with a minimum of expense and effort.
3. The term *glasnost* may be loosely translated as 'openness' or 'publicity'. As suggested throughout this book the Russian tradition of secrecy had led to corruption and misinformation – state lies – which were disrupting the smooth workings of the command economy of the USSR. By the early 1980s the exhaustive repertoire of state secrets and manipulations included: the repression of the media, military secrecy, and the repression of 'official' statistics (which covered entire areas of life in the Soviet Union including crude oil reserves, mortality rates, crime, road accidents, and balance of payments data). From the beginning of his tenure, Gorbachev was committed to a new policy of *glasnost*, or honesty. 'Broad, up-to-date and honest information', he told a conference in December 1984, 'is a sign of trust in people, respect for their intelligence and feelings, and their ability to make sense of developments (White, 1991, p. 110). Perhaps more importantly, *glasnost* was designed to aid in the decentralization of the command economy, reducing both: (1) individual (and enterprise) dependence on the upper echelons of the Soviet bureaucracy; and (2) the number of 'unplanned' errors attributed to misinformation that had been distributed at all levels of the complex planning hierarchy.

    Perestroika (restructuring) has been defined by the *Short Political Dictionary* published in Moscow in the year 1989 in the following manner. 'The strategic course worked out [by] the CPSU at the XXVII Party Congress [of 1986] and the following plenums of the Central Committee, and at the 19th Party Conference [in 1988]'. It is 'a deep renewal of all aspects of Soviet Society, which is revolutionary in character', involving the acquisition by socialism of the 'most up-to-date forms of organization' and the 'fullest disclosure of its merits in all respects: economic, sociopolitical and ideological.' The term, which was not new to the Soviet Union, and had in fact been used quite frequently by Stalin, provided little in the way of practical guidance to state officials and policy makers (White, 1991, p. 189).

    The Soviet term *demokratizatsia* (democratization) 'meant [only] that some choice be allowed in the Soviet system – that in factories, in elections to government bodies, and even in party elections there should be a choice of candidates [not a multi-party political system]. The hope was that *glasnost* and *demokratizatsia*, even in their limited Soviet version, would entice ordinary citizens to pitch in voluntarily to help the reform effort. This was crucial because Gorbachev understood that without active popular support and help, no substantial economic reforms would be possible (Kort, 1993, pp. 296–7).
4. In 1989 the Oil and Gas Ministries of the USSR were merged into a single entity, Minneftegaz, which assumed the responsibility for field development, reservoir management and production.

7. The New Wild West

7.1 TRANSITION TO WHAT?

> Western writers sometimes seem to liken the Soviet system to a sailboat straining against an unfavourable wind. Change the wind, we seem to say, and the boat will quickly right itself. But a more appropriate metaphor for the Soviet economy is that of a gnarled tree that has grown up leaning against the north wind of forced-draft industrialisation. Its past is written into the composition and location of its capital stock, the patterns of its roads and railroads, the size and type of its plants, the distribution of its manpower, the kinds of fuel it burns and ore it uses. Even a perfect leader and a perfect reform, whatever those might be, could not right in a generation what has taken two generations to form. (Gustafson, 1989, pp. 23-24)

For a decade beginning in 1991, the Russian oil economy had to adjust to the dual problem of the dissolution of the Soviet Union and the attempt to establish a market economy. In 1990 the 'USSR Presidential Guidelines for the Stabilisation of the Economy and Transition to a Market Economy' (16 October 1990) described the situation as follows:

> The position of the economy continues to deteriorate. The volume of production is declining. Economic links are being broken. Separatism is on the increase. The consumer market is severely depressed. The budget deficit and the solvency of the government are now at critical levels. Antisocial behaviour and crime are increasing. People are finding life more and more difficult and are losing their interest in work and their belief in the future. (as cited in Granville, 1995, p. 10)

More than ten years later, and despite numerous attempts to implement reform, it is a description that can be applied with disturbing accuracy to the economic situation existing in the Russian Federation. The transition to a market economy has proved far more difficult than anyone imagined in the heady days after the fall of the Berlin Wall (Kerr et al., 1994). Rather than asking 'How long will a full transition to a market economy take?', the more appropriate question may well be 'Transition to what?' (Kerr and MacKay, 1997). Just as there were no 'road maps' for the Bolsheviks to use in establishing a command economy, there were none for the process of transition instituted by Yeltsin in Russia and those in other former command economies in Central and Eastern Europe and the New Independent States of the former Soviet Union. Thus, the attempt to move from a command to a market economy represents the second 'great economic experiment' of the 20th century. It is an experiment that appears likely to consume a considerable portion of the 21st century as well.

7.1.1 Price Reform the Old Way

From 1991 to 1992, the Yeltsin economic reform programme was concentrated on the gradual elimination of price controls, and an attempt to create a solid foundation for the transition to a free market economy. As might have been anticipated, the liberalization of strategic commodities – including energy, agricultural products, precious metals and freight tariffs – was perceived as a threat to industrial development and the general rate of inflation, so that these prices remained under the strict supervision, that is, control, of the government.

The main elements of pricing policy reforms in the Russian Federation in 1991–92 may be summarized briefly as follows:

- *January 1991: A Partial Reform of Producer and Agricultural Prices*: The January 1991 price reform was undertaken in an effort to stabilize the economy and ease the transition to a free market economy. Prices were cautiously 'amended' to improve production incentives, thereby preventing any 'superfluous' deterioration of the fractured Soviet industrial complex. To be specific, according to Koen and Phillips, (1993, pp. 14–15):

    The January 1991 reform shifted many producer prices from the fixed to the contractual category. In theory, enterprises were permitted to negotiate contract prices of so-called new goods within administratively set limits. In practice, these prices reportedly were still heavily regulated and linked to state order prices. After the reform, contractual prices accounted for 40 per cent of the total in light industry, 50 per cent in machine construction, and about 25 per cent in the raw materials, energy and metals sectors. As before the reform, the prices of new products were allocated to be set on a contract basis.

The reform sent the whole industrial price index for the former Soviet Union soaring by 229 per cent in the year 1991. Not surprisingly, the aggregate wholesale price indices for fuels and electrical energy were increased at a slower pace – 134 and 116 per cent, respectively – suggesting a 'real' reduction in the relative price of domestic energy products (IMF, 1992). To further aggravate matters, most official retail prices were held constant until April 1991, so that 'the first effect of the reform was to increase the subsidy bill considerably' (Granville, 1995, p. 15).

- *2 April 1991*: *Partial Retail Price Reform*: The April reform was designed to ease the degree of central control over retail prices gradually, so as not to disturb the delicate balance between controlled deregulation and 'speculative hysteria or chaos. To this end: (1) the aggregate index of administered retail prices was increased by a total of 60 per cent in April 1991; (2) the share of retail prices that were fixed; that is, set by central command, was reduced to 55 per cent; (3) the share of contractual prices – to be negotiated between the producer and the retail unit – was increased to 30 per cent; (4) the share of regulated (loosely controlled) prices was increased to 15 per cent; (5) the fixed prices for a number of 'problem' commodities were increased in an effort to establish 'a more rational pattern of relative prices (e.g. among bread, meat and fish) and [reduce] subsidies' (IMF, 1992, pp. 7–8); and (6) Russian consumers were compensated for 85 per cent of the increase in expenditures incurred as a 'direct' result of the April reform. Unhappily, the growth in household compensation payments surpassed the associated increase in collected 'retail' tax revenue, aggravating an already oppressive, and worsening, government budget deficit.

- *2 January 1992*: *Price Liberalization*: On 2 January 1992 President Yeltsin took the first serious steps towards the elimination of price controls, following the advice of western 'experts', who told him that this was the first step toward creating the basic foundations for a free market economy (Hobbs et al., 1997). In the early days, western economists advised a simple, two pronged approach to a market economy – privatization and price liberalization – forgetting that market economies have a myriad of institutions that underpin their operation. It was a recipe for a disaster (Kerr et al., 1994).

For the first time since the Bolshevik victory of 1917, the prices for most good and services were liberalized, and allowed to find their own levels. Important exceptions were made, however, including energy (all fuels and electricity), agriculture (basic foodstuffs such as milk, bread, vodka, sugar, vegetable oil, salt and baby food), precious metals, and freight tariffs. While still tightly controlled, the prices for crude oil, petroleum products and coal were increased by approximately 400 per cent. The administered price of coking coal was raised by 700 per cent. In a slight concession to avid 'reformers', the share of oil, natural gas and coal that could be sold 'legally' at free prices was increased to 40 per cent in May 1992. Prices in the state distribution sector were partially deregulated, but confined to a 25 per cent ceiling on all mark-up ratios.

As expected, the newly liberated prices leapt to unprecedented highs. Official estimates place the aggregate increase in the producer and consumer price indices at 382 per cent and 296 per cent, respectively, in January 1992. At this point, the forces of supply and demand, of course greatly distorted by the lack of market institutions and limited market intervention by local authorities, began to exert their influence on price levels and product availability. Freely priced goods such as yoghurt and cheese became plentiful, while severe shortages developed for products whose prices were still regulated or fixed by central command.

The maintenance of even this limited number of 'essential' price controls would prove overly expensive for the newly formed Commonwealth of Independent States (CIS) government. Tax revenues were waning, and the government budget had been stretched to the limit by years of acute economic depression. A massive 'unpaid' bill for government subsidies led to the eventual disappearance of important basic food stuffs – milk and baby food – from the Russian domestic market place. Yeltsin was forced to implement a 'second wave' of liberalization policies, and the prices for all remaining foodstuffs were deregulated in March–April 1992. Administered prices were maintained for a few politically sensitive commodities such as rents, utilities, public transportation, and state purchases of wholesale grain. The liberalization of energy prices, which was perceived as a serious threat to the grain harvest, industrial production and the general rate of inflation, was delayed until March 1995.

Table 7.1 illustrates the evolution of domestic crude oil pricing policies in the Russian Federation. The effects of these policies on domestic oil prices may be summarized as follows: in the January 1991 price reform, the average wholesale enterprise price of crude oil was increased to 65 roubles per tonne, a 152 per cent increase over the 25.7 roubles per tonne reported in 1990 (see Table 7.1). As mentioned above, the wholesale industry prices for crude oil were raised at a slower pace – a 130 per cent increase to the 63–74 roubles per tonne range. In stark contrast to the upstream arrangement, the retail prices for petroleum products were increased only hesitantly, and the official retail prices for regular (76 RON) and premium (92 RON) gasoline were actually held constant at 0.3 and 0.4 roubles per litre, respectively, for the years 1990 and 1991 inclusive (IEA, 1993).

Table 7.1: Average producer prices for crude oil in the Russian Federation, 1990–94

(roubles per metric tonne)

Year/Quarter	Nominal Contract Prices (Rbl/t)	GDP Deflator Index (1970=1)	Real Prices (Constant 1970 Rbl/t)	Contract Price [a] (US$/t)
1990	25.7	1.267	20.279	1.329
1991	65.0	2.645	24.575	1.049
1992 Q1	315.0	24.651	12.779	1.817
Q2	860.0	47.741	18.041	6.671
Q3	1,990.0	79.929	24.897	11.557
Q4	7,433.0	142.984	51.985	18.437
1993 Q1	14,233.0	292.792	48.611	24.893
Q2	20,767.0	531.204	39.094	22.595
Q3	32,733.0	1,011.467	.32.362	31.875
Q4	43,347.0	1,704.090	25.437	37.091
1994 Q1	67,670.0	2,620.023	25.828	42.842
Q2	74,666.6	3,451.450	21.633	39.818
Q3	75,333.3	4,265.025	17.663	35.145
Q4	89,966.6	5,726.153	15.712	28.162

Note: [a] Contract prices are calculated as a simple arithmetic average of monthly data. The Moscow Interbank Currency Exchange rate (MICES) was used to calculate the contract price in US dollars per metric tonne.

Source: Khartukov (1995).

It is important to note that administered or planned price increases were implemented at a faster pace in most other industrial sectors of the Russian Federation. Needless to say, the discrepancy in relative price increases was reflected in a dramatic increase in crude oil production costs. While the retail prices for petroleum products languished at 1990 levels, the cost of producing a tonne of crude oil in the former USSR soared to 57.39 roubles per tonne in 1991, a 171 per cent increase over the 21.13 roubles recorded in 1990.

The situation reached crisis proportions in 1991 when average production costs rose to 98.2 per cent of average ex-field gate wholesale enterprise price in the Russian Federation (see Tables 7.2, 7.3 and 7.4). This 'relative' reduction in crude oil prices was functionally unsustainable, and mounting industry subsidies would soon set the stage for yet another substantial price increase. The long years of control and stagnation had finally come to an end. From this point on, the nominal increase in crude oil prices would 'resemble the flight of a champagne cork' (Khartukov, 1995, p. 3) released suddenly from a bottle that had been shaken violently with no practical benefit to its contents.

On 1 January 1992 the average wholesale enterprise price for crude oil was raised to 315 roubles per tonne, a 385 per cent increase over 1991 levels. At the same time, wholesale industry prices were 'unified' and raised to a single, state-controlled level of 350 roubles per tonne, a 400 per cent increase. An inconsequential volume of domestic oil prices was cautiously liberated. To summarize the complex repertoire of tentative price reforms, a two-tiered pricing system was introduced under which a fraction of total Russian Federation production – approximately 30 per cent – could be sold at the free market prices available on the domestic market. As always, approximately 10 per cent of total production was earmarked for exports to the non-CIS nations and sold at the world oil price.

These experimental pricing policies would, once again, prove insufficient. According to Khartukov (1995, p. 3) Yeltsin's grand price liberalization scheme 'caused prices in the other (input-supplying) sectors to catapult by factors of 7 to 13 within a mere four months', and the government was forced to permit another administrative increase in crude oil prices.

Less than one month later, on 18 May 1992, the two-tiered pricing system was abandoned and the 'free market' prices for non-State order oil were replaced by a single fixed wholesale industry price of 1800 roubles per tonne, the designated 'base' price. In theory, all wholesale industry contract prices could be freely negotiated to a ceiling of 2200 roubles per tonne (excluding Value Added Tax). In practice, negotiations for crude oil prices exceeding 1900 roubles per tonne were nonsensical. A progressive 60–90 per cent price equalization tax was levied on all prices negotiated in the 1800–2200 roubles per tonne range. Revenues accruing from the price equalization tax, and indeed all revenues amassed from the sale of crude at prices above the 2200 mark, were diverted to a newly established Price Regulation Fund (PRF). The legislation effectively lowered the wholesale price ceiling to 1980 roubles per tonne – the average contract price negotiated in the Russian Federation in August 1992.

The effects of even these huge administrative increases in domestic oil prices were counter-productive. While the wholesale enterprise price of crude oil had virtually exploded, registering a 2946.15 per cent nominal increase from 65 roubles per tonne in 1991 to 1980 roubles per tonne in August 1992, the oil industry would receive no 'real' benefits from the carefully planned price increments. The oversight has been attributed to deliberate government policy – the maintenance of strict energy price controls in the face of Yeltsin's sweeping (January 1992) price liberalization – and a comprehensive, albeit still highly restrictive, currency reform. A number of factors contributed to the confusion.

The first was the old Soviet aversion to inflation, and a 'mistaken premise that the increase in energy prices would be inflationary' (Granville, 1995, p. 23). Granville (1995, p. 23) suggests that:

> [The] authorities failed to see that the rise in prices entailed by liberalisation would cause only a temporary increase in inflation, which would soon have returned to its previous level. Indeed, one can go further than this. Higher energy prices would have decreased the budget deficit and therefore reduced the need for monetary financing. The result would have been lower inflation, with a one-month jump in the overall price level.

Instead, abstruse fears of hyper-inflation prevailed, and Russian oil producers received only a token nominal increase in crude oil prices. When valued in constant 1970 roubles, the real crude oil price reached levels as low as 18.014 (1970 rbls./t) in the second quarter of 1992; a real 26.70 per cent reduction from the 24.575 (1970 rbls/t) reported in 1991 (see Table 7.1). This must be viewed against the background that in the years following the January 1991 price reform, inflation ran rampant throughout the Russian Federation sending the Gross Domestic Product price index (1970=1.00) soaring from an annual average of 2.645 in 1991 to 79.929 in the third quarter of 1992; a 2922 per cent increase in less than two years.

A second factor was the gradual liberalisation of official exchange rates, and resulting depreciation of the rouble on international capital markets. Despite massive intervention by the government and Central Bank of Russia, the liberalisation of Russian exchange rates resulted in what can only be described as a precipitous depreciation of the rouble on the

Table 7.2: *Taxes and costs of crude oil in the Russian Federation – crude oil produced by Russian enterprises for domestic use, 1991–97*

	1993 1018.00		1994 2212.00		1995 4560.00		1996 5114.83		1997 5784.90	
Exchange Rate Rbls/US$	US$/tonne	Rbls/tonne	US$/tonne	Rbls/tonne	US$/tonne	Rbls/tonne	US$/tonne	Rbls/tonne	US$/tonne	Rbls/tonne
International Oil Price:[a]	128.99	131,315.15	119.97	265,380.89	144.08	838,697.32	128.99	588,209.33	157.01	803,062.04
Domestic Transportation Costs[b]	0.27	274.86	1.62	3,583.44	5.60	32,395.55	4.00	18,240.00	5.80	29,666.01
VAT[c]	5.98	6,086.00	6.78	15,000.00	14.74	85,260.00	11.93	54,390.00	15.81	80,861.28
Wholesale Enterprise Price[d]	29.89	30,430.00	33.91	75,000.00	70.18	406,000.00	56.80	259,000.00	75.28	385,053.71
Excise Levy	6.28	6,390.30	6.68	14,775.00	9.51	55,000.00	9.09	41,440.00	12.22	62,499.99
Price Regulation Fund	4.28	4,361.91								
Suppliers Price: Gross Wellhead	19.33	19,677.79	27.23	60,225.00	60.67	351,000.00	47.71	217,560.00	63.06	322,553.72
Royalty[e]	1.88	1913.56	2.17	4,793.91	4.83	27,939.6	3.80	17,317.78	5.02	25,675.28
Geology Fee	2.36	2,403.97	2.72	6,022.50	6.07	35,100.00	4.77	21,756.00	6.31	32,255.37
Special Purpose Levies[f]	7.93	8,069.72	9.50	21,012.55	3.09	17,869.63	15.97	72,821.67	10.8	55,231.72
Sub-Total:	12.17	12,387.25	14.39	31,828.96	13.99	80,938.23	24.54	111,895.45	22.12	113,162.37
Government Funds and Charges										
Production Costs[g]	9.39	9,559.02	13.38	29,596.56	34.57	199,984.68	20.42	93,115.20	28.95	148,074.33
Including workover and maintenance charges[h]	4.70	4,779.51	6.69	14,798.28	11.22	64,887.46	11.22	51,147.95	11.22	57,371.29
Producer Balance	-2.23	-2,268.48	-0.54	-1,200.52	12.11	70,077.08	2.75	12,549.35	11.99	61,317.02
Benefits for the Profit Tax[i]					2.42	14,015.42	0.55	2,509.87	2.40	12,263.40
Taxable Profit					9.69	56,061.66	2.20	10,039.48	9.59	49,053.61
Profit Tax					3.39	19,621.58	0.77	3,513.82	3.36	17,168.77
Dividend Tax					0.47	2,733.01	0.11	489.42	0.47	2,391.36
Profit for Payment of Dividends					2.68	15,487.03	0.61	2,773.41	2.65	13,551.06

Notes:

a: U.K. Dated Brent f.o.b. Sullom Voe.

b: Domestic transportation costs, average for the Russian Federation

c: The VAT tax rate is equal to 20 per cent of the wholesale enterprise price in the years 1993–94, and 21 per cent of the wholesale enterprise price in the years 1995–87.

d: Wholesale enterprise price, ex-field gate (excluding VAT/ST).

e: 8% of wholesale enterprise price minus the excise levy and allocations to the price regulation fund.

f: Special purpose levies include: The Ministry of Fuel and Energy Investment Fund (until 1996), Science, Land Use and Environment Funds, asset tax, the excess wage tax (until 1996) and road users tax.

g: Production costs include exploration, development and lifting costs, depreciation, and current expenses (all variable operating costs, and payments to the social reserve fund [approximately 37.5 per cent of salaries in 1993]).

h: Minor workover, and maintenance charges that must be paid every three years to prevent shut-ins. The World Bank, *Staff Appraisal Report, Russian Federation Oil Rehabilitation Project,* 26 May 2 1993, p. 142.

i: Benefits for the profit tax are assumed to equal 20 per cent of the 'producer balance profit'. The rate may be an underestimate. Under existing legislation, oil and gas producers are permitted to reduce their taxable base by up to 50 per cent if the profit is allocated to modernization, reconstruction, expansion and development. Professor Alexander G. Kemp, The Russian Petroleum Tax System: Evolution, Effects and Prospects, Paper presented to the 15th CERI International Oil and Gas Markets Conference, 30 Sept – 1 Oct 1996.

Sources: World Bank (1993); World Bank (1994); Khartukov (1998); Khartukov (1995); Smith (1997).

243

Table 7.3: Taxes and costs of crude oil in the Russian Federation – crude oil produced by Russian enterprises for export to destinations outside the CIS, 1991–97

	1993 1018.00		1994 2212.00		1995 4560.00		1996 5114.83		1997 5784.90	
Exchange Rate Rbls/US$	US$/tonne	Rbls/tonne	US$/tonne	Rbls/tonne	US$/tonne	Rbls/tonne	US$/tonne	Rbls/tonne	US$/tonne	Rbls/tonne
Export Price:[a]	108.73	110,687.14	107.13	236,971.56	129.60	749,725.63	114.40	521,664.00	138.40	707,892.47
Export Duty[b]	0	0	0	0	0	0	0	0	0	0
Transportation Costs[c]	12.00	12,216.00	15.60	34,507.20	22.25	128,714.47	18.00	82,080.00	22.25	113,804.97
VAT[d]	0	0	0	0	0	0	0	0	0	0
Wholesale Enterprise Price[e]	29.89	30,430.00	33.91	75,000.00	70.18	406,000.00	56.80	259,000.00	75.28	385,053.71
Excise Levy	6.28	6,390.30	6.68	14,775.00	9.51	55,000.00	9.09	41,440.00	12.22	62,499.99
Suppliers Price: Gross Wellhead	90.45	92,080.84	84.85	187,689.36	97.84	566,011.15	87.31	398,144.00	103.93	531,587.51
Royalty[f]	8.65	8,810.70	8.53	18,862.94	7.79	45,054.49	9.11	41,524.45	8.27	42,314.37
Geology Fee	2.36	2,403.97	2.72	6,022.50	6.07	35,100.00	4.77	21,756.00	6.31	32,255.37
Special Purpose Levies[g]	7.93	8,069.72	9.50	21,012.55	3.09	17,898.63	15.97	72,821.67	10.80	55,231.72
Sub-Total: Government Funds and Charges	18.94	19,284.39	20.75	45,897.99	16.95	98,053.12	29.85	136,102.13	25.38	129,801.46
Production Costs[h]	9.39	9,559.02	13.38	29,596.56	34.57	199,984.68	20.42	93,115.20	28.95	148,074.33
Including workover and maintenance charges[i]	4.70	4,779.51	6.69	14,798.28	11.22	64,887.46	11.22	51,147.95	11.22	57,371.29
Producer Balance Profit	62.12	63,237.43	50.72	112,194.81	46.32	267,973.35	37.05	168,926.67	49.60	253,711.72
Benefits for the Profit Tax	12.42	12,647.49	10.14	22,438.96	9.26	53,594.67	7.41	33,785.33	9.92	50,742.34
Taxable Profit	49.70	50,589.94	40.58	89,755.85	37.06	214,378.68	29.64	135,141.34	39.68	202,969.38
Profit Tax	15.90	16,188.78	14.20	31,414.55	12.97	75,032.54	10.37	47,299.47	13.89	71,039.28

Dividend Tax	2.53	1.98	1.81	1.44	1.93
	14.36	11.21	10.24	8.19	10.96
Profit for Payment of Dividends	2,580.09	4,375.60	10,450.96	6,588.14	9,894.76
	14,620.49	24,795.05	59,222.11	37,332.79	56,070.29

Notes:

a: Urals Blend at Russian export terminal.

b: The excise duty was set at approximately 30 ECU/tonne of the sales price in the years 1993–94, and 21.5 ECU/tonne in 1995. The tax was abolished on 1 July 1996. The effective tax rate is assumed to be equal to zero. In the years 1993–96, producers were granted a credit for this tax wherever equivalent investment in oil field development can be proved.

c: Pipeline transportation costs from West Siberia to the Black Sea:

d: The VAT tax rate on exports of crude oil to non-CIS countries is 'zero' rated, i.e. producers are permitted a refund on VAT expenditures. The rate is assumed to equal zero, despite the fact that producers have had difficulty in obtaining refunds. Professor Alexander G. Kemp, 'The Russian Petroleum Tax System: Evolution, Effects and Prospects', Paper presented to the 15th CERI International Oil and Gas Markets Conference, Calgary, Alberta, 30 September – 1 October 1996.

e: Wholesale enterprise price, ex-field gate (excluding VAT/ST).

f: The Royalty is assumed to be equal to approximately 8% of the export price (the sales price) in the years 1993–95, and 8% of the sales price minus transportation charges and the excise tax in the years 1996–97.

g: Special purpose levies include: The Ministry of Fuel and Energy Investment Fund (until 1996), Science, Land Use and Environment Funds, asset tax, the excess wage tax (until 1996) and road users' tax.

h: Production costs include exploration, development and lifting costs, depreciation, and current expenses (all variable operating costs, and payments to the social reserve fund [approximately 37.5 per cent of salaries in 1993]).

i: Minor workover, and maintenance charges that must be paid every three years to prevent shut-ins. The World Bank, *Staff Appraisal Report, Russian Federation Oil Rehabilitation Project*, 26 May 1993, p. 142.

Sources: Sipovsky (1996); Grey (1998); World Bank (1993); World Bank (1994); Khartukov (1995); Khartukov (1998); Smith (1997).

Table 7.4: Taxes and costs of crude oil in the Russian Federation – joint venture exports, 1991–97

| Exchange Rate Rbls/US$ | 1993 | | 1994 | | 1995 | | 1996 | | 1997 | |
| | 1018.00 | | 2212.00 | | 4560.00 | | 5114.83 | | 5784.90 | |
	US$/tonne	Rbls/tonne	US$/tonne	Rbls/tonne	US$/tonne	Rbls/tonne	US$/tonne	Rbls/tonne	US$/tonne	Rbls/tonne
Export Price[a]	108.73	110,687.14	107.13	236,971.56	114.40	521,664.00	138.40	707,892.47	129.60	749,725.63
Export Duty[b]	0	0	0	0	0	0	0	0	0	0
Transportation Costs[c]	12.00	12,216.00	15.60	34,507.20	18.00	82,080.00	22.25	113,804.97	22.25	128,714.47
VAT[d]	0	0	0	0	0	0	0	0	0	0
Wholesale Enterprise Price[e]	29.89	30,430.00	33.91	75,000.00	56.80	259,000.00	75.28	385,053.71	70.18	406,000.00
Excise Levy	6.28	6,390.30	6.68	14,775.00	9.09	41,440.00	12.22	62,499.99	9.51	55,000.00
Suppliers Price: Gross Wellhead	90.45	92,080.84	84.85	187,689.36	87.31	398,144.00	103.93	531,587.51	97.84	566,011.15
Royalty[f]	8.65	8,810.70	8.53	18,862.94	9.11	41,524.45	8.27	42,314.37	7.79	45,054.49
Geology Fee	2.36	2,403.97	2.72	6,022.50	4.77	21,756.00	6.31	32,255.37	6.07	35,100.00
Special Purpose Levies[g]	7.93	8,069.72	9.50	21,012.55	15.97	72,821.67	10.80	55,231.72	3.09	17,898.63
Sub-Total:	18.94	19,284.39	20.75	45,897.99	29.85	136,102.13	25.38	129,801.46	16.95	98,053.12
Government Funds and Charges										
Production Costs[h]	20.00	20,360.00	20.00	44,240.00	21.12	96,307.20	28.95	148,074.33	34.57	199,984.68
Including workover and maintenance charges[i]	10.00	1,018.00	10.00	2,212.00	11.22	51,147.95	11.22	57,371.29	11.22	64,887.46
Producer Balance Profit	51.51	52,436.45	44.10	97,551.37	36.35	165,734.67	49.60	253,711.72	46.32	267,973.35
Benefits for the Profit Tax	10.30	10,487.29	8.82	19,510.27	7.27	33,146.93	9.92	50,742.34	9.26	53,594.67
Taxable Profit	41.21	41,949.16	35.28	78,041.10	29.08	132,587.74	39.68	202,969.38	37.06	214,378.68
Profit Tax	13.19	13,423.73	12.35	27,314.38	10.18	46,405.71	13.89	71,039.28	12.97	75,032.54
Dividend Tax	2.10	2,139.41	1.72	3,804.50	1.42	6,463.65	1.93	9,894.76	1.81	10,450.96
Profit for Payment of Dividends	11.91	12,123.31	9.75	21,558.85	8.03	36,627.36	10.96	56,070.29	10.24	59,222.11

Notes:

a: Urals Blend at Russian export terminal.

b: The excise duty was set at approximately 30 ECU/tonne of the sales price in the years 1993–94, and 21.5 ECU/tonne in 1995. The tax was abolished on 1 July 1996. The effective tax rate is assumed to be equal to zero. In the years 1993–96, producers were granted a credit for this tax wherever equivalent investment in oil field development can be proved.

c: Pipeline transportation costs from West Siberia to the Black Sea.

d: The VAT tax rate on exports of crude oil to non-CIS countries is 'zero' rated, i.e. producers are permitted a refund on VAT expenditures. The rate is assumed to equal zero, despite the fact that producers have had difficulty in obtaining refunds.

e: Wholesale enterprise price, ex-field gate (excluding VAT/ST).

f: The Royalty is assumed to be equal to approximately 8% of the export price (the sales price) in the years 1993–95, and 8% of the sales price minus transportation charges and the excise tax in the years 1996–97.

g: Special purpose levies include: The Ministry of Fuel and Energy Investment Fund (until 1996), Science, Land Use and Environment Funds, asset tax, the excess wage tax (until 1996) and road users' tax.

h: Production costs include exploration, development and lifting costs, depreciation, and current expenses (all variable operating costs, and payments to the social reserve fund [approximately 37.5 per cent of salaries in 1993]).

i: Minor workover, and maintenance charges that must be paid every three years to prevent shut-ins.

Sources: Sipovsky (1996); Grey (1998); World Bank (1993); World Bank (1994); Khartukov (1995); Khartukov (1998); Smith (1997).

international exchange markets. By the end of 1992, the Moscow inter-bank foreign exchange rate had risen to 414.5 roubles per US dollar; 18 times higher than the 22.88 roubles per US dollar reported in 1990.

The effects of even these huge administrative increases in domestic oil prices were counter-productive. While the wholesale enterprise price of crude oil had virtually exploded, registering a 2946.15 per cent nominal increase from 65 roubles per tonne in 1991 to 1980 roubles per tonne in August 1992, the oil industry would receive no 'real' benefits from the carefully planned price increments. The oversight has been attributed to deliberate government policy – the maintenance of strict energy price controls in the face of Yeltsin's sweeping (January 1992) price liberalization – and a comprehensive, albeit still highly restrictive, currency reform. A number of factors contributed to the confusion.

The first was the old Soviet aversion to inflation, and a 'mistaken premise that the increase in energy prices would be inflationary' (Granville, 1995, p. 23). Granville (1995, p. 23) suggests that:

> [The] authorities failed to see that the rise in prices entailed by liberalisation would cause only a temporary increase in inflation, which would soon have returned to its previous level. Indeed, one can go further than this. Higher energy prices would have decreased the budget deficit and therefore reduced the need for monetary financing. The result would have been lower inflation, with a one-month jump in the overall price level.

Instead, abstruse fears of hyper-inflation prevailed, and Russian oil producers received only a token nominal increase in crude oil prices. When valued in constant 1970 roubles, the real crude oil price reached levels as low as 18.014 (1970 rbls/t) in the second quarter of 1992, a real 26.70 per cent reduction from the 24.575 (1970 rbls/t) reported in 1991 (see Table 7.1). This must be viewed against the background that in the years following the January 1991 price reform, inflation ran rampant throughout the Russian Federation sending the Gross Domestic Product price index (1970=1.00) soaring from an annual average of 2.645 in 1991 to 79.929 in the third quarter of 1992, a 2922 per cent increase in less than two years.

A second factor was the gradual liberalization of official exchange rates, and resulting depreciation of the rouble on international capital markets. Despite massive intervention by the government and Central Bank of Russia, the liberalisation of Russian exchange rates resulted in what can only be described as a precipitous depreciation of the rouble on the international exchange markets. By the end of 1992, the Moscow inter-bank foreign exchange rate had risen to 414.5 roubles per US dollar, 18 times higher than the 22.88 roubles per US dollar reported in 1990. The devaluation effectively undermined most of the gains accruing from what would subsequently prove to be minor, nominal increases in the level of domestic Russian crude oil prices. As a result, producer contract prices in the Russian Federation averaged US\$0.91 per barrel in the second quarter of 1992 – a 405 per cent increase over the US\$0.18 reported in 1990 – but still only 5 per cent of the world oil price (approximately US\$18.00 per barrel). The net effect was to discourage any interest in investing in the sector. Substantial real increments in the level of crude oil prices would be required to meet even the Spartan targets of the Yeltsin government, that is, raising domestic energy prices to one-third of world prices in September 1992 (Watkins, 1993).

The separation of domestic and export markets for crude oil and petroleum products also contributed to industry confusion. The state monopoly over foreign trade, which had been forged gradually through the isolation of Stalinism, and expanded and consolidated in the Brezhnev years, was, in the manner of price controls, rigorously maintained, and politically guarded after the end of the command era. As always, aggregate oil export quotas were determined by the appropriate authorities, that is, the Ministry of Economics. The sole indication of Russia's revolutionary attempt at transition to a free market economy was an insignificant, barely perceptible, change in administrative procedure. In the new world of the 1990s: '*Special exporters* (usually producers or trading companies) were the only entities legally authorised to export, acting as agents for producers in realizing their export quota's' (IEA, 1995 p. 131). An age-old, and irreproachable Soviet methodology set crude oil and product exports as the residual of 'anticipated' production volumes minus domestic consumption (OECD/IEA, 1995). Not surprisingly, aggregate oil export quotas remained constant at approximately 10–15 per cent of total Russian production in the years 1991–93 inclusive. As in the past, numerous administrative imperfections, including uncertainty and inaccuracy in the forecasts for domestic consumption, led to frequent adjustments and haphazard, ad-lib revision (IEA, 1993).

However flawed, this rigid system of export quotas and export taxes was crucial to the maintenance of domestic price controls, and a low internal price level. With domestic crude oil prices fixed at a mere 5 per cent of the world oil price in the second quarter of 1992, the abolition of export quotas would have diverted most of the oil production in the Russian Federation to international non-CIS markets, leaving the domestic consumers bereft of supplies.

The first steps towards the liberalization of domestic oil prices were taken on 18 September 1992. Decree number 1098 abolished the practices of both 'allocated sales', and zone-based pricing. The concept of cost-plus pricing was maintained but, for the first time ever, was to be implemented on an individual basis. Under the new regime, the wholesale enterprise price was defined as the sum of production costs (for each individual enterprise), taxes (including royalties and other government charges) and a standard 50 per cent profit margin. Prices could be 'freely' negotiated between producers and consumers, but were subject to a strict limit, or absolute ceiling, of 1.5 times production costs. As before, a base wellhead price was defined, and fixed at 4000 roubles per tonne (excluding VAT and excise taxes). An enterprise could exceed this base price, but only by an amount equal to, or less than, 1.5 times the firm's individual production costs. Prices negotiated in excess of the base price were subject to a subdued but still highly progressive price regulation tax.

The September oil price reform was aggressive by Russian standards, and as a result, carried real implications for the level of domestic oil prices. When valued in 1970 roubles, wholesale enterprise contract prices reached 51.985 (1970 rbls/t) in the fourth quarter of 1992, a 189 per cent increase over the 18.014 (1970 rbls/t) reported in the second quarter (see Table 7.1). Remarkably, the nominal increase in producer prices was sufficient to offset even a 1724 per cent annual increase in production costs.

Exploration, development and lifting costs, including current expenses, reached only 1120 roubles per tonne in 1992, an encouraging 43.6 of the ex-field gate wholesale enterprise price (see Tables 7.2, 7.3 and 7.4).

7.1.2 The Effective Elimination of Industrial Profitability Through Taxation

The 'real' gains in gross profitability – price minus production costs – would never be realized by the ailing Russian oil industry. Domestic producers, long accustomed to paying only token charges for centralized assets, including the free use of land and mineral rights, were 'suddenly' subjected to a plethora of taxes and levies and duties imposed by federal, regional and local governments and authorities (see Table 7.5).

To cite only a few of the new, and unduly burdensome, levies: (1) a 26 ECU per tonne export tax was introduced in December 1991, and increased to 44 ECU per tonne in mid-1992; (2) mandatory currency conversion or surrender requirements were introduced in December, 1991; (3) a 6–16 per cent royalty payment for the use of subsoil resources was introduced on

21 February 1992; (4) a 10 per cent geology fee (revenue tax) for the use of subsoil resources was introduced on 21 February 1992; (5) an 18 per cent excise tax was introduced on 17 September 1992; (6) a 32 per cent profits tax was introduced in 1992; (7) a 28 per cent value added tax replaced the turnover tax in January, 1992; (8) deductions to the Federal Oil Investment Fund averaging approximately 28 per cent of gross proceeds excluding the excise tax were introduced in 1992; and (9) a host of additional taxes was introduced in 1992, including the road tax, the excess-wage tax, property (assets) taxes, the land tax, environmental taxes, and a wide variety of colourful, and diverse 'local' levies. Often these taxes were thinly disguised mechanisms for allowing local bureaucrats, still in place but freed from control at the centre, to 'feather their own nests'. Corruption exercised through the ability to license (tax) firms and other economic actors became the rule in former command economies (Kerr and MacKay, 1997).

By the end of 1992, fiscal charges and payments had reached 69 per cent of the ex-field gate wholesale enterprise price. The 'new' payments to the government effectively eliminated gross profits, forcing the typical domestic enterprise to operate at a loss of approximately 327 roubles per tonne (see Tables 7.2, 7.3 and 7.4).

Adding insult to injury, Russian domestic oil prices had yet to reach the government target of 30 per cent of the world oil price by September 1992. Producer prices for crude oil averaged only US$2.51 per barrel in the fourth quarter of 1992, a disappointing 15 per cent of the spot market price for Arabian Light of US$17.12 per barrel. The urgent necessity for sensible price and tax reform was indisputable. Less than two months later, in February 1993, both the price cap and progressive taxation schemes were relaxed, and the base price was raised to 9000 roubles per tonne.

The reduction of the price regulation (equalization) tax from 60–90 per cent in May–September 1992 to 10–30 per cent in February 1993 had significant implications for the nominal rate of increase in domestic oil prices. Wholesale enterprise contract prices rose steadily from a national average of 2050 roubles per tonne in September 1992 to 32 500 in July 1993, a 1485 per cent increase in less than twelve months. At this point, the exhaustive series of administered oil price increases began, finally, to take its toll on demand. The wholesale enterprise contract price stabilised at approximately 32 500–32 700 roubles per tonne between the months of July and August 1993. According to the World Bank (1994, p. 142), in effect:

Table 7.5: The evolution of the petroleum fiscal regime in the Russian Federation, 1991–97

Month/Year	Form of Taxation (Policy)	Tax Rate
	Value Added Tax	
January 1992	The Russian Federation introduces a general value added tax (VAT) to replace the turnover tax.	28%
1993	The VAT is reduced to 20% plus an effective surcharge of 3%. At the same time VAT charges are levied on imports from non-CIS countries. Oil producers are permitted to offset the VAT paid on imports with the VAT levied on their sales. An actual refund may be received if the volumes in question are exported to the non-CIS nations. Firms are 'reimbursed by invoice to the government with up to a two year delay'.	20%, plus a special 3% tax levied on the sales of all goods.
	Export Taxes	
31 December 1991	Decree #91: An export tax is introduced for crude oil exports. The federal ad valorem customs tariff is to be paid in Roubles no later than 60 days after the oil is presented to customs for inspection. Note: 'In calculating this a tax credit is provided based upon the losses arising from the 40% currency conversion described (below)'.	26 ECU/tonne (approximately \$35/tonne US)
Mid-1992	The export tax is increased.	44 ECU/tonne (approximately \$60/tonne US)
September 1992	The export tax is reduced.	21 ECU/tonne (approximately \$28/tonne US)
1993	(Law on the Customs Schedule—21 May 1993, and Government Ordinance # 1103–30 October 1993) The export tax is increased to 30 ECU/tonne at the beginning of 1993. Exemptions were provided for: (i) Centralized exports to meet State needs; (ii) Exports under intergovernmental agreements within the CIS; (iii) 100 per cent of the incremental production from rehabilitation projects – until such time as the costs have been fully recovered (Resolution #179 March 1993); (iv) 60 per cent of production from a wide range of new field developments – for five years commencing with the startup of production (Resolutions #180 and #218); (v) Joint ventures formed before 1 January 1992 – until such time as costs have been fully recovered. In short, producers were 'granted a credit for this tax where equivalent investment in oilfield development can be demonstrated (See for example Resolution #179), and an exemption until project payout for major rehabilitation and new field investments'	30 ECU/tonne

1 January 1995	The export tax is reduced to 23 ECU/tonne.	23 ECU/tonne.
1 April 1995	The export tax is to be reduced gradually in line with corresponding and complimentary increases in the excise tax.	20 ECU/tonne
April 1996	The export tax is reduced to 10 ECU/tonne.	10 ECU/tonne
1 July 1996	The export tax is abolished.	

Tax on Centralized Exports

	The introduction of an export duty on centralized exports, or exports for State needs. The tax applies to all oil and petroleum products sold using an export licence that has been obtained through an 'official' auction.	The effective tax rate is varied on a case by case basis, and may be approximated by the 'difference between the domestic purchasing price (adjusted for transport costs and margin) and the world market' price.

Excise Taxes

17 September 1992	Decrees # 893 August 1992, # 724 September 1992, # 847 November 1992: Law on the Government Regulation of Prices on Some Types of Fuel and Energy: The Russian Federation introduces an excise tax – a percentage ad valorem tax based on the domestic price (even when the volumes in question were to be exported).	Revenue Tax: Initially set at 18% of gross product sold by an enterprise.
1 November 1992	Decree #847: The excise tax is differentiated to reflect differences in production costs. Exemptions were granted to companies employing more than 30 per cent of foreign capital if they had registered before the end of 1991, and were engaged in export activities.	Revenue tax: Rates varied from 5–30% according to the ease of mineral extraction from named production associations. A rate of 18% was applied to 'unnamed' companies. For Example: Under decree #847 1 November 1992, KNG was charged 25% while PNG and VNG were exempt from the excise tax.

253

Month/Year	Form of Taxation (Policy)	Tax Rate
July 1993	Prime Minister Chernomyrdin signs a decree increasing the average excise tax from 18% to 30%.	The average excise tax rate is set at 30%
13 September 1993	Decree # 910: The excise tax is reformed.	'Rates varying from nothing to 42% apply to a number of producers with extremely high or low natural production costs. Those in the middle range of costs pay 24%... All joint ventures currently operating have to pay the 24% rate even though their production costs vary greatly'.
14 April 1994	Government Ordinance #320 on Excises on Oil Produced in Russia: The excise tax default rate is set at 14 750 Rbls/t. The excise tax is changed from a percentage ad valorem tax to an equivalent Rbl/t tax to be indexed to the Rbl/US$ exchange rate and adjusted monthly.	The average default tax rate is set at 14 750 Rbls/t (approximately US$10/t). The actual excise tax rates are varied by field according to production costs, and range from 0–36 000 Rbls/t. i.e. 11 000 Rbls/t (US$7.30/t) for MNG, and 8500 Rbls/t (US$5.70/t) for TN and YNG.
1 April 1995	Decree #590: The excise tax is to be increased gradually in line with corresponding and complementary reductions in the export tax.	The average default rate is set at 39,000 Rbls/tonne. Actual rates can vary between 20,000 and 50,000 roubles per tonne as determined separately for each entity by the Russian Ministry of Fuel and Energy (MF&E).
June 1995	The maximum rate of the excise tax is raised to '53 040 Rbls/tonne ($US11.3/tonne), with the weighted average of its differential rates reaching 41 400 Rbls/tonne ($US8.8/tonne)'.	The average default rate reaches 41,400 Rbls/tonne.

19 June 1996	The excise tax is reformed and differentiated further according to the following new field categories: (A) Older fields utilizing enhanced recovery techniques; (B) Mature fields with production less than 2.65 tonnes/day, and water encroachment rates of 70-90%, or higher; (C) New fields coming on stream in the years 1996–2000; and (D) Fields with 'abnormal' properties, i.e. heavy fractions, well depths greater than 4500 meters, a gas content greater than 1000 cm/t, and foundation temperatures higher than100 degrees C.	Fields in Groups (A) and (D): The excise tax is set at 50% of the maximum rate. Fields in Group (B): The excise tax is reduced to 75% (of the rate which would otherwise apply) if water encroachment falls in the range of 70-90%, and 25% if water encroachment exceeds 90%.. Fields in Group (C): Receive a full exemption from the excise tax until capital investment has been fully recovered. The exemption is limited to a period of 5 years following the initial development date, regardless of the status of 'recovered' capital investment costs.
11 July 1996	The Ministry of Finance and State Tax Service issue an ordinance raising the average default rate from 55 000 Rbls/t to 70 000 Rbls/t.	The average default rate is set at 70 000 Rbls/t. The maximum default rate – applied to a significant number of Joint Ventures – reaches 82 536 Rbls/t.
8 April 1997	The maximum rate of the excise tax is raised to 84,005 Rbls/tonne ($US14.52/tonne), with the weighted average of its differential rates reaching approximately 55, 000 Rbls/tonne ($US9.5/tonne).	The average default rate reaches 55 000 Rbls/tonne.

255

Month/Year	Form of Taxation (Policy)	Tax Rate
	Royalty Payments	
21 February 1992	Law on Subsoil Resources (Government ordinances #318 – 18 May 1992; #478 – 9 July 9 1992; and #828 – 28 October 1992): Royalty payments are introduced for use of the subsoil. The royalty consists of a two-tier payment; (i) an initial downpayment; and (ii) periodic payments following the initiation of mining operations. Important exemptions include: (i) Oil production requiring thermal, chemical, physical, or micro biological recovery technology; and (ii) 'approved' or specified mining research efforts. The federal government receives 40 per cent of the royalty tax revenue, 30 per cent is earmarked for the regional government, and the remaining 30 per cent is allocated to the local authorities.	Revenue Tax: Varies from 6–16% of the value of gross wellhead prices (output). The rate is determined by the type, quantity and quality of the sub-soil mineral, the geographical, economic and technical environment in which the mineral in question is to be prospected, and the financial risks undertaken by the prospector. The exact value is determined by negotiation, or in the case of new oil fields, bidding. In practice, the rate of royalty… is deductible from taxable profit as production expenses… [and] is usually negotiated at 8%.
21 February 1992	Law on Subsoil Resources (Government ordinances #318 – 18 May 1992; #478 – 9 July 1992, Resolution #4546-1 of the Supreme Soviet of the Russian Federation – 25 February 1993): A geology fee is introduced for all fields that fall under the jurisdiction of the Geology Committee (Roskomnedra). Oil producers are permitted to reduce their payments to the Geology Fund if an equivalent amount is spent on new field development. New fields are exempt from the levy in all cases where the producer has incurred the entire exploration costs. Other important exemptions include: Oil production requiring thermal, chemical, physical, or micro biological recovery technology and 'approved' or specified mining research efforts. Revenues are allocated to the federal non-budgetary Fund for the Reproduction of Mineral and Raw Material Stocks.	Revenue Tax: 10% of the value of gross well head prices (output). 'Large' taxpayers are required to make an advance payment every 10 days.
30 December 1993	Government Resolution #1359: The geology fee is disaggregated (differentiated) to reflect differences in geological exploration costs.	Revenue Tax: Varies from 0 in the older producing regions to 10 per cent of 'equivalent domestic revenues' in regions where the exploration effort is more pronounced.

256

Price Regulation Fund (PRF)

18 May 1992	Decree $318: A progressive price regulation tax is levied on all prices negotiated in the 1800–2200 roubles per tonne range.	Progressive tax varying from 60–90%.
18 September 1992	The price regulation tax is reduced.	Progressive tax varying from 30–50%.
1 February 1993	The price regulation tax is reduced.	Progressive tax varying from 10-30%.
January 1993	The Minister of Finance broadens the domain of the tax to include exports as well as domestic sales.	
July 1993	The price regulation tax is abolished.	

Profits Tax

18 September 1992	A profitability ceiling is introduced limiting prices to exactly 1.5 per cent of production cost. Note: World Bank Report claims this tax was abolished in late 1993 (WBSAP June 13, 1994 p. 143)	Gross (or Balance) Profit – defined as net revenue minus production costs – must be less than, or equal to, 50% of production costs.
1992	A tax is levied on gross profits – defined as 'the proceeds received from the sales of services, goods, fixed assets and other property and/or proceeds from non sale operations less allowed expenditures'. Benefits for the profit tax: oil producers are permitted 'to reduce their taxable base by the amount of profit allocated to modernization, reconstruction, expansion and development of production. The taxable base cannot be reduced by more than 50%'.	32% of Gross Profits
22 December 1993	Law of the Russian Federation Concerning Tax on the Profit of Enterprises and Organizations, as amended. Decree No. 2270 of the President of the Russian Federation of 22 December 1993 Concerning Certain Changes in Relation to Taxation and the Relations between Budgets at Various Levels: The law on profit tax is amended to include an excess wage tax – a tax on wages paid to employees in excess of certain norms.	38% of salaries paid in excess of 6 times the monthly minimum wage. (Rbls. 123,000 or US$41.00 as of 1 July 1994)

257

Month/Year	Form of Taxation (Policy)	Tax Rate
January 1994	Law of the Russian Federation Concerning Tax on the Profit of Enterprises and Organizations, as amended. Decree No. 2270 of the President of the Russian Federation of 22 December 1993 Concerning Certain Changes in Relation to Taxation and the Relations between Budgets at Various Levels: The profit tax is increased to a maximum of 38% of 'balance' (operating) profit.	35% of Gross Profits with regional increases of up to a maximum of 3%. The 38% maximum rate is allocated between the Federal government (13%) and local authorities (25%).
	Withholding Tax: External profit distribution (dividends) is subject to a 15% withholding tax. The tax is not applied to domestic producers.	15%
March 1995	The profitability ceiling is abolished for crude oil and petroleum products.	
	Mandatory Currency Conversions	
30 December 1991	Edict #335 (& Instruction #3 from the Central Bank – 22 January 1992) outlined severe hard-currency stabilization requirements. The mandatory sale of 40 per cent of all hard currency export earnings to the 'state' hard currency reserve at the special commercial exchange rate of 55 Rbls/t plus the mandatory sale of an additional 10 per cent of export earnings to the hard currency stabilization fund at the CBR market exchange rate (approximately 90–110 Rbls/US$).	Approximately 30 per cent of export earnings. NOTE: Estimates of the implied tax rate vary significantly according to time of estimation and method of analysis. According to Linda S. Goldberg, in the months from January through April 1992, the foreign exchange surrender requirements 'effectively taxed exporters at a rate of at least 30 percent of export earnings'. World Bank analysts ignore the implied tax rate (opportunity cost) but notice that 'currency conversions at market rates result in an approximate 1% tax on revenues due to transactions costs'.

November 1991	Foreign Investment Act – Article #25: Joint ventures with more than 30% foreign participation receive an exemption from the mandatory currency conversion requirements.	
March 1993	Resolution #374: Oil producers are granted an exemption from all mandatory currency conversions.	

Special Purpose Levies

May 1992	Crude oil producers in the Russian Federation are required to pay contributions (taxes) for a number of special purpose funds. By year-end 1992, the special purpose levies include:	
	The Ministry of Fuel and Energy Investment Fund	2 % Royalty
	Note: Payments to the Federal Oil Investment Fund: 'Investment Fund deductions were designed for intrasectoral redistribution of differential rent and represent the largest immediate withdrawal from oil producers' revenues: on the average, 28% of gross proceeds (excluding excise tax). Differentiated between 5% and 40% for individual producers and supposed to be returned with the "life-giving rain" of centralized investments, those deductions, however, have tended to vanish into bottomless government coffers, as the state practically ceased to finance the industry by 1993'.	
	Science, Land Use, and Environment Funds	5% of the costs of production
	Insurance Fund	2% of Gross Revenues
	Asset Tax	2% of Net Fixed Assets
	Road Users' Tax	0.4% of gross revenues from Jan. 1992 – May 1993 1.25 % of gross revenues less excise tax from June 1993–Dec. 1993 2% of gross revenues less excise tax from Jan 1994–1995
February 1993	Social Reserve Fund – levied on total wages for the purpose of rebuilding social infrastructure.	37.5% of total wages

259

Month/Year	Form of Taxation (Policy)	Tax Rate
June 1994	By June 1994, the list of special purpose levies has been expanded to include: Road users tax – 1.54% of products sold. Land tax – roubles per hectare. Employment Fund – 2% of wages (or the Remuneration Fund). Social Insurance Fund – 5.4% of wages (or the Remuneration Fund).. Pension Fund – 28% of wages (or the Remuneration Fund). Medical Insurance – 3.6% (or the Remuneration Fund). Transportation tax – 1% of the Remuneration Fund. Educational charges – 4.5% of products sold. Environmental charges – forests 0.02% of the Remuneration Fund. Environmental charges – territorial cleaning 0.5% of the Remuneration Fund. Environmental charges – water fees 0.05% of product sales. Vehicle taxes – roubles per horsepower for the entire fleet Vehicle taxes – a percentage of the value of new vehicles. Militia tax – 3% of the minimum wage multiplied by employment. Urban transport tax – 2% of the remuneration fund. Advertisement tax – 0.005% of product sold. Fund for the Support of Agriculture – 3% of production costs.	
May 1995	Deductions to investment funds are officially abolished.	
June 1995	By June 1995, the list of special purpose levies and local charges had risen to approximately 72 810 roubles per tonne (28.1% of the wholesale enterprise price for crude oil in the Russian Federation). The list of charges included deductions for the industry investment fund, the road-use tax, transport tax, property (assets) tax, environmental levies, land tax and other local taxes.	
June 1996	By June 1996, the list of special purpose levies has been modified, and included the following federal, regional and local charges: Road users' tax (1 to 2% of gross revenues), Assets Tax (2% of net fixed assets), Land tax (roubles per hectare), Pension fund, Social Security fund, Medical Insurance fund, State Unemployment fund, Pollution fee, Social and Accommodation tax, Municipal tas, Forest fee, Transport police fee, Transport tax, Licence fee, and Education tax.	

260

June 1997	By June 1997, the list of special purpose levies and local charges had fallen to approximately 17 900 roubles per tonne (4.4% of the wholesale enterprise price for crude oil in the Russian Federation). The list of charges included the road-use tax, property (assets) tax, environmental levies, land tax, and other local taxes.	

Miscellaneous

December 1991	Decree #213: All import duties are abolished.	
September 1992	Import duties reestablished at 15%. Exemptions are possible for imports of oilfield equipment.	
8 February 1993	A Repatriation Tax is levied on U.S. investors.	15% of remitted dividends
March 1993	Decree #441: Upstream oil field projects – and imports of oil field equipment – are 'officially' exempt from import duties.	
March 1993	Decree #10R, and Resolutions # 218, # 179, and # 180: Exports of incremental oil are declared 'duty free' for rehabilitation projects.	
1994	Wage penalties – Production Associations are required to add back all wages paid in excess of 6 times the minimum wage.	The penalty results in a 10% increase in taxes payable.
8 April 1994	Presidential Decree #307: Oil producers are granted the right to cut off supplies to delinquent consumers.	
June 1994	A withholding tax is levied on external profit distribution – This levy does not affect domestic producers.	5%
April 1996	A transport fee surcharge is established at $3.1 per tonne (average).	
July 1996	The transport fee surcharge is raised to $6.2 per tonne (average).	

Export Quotas

23 May 1994	Decree #1007 'On the Cancellation of Quota's and Licensing for the Export of Goods and Services': The decree stipulates the elimination of all export quotas – including those imposed on 'strategically' important commodities – by 1 July 1994. The official listing of strategically important goods was maintained, and the complex licensing requirement for strategic goods was replaced by a simple registration requirement.	

261

Month/Year	Form of Taxation (Policy)	Tax Rate
1 July 1994	A 'supplementary' decree delays the abolition of export quotas on crude oil and petroleum products until 1 January 1995.	
May— October 1994	Mazut exports are declared to be duty free (given an export tax holiday from May to October 1994).	
31 December 1994	Prime Minister Chernomyrdin issues a Resolution calling for the abolition of export quotas, while retaining a measure of control over the oil industry through the maintenance of the 'special exporter' status. The resolution requires that 'the volumes of exportable oil envisaged for each producer should vary with the amount of oil produced'.	Domestic producers are permitted to export approximately 30% of their 'own' production. Joint Ventures are permitted to export approximately 60% of their 'own' production in January and February. This share was subsequently raised to 100% for the March export schedule.
December 1994–March 1995	A ban is imposed on Mazut Exports from December 1994–March 1995	
28 February 1995	Government Decree #209: The principal of equal access to pipelines is established, 'with priority given to producers and suppliers that exported according to Russia's few remaining obligations (i.e. inter-governmental agreements for deliveries of crude to Slovakia, the Czech Republic and Cuba). Domestic producers are permitted to export the maximum of: (a) all requested volumes; or (b) an amount proportional to production in the previous quarter. Thirteen Joint Ventures are granted priority access, and permitted to export their 'own' product without restriction'.	
6 March 1995	Presidential Decree 'On Measures for State Regulation of Natural Monopolies in the Russian Federation': The 'Federal Service' is granted the authority to supervise access to export pipelines in the Russian Federation.	
25 March 1995	President Yeltsin signs a decree disbanding the institution of special exporters, and prohibiting the manipulation of exports through mandatory domestic sales targets.	

262

1996	The centralized export system is replaced by a system of 'exports for State needs'. Under export for state needs the government has controlled around one-fourth of the export capacity and uses it to make money by buying at the lower wellhead price and reselling it at the higher export price. The revenues have gone off-budget to various projects (such as a rehabilitation of the Kremlin dome).
1997	The system of 'exports for State needs' is abolished. A new system is implemented in which the allocation of scarce export facilities is determined on the basis of production shares (with consideration for the level of tax arrears). Federal programme exports reaches 15 248 000 tonnes in 1997, approximately 16.5% of total crude oil exports from the Russian Federation (excluding exports to other CIS member countries), and 19% of total exports to non-CIS/non-Eastern European nations.

Producers were unable to sell oil to the domestic market at a full (allowed) markup because of consumer price resistance. As a consequence, and due to mounting payment arrears, producers began to shut-in production and pull oil from the domestic market. By some estimates 20 million tonnes of production, representing 6% of total production, was voluntarily shut-in during 1993.

The progressive price regulation tax – a sliding scale price cap – was considered to be redundant and ineffective and, consequently, was abandoned in July 1993. The cost-plus rule – restricting prices to production costs plus taxes and a strict 50 per cent profit margin – was maintained as the last vestige of oil price control in the Russian Federation (IEA, 1993).

At the same time, the Yeltsin government was beginning to recognize the appalling implications of the onerous tax burden that had been forced, all too hastily, on the nation's bedraggled crude oil industry. Of particular concern was the disproportionate number of taxes that were based on a firm's gross revenues. These included the excise tax, royalty payments for the use of the subsoil, the geology fee, and payments to the Federal Oil Investment Fund (approximately 28 per cent of gross proceeds excluding the excise tax). As a general rule of thumb, a revenue-based tax is insensitive to the costs and profitability of an individual enterprise, and as a result, tends to discourage the high cost producers thereby distorting otherwise straightforward investment decisions. In a worst case scenario, it can drive them out of business altogether.

In the case of the Russian Federation's oil industry where costs were largely determined by natural geological and environmental conditions, and not the inefficiency of management, the long-term implications for the economic viability of domestic production associations would have been devastating. For example: while a corporate profits tax was introduced in 1992, it was never fully enforced for the simple reason that only a few firms had any taxable income (profits) left to tax.

Tentative measures to alleviate the 'revenue situation' were inaugurated in November 1992, at which time the excise tax was reformed (modified) to reflect differences in the profitability of an individual Russian project and/or production association. The new, and 'improved', excise tax rate varied from 5–30 per cent according to the ease (cost) of mineral extraction from named production associations (World Bank, 1994). An average rate of 18 per cent was applied to 'unnamed' companies.

Recommendations that revenue-based taxes be abolished altogether

The New Wild West 265

were greeted with a surprising degree of enthusiasm. As mentioned above, the price equalization tax sliding scale price cap was abandoned in July 1993. According to the IEA (1993, p. 16) by:

> early 1993 it looked as if the next revenue tax to be abolished may be the excise tax. At an oil conference in April 1993, Russian Deputy Finance Minister Molchanov hinted in that direction. But in July 1993 Prime Minister Chernomyrdin signed a decree increasing the average excise tax from 18 per cent to 30 per cent.

The retrenchment was short-lived, and less than two months later, on 13 September 1993, the average excise tax was reduced from 30 per cent to 24 per cent (see Table 7.5). In a related development, the government issued Presidential decree number 2285 'Concerning Issues Related to Production Sharing' on 24 December 1993. The new decree was designed to simplify the complex Russian tax regime, shifting the entire structure 'towards profit-based taxes for new licenses structured as production sharing agreements' (World Bank, 1994, p. 12). A number of major new projects, involving capital investments estimated in excess of US$35 billion and corresponding increases in production, would come under the new tax system. International, as well as domestic oil companies, would be affected. Among the international companies are Amoco, Elf, Exxon, Marathon, Mobil, Shell, Texaco and Total. Finally, on 30 December 1993, the geology fee was disaggregated to reflect 'regional' differences in geological exploration costs (Government Resolution Number 1359).

Even this limited attempt at tax reform would have minor positive implications for the profitability of the Russian oil industry. By the end of 1993, fiscal charges and payments had fallen to 66.3 per cent of the ex-field gate wholesale enterprise price. Under the new petroleum fiscal regime, a typical domestic enterprise could operate at a slight net gain of approximately 694 roubles per tonne. Somewhat ironically, the tax break was, however, accompanied by a subtle but disturbing reduction in direct government subsidies to the oil industry (IEA, 1993), and the 'state practically ceased to finance the industry by 1993' (Khartukov, 1995, p. 16).

The 'real' implications of the July 1993 price reform were far from encouraging. Once again, progress on this front would be delayed by deliberate (stubborn) government policy. To cite the usual examples: (1) the maintenance of strict energy price controls – the 1.5 times cost plus rule – in the face of rapid price liberalisation, and rampant inflation, in most other sectors of the economy; (2) the continued separation of the domestic and international (export) markets for crude oil and petroleum products; and (3) the liberalization of the exchange rate, and resulting depreciation of the rouble on international currency markets.

Over the 1992–94 period, the government had been attempting to engineer a controlled (and gradual) liberalization of the new 'unified' Russian exchange rate. Despite massive intervention by the CBR, the cautious exchange 'stabilization' programme would be undermined by years of loose monetary policy, and a weak fiscal policy; that is, the attempt to finance a large and growing budget deficit, while simultaneously providing substantial credits to the agricultural industry. By the end of September 1994, the Government's hard currency reserves had reached record lows, and the CBR was forced to relax the level of direct intervention on international currency markets. On *Black Tuesday* (11 October 1994) the exchange rate collapsed to 3962 roubles/US dollar, a 27 per cent devaluation in less than 24 hours.

In short, while nominal domestic crude oil prices had risen to 76 909 roubles per barrel in 1994 representing a 177 per cent increase over the 27 770 roubles per tonne reported in 1993, they were still less than 30 per cent of the world oil price (see Table 7.6). Given the official Moscow inter-bank auction exchange rate for 1994, approximately 2212 rbls/US$, Russian Federation producer prices averaged only US$4.74 per barrel, only 28.97 per cent of the US$16.36 reported for UK Brent Blend (dated, f.o.b. Sullom Voe). Given the chaotic state of the Russian market, however, direct comparisons of Russian prices with those in the international market must be viewed with considerable caution.

A rough estimate of the implied opportunity cost of Russian Federation pricing policies to the domestic crude oil industry is illustrated in Table 7.7. As a general rule, the opportunity cost to domestic producers is simply the foregone revenue from crude oil supplies, which, in a free market economy, would undoubtedly have been sold at the world oil price. According to International Energy Agency (IEA, 1993), this value can be determined by a simple mathematical formula. To be specific: the opportunity cost to domestic producers is equal to the world oil price minus the Russian Federation producer price times the amount of oil consumed in the domestic market.

Discounting the questions raised by the various elasticities of the demand and supply curves for oil including: (1) the inevitable reduction in the quantity of domestic oil consumption due to an increase in the domestic oil price; and (2) reductions in the world oil price as a result of increases in

Table 7.6: Russian Federation crude oil prices vs. the world oil price, 1990–97

Year	Russian Federation Producer Prices Contract (field gate) Rbls/t	Moscow Inter-Bank Auction Exchange Rates Rbls/US$	Russian Federation Producer Prices Contract (field gate) US$/Bbl	United Kingdom Brent Blend (Dated) f.o.b. Sullom Voe US$/Bbl	Russian Federation Producer Prices Expressed as a Percent of Brent (%)
1990	25.70	19.34	$0.18	$24.55	0.74
1991	65.00	61.95	$0.14	$20.73	0.69
1992	2,649.50	227.90	$1.59	$20.05	7.91
1993	27,770.00	1,018.00	$3.72	$17.59	21.15
1994	76,909.17	2,212.00	$4.74	$16.36	28.97
1995	229,368.00	4,560.00	$6.86	$17.59	38.99
1996	341,159.16	5,114.83	$9.10	$21.41	42.48
1997	402,051.94	5,784.92	$9.48	$19.77	47.93

Note: Russian Federation producer prices are ex-field gate or ex-refinery.

Sources: Khartukov (1995); IEA (various issues); Barton (1998); Khartukov (1998).

Table 7.7: The implied opportunity cost of Russian Federation pricing policies to the domestic oil industry, 1990–94

Year	Final Consumption of Crude Oil and Petroleum Products Russian Federation (million tonnes of oil equivalent)	Final Consumption of Crude Oil and Petroleum Products Russian Federation (million barrels of oil equivalent)	Russian Federation Producer Prices Contract US$/Bbl	United Kingdom Brent Blend 38 deg. Posted US$/Bbl	United Kingdom Brent minus Russian Federation Producer Price	Implied Opportunity Cost (billions of $US)
1991	117.0	858.0	$0.14	$17.75	$17.61	$15.109
1992	112.9	827.9	$1.59	$18.30	$16.71	$13.834
1993	106.3	779.5	$3.72	$13.15	$9.43	$7.350
1994	98.8	724.5	$4.74	$16.20	$11.46	$8.303

Sources: World Bank (1997); Khartukov (1995); Khartukov (1998); EIA (various issues).

Russian Federation oil exports; the IEA estimate is, in fact, a very good first approximation. Assuming the validity of the IEA methodology as a crude benchmark, or useful rule of thumb, the foregone revenues from oil supplies which could have been sold at world oil prices reached 8.30 billion US dollars in 1994, a US$950 million increase over the $7.35 billion estimated for 1993.

Nevertheless, calls for complete oil price liberalization, and the removal of the state export quota would continue to fall on deaf ears. Remarkably, the refusal to liberate the export market for strategically important goods and services would be rigorously applied to only one group of commodities – crude oil and petroleum products. As early as June and December 1993, the government had removed the export quotas on timber, fertilizers, coal and meat products, due to declining international competitiveness and a growing need for hard currency supplies. The 'free' export of strategically important commodities was permitted less than a year later, in July 1994.

At the eleventh hour, crude oil and petroleum products would be 'judiciously' excluded from this list. As reported by Granville (1995, p. 136):

> A decree (issued on 23 May 1994) stipulated the elimination of all export quotas as of July 1, 1994. The list of strategically important goods was maintained even for goods whose exports no longer require a quota, and the licensing requirement for these goods was replaced by a registration requirement. A second decree published on 1 July 1994 postponed the elimination of export quotas on oil and oil products until 1 January 1995.

In reality, the full liberalisation of crude oil exports would be delayed until March 1995. On 31 December 1994 Prime Minister Chernomyrdin signed a Government Resolution replacing the archaic 'Soviet' quota system with 'quarterly export allocations' to be approved by the government of the Russian Federation (Granville, 1995). A strict measure of control over oil exports was guaranteed by the retention of *special exporter* status, and the prohibition of Mazut fuel oil exports from December 1994 to March 1995 (see Table 7.5).

At the same time, years of overly aggressive, and unrestrained monetary policies, and a corresponding increase in domestic inflation, had elevated both the GDP deflator and crude oil production costs to new, and disturbing highs. When valued in constant 1970 roubles, the 'real' crude oil price had fallen to levels as low as 15.71 (1970 rbls/tonne) by the fourth quarter of 1994, a 59.8 per cent 'real' reduction from the 39.09 (1970 rbls/tonne) reported in the second quarter of 1993. In stark contrast, nominal crude oil production costs reached 29 600 roubles per tonne in 1994, a 210 per cent increase over the 9560 reported in 1993. By the end of 1994, the exploration, development and lifting costs for a typical Russian enterprise accounted for a full 39.5 per cent of nominal crude oil prices, a 25.8 per cent increase over the 31.4 reported in 1993. The average Russian oil producer was, once again, squarely in the red, recording a net loss of over 1200 rbls/t.

7.1.3 Price Liberalization and the Removal of the State Export Quotas: 1995–97

By January 1995 the Yeltsin government was forced to consider a serious reform programme. Prices were fully liberalized, and all forms of export allocations, including the manipulation of exports through mandatory domestic sales targets and *special exporter* status, were abandoned/ prohibited on 25 March 25 1995.

Less than one month later, on 1 April 1995, the Russian government unveiled a plan under which the (20 ECU/tonne) export tax – the last administrative barrier to the equalization of domestic (Russian) and free market (international) crude oil prices – was to be gradually reduced (abolished). According to a series of 'official' statements by the Russian Duma, the shortfall in government revenue was to be made up (replaced) by a series of complementary increases in the excise tax.

Indeed, Russia appeared determined to take decisive measures to revitalize the ailing oil industry. According to the OECD/IEA (1995, pp. 135): 'Care was taken to avoid increasing the statutory burden of the [excise] tax and to maintain its variable, profit-sensitive character'. By 1 July 1996, the export duty had been abolished, and the average excise tax had been raised to 55 000 roubles per tonne, a 41 per cent increase over the 39 000 roubles per tonne levied in April 1995.

Tables 7.6, 7.8, 7.9 and 7.10 illustrate the effects of these policies on crude oil and petroleum product prices in the Russian Federation. Despite the best efforts of the Russian Duma (Parliament), full price liberalization, that is, the achievement of 'parity' between export and domestic prices, has yet to be achieved. According to Kemp and Jones (1996, p.2): 'Parity is defined as the situation where the export price minus any export tax equals the domestic price plus the excise tax plus transportation costs to the export border'.

Table 7.8: Russian Federation gasoline prices vs. international gasoline prices, 1990–97

Year	Russian Federation Regular Automotive A-76 Gasoline Ex-Refinery Price Rbls/t	Moscow Inter-Bank Auction Exchange Rates Rbls/US$	Russian Federation Regular Automotive A-76 Gasoline Ex-Refinery Price US$/Bbl	Spot Market Product Prices Unleaded Gasoline f.o.b. Rotterdam US$/Bbl	Russian Federation Gasoline Prices Expressed as a Percent of Rotterdam (%)
1990	73.49	19.34	$0.52	$34.64	1.50
1991	192.05	61.95	$0.42	$30.68	1.38
1992	5,013.80	227.90	$3.00	$27.41	10.95
1993	53,648.60	1,018.00	$7.19	$23.86	30.11
1994	180,499.20	2,212.00	$11.13	$21.55	51.65
1995	579,120.00	4,560.00	$17.32	$22.77	76.05
1996	787,683.82	5,114.83	$21.00	$26.45	79.38
1997	919,802.28	5,784.92	$21.68	$26.59	81.54

Note: Russian Federation product prices are for regular unleaded gasoline (70% A-76 + 30% A-92). All product prices are ex-refinery prices excluding VAT.

Sources: OECD.IEA (1995); Khartukov (1995); IEA (various issues); Barton (1998); Khartukov (1998).

Table 7.9: Russian Federation heavy fuel oil prices vs. international heavy fuel oil prices, 1990–97

Year	Russian Federation Heavy Fuel Oil Ex-Refinery Price Rbls/t	Moscow Inter-Bank Auction Exchange Rates Rbls/US$	Russian Federation Heavy Fuel Oil Ex-Refinery Price US$/Bbl	Spot Market Product Prices Heavy Fuel Oil f.o.b. Rotterdam US$/Bbl	Russian Federation Heavy Fuel Oil Price Expressed as a Percent of Rotterdam (%)
1990	32.88	19.34	$0.23	$13.54	1.71
1991	99.12	61.95	$0.22	$10.54	2.07
1992	2,392.95	227.90	$1.43	$11.11	12.88
1993	25,755.40	1,018.00	$3.45	$8.80	39.22
1994	62,157.20	2,212.00	$3.83	$11.20	34.23
1995	226,176.00	4,560.00	$6.76	$12.75	53.05
1996	340,136.20	5,114.83	$9.07	$14.05	64.56
1997	426,927.10	5,784.92	$10.06	$12.56	80.13

Note: Russian Federation heavy fuel oil prices are listed as an average for the years 1990–95, and M-40 grade in the years 1996–97. Rotterdam prices are listed for heavy fuel oil with a sulphur content of 3.5%. All product prices are ex-refinery prices excluding VAT.

Sources: OECD/IEA (1995); Khartukov (1995); IEA (various issues); Barton (1998); Khartukov (1998).

Table 7.10: Russian Federation diesel prices vs. international diesel prices, 1990–95

Year	Russian Federation Diesel Fuel Oil Ex-Refinery Price Rbls/t	Moscow Inter-Bank Auction Exchange Rates Rbls/US$	Russian Federation Diesel Fuel Oil Ex-Refinery Price US$/Bbl	Spot Market Product Prices Diesel Fuel f.o.b. Rotterdam US$/Bbl	Russian Federation Diesel Fuel Oil Prices Expressed as a Percent of Rotterdam (%)
1990	58.02	19.34	$0.41	$29.05	1.41
1991	179.66	61.95	$0.40	$27.41	1.44
1992	4,626.37	227.90	$2.77	$24.14	11.47
1993	44,283.00	1,018.00	$5.93	$22.64	26.20
1994	159,042.80	2,212.00	$9.80	$20.18	48.58
1995	510,720.00	4,560.00	$15.27	$20.86	73.20
1996	736,535.52	5,114.83	$19.64	$26.32	74.61
1997	966,081.64	5,784.92	$22.77	$23.73	95.98

Note: Russian Federation diesel prices are listed as an average for the years 1990–95, and with 0.2% sulphur in the years 1996–97. All product prices are ex-refinery prices excluding VAT.

Sources: OECD/IEA (1995); Khartukov (1995); IEA (various issues); Barton (1998); Khartukov (1998).

The reforms of the Yeltsin period failed to fully integrate the oil economy of the Russian Federation into the international petroleum market. Prices remain, to a considerable degree, independent of international market forces. This raises the question: What has the process of transition led to? If, after a decade of reform, the Russian oil economy is not a market economy, then what determines both prices and resource allocation? This becomes the central question in the post-command era. If markets don't lead the Russian oil economy, then how must the sector be analysed and what criteria must be used in making investment decisions?

7.2 A GAME WITHOUT RULES

The failure to realize the full benefits of price liberalization including parity between domestic and export prices, industrial profitability, and the attraction of foreign investment funds, has been attributed to a number of constraints and economic disequilibria.

7.2.1 Binding Physical and Administrative Constraints on Export Capacity

The crude oil pipeline network of the Russian Federation boasts an aggregate capacity, defined across all the delivery points in the Russian Federation, of approximately 13 MMB/d. According to estimates provided by the United States Energy Information Agency (USEIA), the system has been operating at a mere 60 per cent of full capacity (USEIA, 1997). At the same time, despite the existence of significant excess pipeline capacity, regional bottlenecks and the gradual erosion of capital and infrastructure have resulted in binding physical constraints on the export transportation network. Sporadic incidents of constraints and overcrowding have been reported since 1993.

The rationale for this apparent pipeline paradox may be summarized briefly as follows. The dissolution of the USSR was accompanied by a dramatic reduction in the demand for Russian oil in Central and Eastern Europe, and the former Soviet Republics. In short, while Russian oil exports to non-CIS nations fell steadily starting in 1988, from 2.89 MMb/d in 1988 to approximately 1.86 mmb/d in 1997, the diversion of volume from Central and Eastern Europe and the non-CIS nations has resulted in bottlenecks at the major export ports to the Western market (Barton, 1998). Specifically, Transneft issued a number of official statements claiming that there 'is little excess capacity' at the major Western export terminals in the Baltic and Black Seas. Ports such as Novorossiysk (Russia's major export terminal on the Black Sea) have been pushed to full capacity, limiting access to crowded terminals in several key regions of the Russian Federation (OECD/IEA, 1995).

The bottlenecks have been aggravated by discriminatory pricing policies, excessive pipeline tariffs, and the Duma's reluctance to relinquish the centralized export system. According to USEIA (1997, p. 134):

> One of the major reasons domestic crude prices have been below world market ones has been the strong downward pressure exerted by government control over access to the export market through its system of quotas and licenses. By restricting the volume of exports, the Government has kept the domestic oil market relatively slack. As a consequence (though also due to mounting payment arrears) producers actually shut-in production and pulled crude from the domestic market, since they had output for which they had neither export quotas, solvent domestic buyers or storage space.

While the rigid system of export quotas was abolished in 1995, it was replaced less than a year later in 1996 by a new system of 'exports for state needs'. Under the new system, the Russian government was permitted to fund a number of 'special federal programmes' by purchasing crude oil supplies at domestic prices, and exporting the oil at the international oil price. This 'taxation by another name' prevents a fully functioning market from developing and inhibits foreign investment due to its ad hoc nature.

In August 1997, for example, the list of special programmes included the reconstruction of the Kremlin dome, the provision of fuel supplies to Chukotka and Kamchatka, the construction of the Tu-324 jet plant, and the construction of the Yelabuga car plant (USEIA, 1997). According to Barton (1998), Federal Programme Exports reached 15.284 million tonnes in the year 1997, approximately 16.5 per cent of total exports from the Russian Federation excluding exports to other CIS member nations (see Table 7.5).

The Federal Programme Export system was eventually abolished, but with a caveat. According to Boris Nemtsov, the First Deputy Prime Minister and (former) Fuel and Energy Minister, the 'State oil export programme will be discontinued, with the proviso that a portion of oil company export receipts be used to pay salary arrears. On 9 July 1997 President Yeltsin signed a decree ending state exports of crude oil in the future; the effective date could be in 1998' (USEIA, 1997, p. 5).

While the abolition of the Federal Export System undoubtedly freed a portion of export capacity for domestic and joint venture producers, the continued scarcity of capacity at key export terminals can be expected to create sporadic bottlenecks and disturbances until such time as new facilities have been constructed and are fully operational.

In the meantime the shortage of export capacity has resulted in the creation of 'economic' rents, as producers have been forced to compete for access to crowded facilities. The challenge to the Russian Duma has been to devise an appropriate system for the allocation of scarce export capacity, and to ensure that the economic rents are collected in a manner that 'does not distort incentives for oil producers, and does not discriminate among producers, including between domestic and foreign producers' (Grey, 1998, p. 31). The reality has been that these rents have been a prime target for capture through corruption given the access to hard currency they provide.

Over the period 1991–97, the allocation of scarce export capacity was directed by the following official rules and procedures:

1. 1991–95: oil exports were subjected to a preferential quota system. A number of Russian (and JV) oil companies were granted tax breaks and preferential 'access' to pipelines; that is, 'roughly 20 oil companies were provided tax breaks and special pipeline access via licensed middlemen' (USEIA, 1997, p. 5);

2. February 1995: decree number 209 established the principal of 'equal access' to pipelines. Priority was given to producers and suppliers that exported crude oil according to Russia's few remaining obligations (inter-governmental agreements for deliveries of crude oil to Slovakia, the Czech Republic and Cuba). Domestic producers were permitted direct access to the Transneft system according to established throughput allocations (i.e. an amount equal to the volume of crude oil produced in the previous quarter). Thirteen Joint Ventures were granted priority access and permitted to export their 'own' production without restriction (see Table 7.5);

3. July 1995: producers were granted the right to 'resell' allocated pipeline space to a number of trading companies. The co-ordination of export terminal operations was placed under the supervision of the seven largest independent Russian oil companies (USEIA, 1997);

4. October 1995: the Federal Energy Commission (FEC) was established to regulate the export system. In theory, the top priority was to be granted to non-CIS crude oil exports (including JV exports). The major integrated Russian oil companies were granted 'second' priority, and permitted to export up to 35 per cent of the volume of crude oil production. All remaining capacity was to be auctioned, and granted to the highest bidder. In reality, the pipeline access granted to JV

producers was strictly limited. In the year 1996 an increase in international oil prices resulted in a significant increase in the volume of crude oil exports so that the JVs were forced to compete for space with major Russian oil companies (USEIA, 1997);

5. 1996: the Duma established a transport fee surcharge (approximately US$6.2 per tonne) in an effort to capture a portion of the economic rents accruing from the scarcity of export capacity (see Table 7.5);

6. 1997: export capacity is allocated according to production shares, 'with consideration for the level of tax arrears' (Grey, 1998, p. 31).

The effect of these many regulatory changes was to inhibit investment in oil field development and, more importantly, investment in pipelines and other transportation related facilities.

As late as November 1998 there were still three administrative restrictions on crude oil exports: (1) the centralized allocation of export capacity according to production shares (Grey 1998); (2) Federal Programme Exports, and (3) barriers concerning the level of tax arrears. As of 1 July 1998, the State tax service had been 'authorised to confiscate hard-currency revenues from pipeline exports, which will be used primarily to settle exporters' overdue taxes. This may further undermine oil producers' ability to pay Western bank loans, which also are secured against oil export' (Khartukov, 1998, p. 2).

As mentioned above, sporadic shortages of export capacity, and the remaining 'centralized' restrictions on crude oil exports, will continue to affect Russian oil markets until such time as new export facilities have been completed. In the meantime, the physical and administrative barriers to exports can be expected to result in a situation where an excess volume of crude oil is supplied to the domestic oil market, placing downward pressure on the domestic oil prices, and preventing parity between domestic and international oil prices.

7.2.2 An Onerous and Complex Upstream Petroleum Taxation Regime

As mentioned frequently throughout this book, the benefits of price liberalization – industry profitability and the attraction of foreign investment funds – have been undermined by the plethora of taxes, levies and duties imposed by federal, regional and local governments. By year-end 1994, the height of the post Soviet taxation era, the sum of fiscal charges, taxes, levies and payments had reached levels as high as 46 610 rbls/tonne, approximately 62.5 per cent of the ex-field gate wholesale enterprise price (excluding VAT) for the average Russian producer. As reported by

Khartukov (1998, p. 3): 'According to official data, in 1994, the after-tax profitability (profit-to-cost-ratio) of the country's oil producing industry dropped to a meagre 7 per cent, compared with 50 per cent in early 1992'.

The onerous tax burden, which was clearly unsustainable, has been reduced significantly since 1994. To cite only a few examples: (1) deductions for the Industry Investment Fund were abolished in May 1995, (2) the excise duty was abolished in July 1996, and (3) the exhaustive repertoire of special purpose levies and local charges was reduced significantly in the years 1994–97. By June 1997 the total (national average) of special purpose levies and local charges had fallen to approximately 17 900 roubles per tonne (4.4 per cent of the average wholesale enterprise price for crude oil in the Russian Federation) and a significant reduction from the 72 810 Rbls/tonne (28.1 per cent of the wholesale enterprise price) reported in June 1995.

By year-end 1996, the average Russian producer was beginning to show a clear profit (see Tables 7.2, 7.3 and 7.4). The net profit for a typical Russian oil company reached 50 700 rbls/tonne in 1997, approximately 12.5 per cent of the wholesale enterprise price (406 000 rbls/tonne). The benefits were reflected, almost instantaneously, in a slight increase in industrial production. The production of crude oil and condensate in the Russian Federation reached 305 476 million tonnes/year in 1997, an increase of over 1 per cent over the 301 283 million tonnes reported in 1996 (Barton, 1998). While modest in comparison to the large gains in crude oil flows reported in the Stalin era, this was the first increase in Russian crude oil flows to be reported since 1988.

7.2.3 Ongoing Difficulties

The expected gains arising from the improvements to the tax regime have, however, been undermined by a number of continuing difficulties, only some of which Russia has the ability to control.

The first was the collapse of international oil prices in the late 1990s. Increasing supplies from the non-OPEC producing nations and the recession in the Asia Pacific region resulted in a significant reduction in the level of world oil prices. For example, the price of Saudi Arabia Light 34° fell from US$17.78 per barrel in 1995 to only US$8.70 per barrel on 9 July 1998. The collapse of world oil prices resulted in a significant reduction of the profitability of Russian oil exports. According to Khartukov (1998, p. 2):

The collapse of world oil prices was bound to result in the crucial deterioration of Russia's oil export economics. The crunch was felt most . . . [in February 1998], when the country's largest oil company, Lukoil, claimed an average loss of $9/metric ton of crude shipped via the port of Novorossiysk. This oil sold for only $85/ton, and the company's export-related expenses exceeded $94/ton.

This situation was alleviated slightly, and for a limited number of oil companies, by the Russian economic crisis of August 1998 culminating in the sudden devaluation of the rouble on 17 August 1998. According to the USEIA (1998, p. 3):

> Paradoxically, Russia's financial crisis has led to increased profits for some oil companies because of the devaluation of the rouble. The reason is that most oil company expenses in Russia are rouble-based, while oil exports to world markets are traded in dollars. By one estimate, 95% of exploration costs are in roubles, while about half of oil company income is received in dollars.

As a general rule, however, the combination of low oil prices and rising production costs reduced both gross revenues, and the profitability of the Russian oil industry. The poor oil prices inhibited the investment necessary to revitalize the industry's infrastructure. Of course, the subsequent rise in international oil prices early in the new millennium will increase profitability but, to a considerable degree, the industry is perceived to be particularly vulnerable to international markets due to the absence of domestic investment funds.

As suggested above, a second problem is the Russian preference for taxes based on volume and gross revenues. These forms of taxation are preferred by governments in Russia because they are easy to monitor and, hence, difficult to avoid. While the onerous Russian tax system has been improved significantly since 1994, the potential benefits have been undermined by the predominance of taxes and levies which are based on the volume of production, and gross revenues. According to information provided by the World Bank, the Russian preference for volume (rather than profit) taxes is also based on familiarity. According to the World Bank (1997, p. 20): 'past practice, the perceived administrative simplicity of excise taxes and the expectation of higher compliance'.

As a general rule, volume (gross revenue) based taxes tend to be highly distortionary, that is, non-neutral, and regressive in nature. According to Kemp and Jones (1996, p. 3):

> The royalty, Mineral Restoration Tax (geological fee) and excise tax are all regressive in nature. They comprise a higher share of profits when these are

reduced whether from oil price falls or cost increases. The current Russian tax and royalty system is of course dominated by impositions based on gross revenues rather than profits.

The following examples illustrate the distortionary nature of these forms of taxation. Examples of taxes based on gross revenue is the royalty (6–16 per cent of the gross wellhead price of crude oil) and the geology fee (10 per cent of the value of gross well head price). These taxes are regressive, and distortionary, in nature; that is, the rate of tax falls as income rises. According to Kemp (1987, p. 91):

> As profitability of exploitation falls due to lower oil prices or higher exploitation costs, a traditional flat-rate royalty or production tax will take an increasing share of gross profits. In an extreme situation, payments have to be made when zero or negative profits are being earned . . . When profitability of exploitation increases due to rising oil prices or unexpectedly low costs, the share of a flat-rate royalty or production tax in gross profits falls. Similarly, the share of economic rent extracted falls. Traditional royalties are thus crude devices which are comparatively insensitive to changes in realised economic rents emanating from oil-price movements or cost variations.

The excise tax (55 000 rbls/tonne) is a tax based on volume. This tax is regressive in nature and as stated by Kemp (1987, p. 92):

> Production-based levies form a wedge between gross revenue and exploitation costs. The requirement of an additional payment based on revenues means that marginal cost exceeds the net price at a higher level of output. The field will be abandoned earlier than would otherwise be the case and resources, the exploitation of which is in the national interest, will remain in the ground.

Given the high proportion of volume and gross revenue based taxes, the bulk of the risk would appear to lie with the investor. According to estimates provided by Kemp (1987, p. 92): 'Under the conventional royalty, cost risks are borne almost entirely by the investor. In particular, the risks of development cost overruns . . . would be borne by the investor under a conventional scheme'.

In short: revenue and volume based taxes are insensitive to the costs and profitability of an enterprise, and as a result tend to discourage the high cost producers and, in a worst case scenario drive them out of business altogether, while simultaneously increasing the risk to be borne by both Russian oil companies and foreign investors.

At the same time, industry profitability, and the ability to attract foreign investment funds, has been greatly inhibited by the high level of uncertainty inherent in the Russian petroleum taxation regime, in particular, frequent revisions to the upstream petroleum taxation regime, and the instability of the generic tax system. In the opinion of World Bank (1997, p. 20):

> the introduction of new taxes was, especially at the outset (1992–1994), largely uncoordinated and introduced instability in the tax system. Changes in the system were numerous and frequent. 'Grandfathering' or stabilisation of tax terms current at the time of an investment commitment, to the extent it has been provided at all, has been hard to win and harder still to hold on to.

Further, there has been considerable and ongoing confusion surrounding the implementation of petroleum taxes. The precise base of the royalty has been the source of considerable uncertainty. In 1992 the Ministry of Finance claimed that the tax base for the royalty was the average domestic selling price minus the excise tax and allocations to the Price Regulation Fund. At the same time, local tax inspectors claimed payment of the royalty based on the sales price. The situation was eventually clarified, and in 1995 the tax base for royalties on crude oil exports was defined as the actual sales price minus the excise tax and transportation costs. According to the World Bank (1997, p. 20), to further aggravate the situation there: 'have been substantial delays in issuing excise tax schedules due to uncertainties over the appropriate differentiation of taxes among producers. There has also been considerable debate over the mechanics of applying tax incentives awarded to foreign joint ventures'.

There has never been a comprehensive and stable legal framework for petroleum taxation and the licensing and operations of Russian and international oil companies. The evolution of the legal framework governing the Russian oil industry is summarized briefly in Table 7.11. The principal components of petroleum taxation in the Russian Federation still remain incomplete. In short, the progress has been delayed by political instability, and economic uncertainty accompanying the dissolution of the USSR, and controversy surrounding Russia's attempt to effect a transformation to a market based economy. While the Supreme Soviet authorized drafting procedures for a comprehensive Oil and Gas Law as early as 1991, the legislation has yet to be either fully completed and/or passed by the Russian Duma.

The plight of the draft tax code provides one example of the difficulties. A draft tax code was submitted to the Duma during the spring of 1997. The

Table 7.11: The legal framework governing the evolution of the Russian oil industry, 1991–97

Legislation	Purpose	Status
Law on Foreign Investment	Permits the establishment of joint ventures or wholly owned subsidiaries by foreign investors. 'Foreign investors are protected from adverse changes in legislation for a period of 10 years after their incorporation (with the exception of laws concerning national security, taxation, environmental protection and health).' 'Investments may not be nationalized except by law and in exceptional circumstances. In the event prompt and full compensation will be paid.' 'Investment disputes are to be settled in national courts by arbitration unless otherwise provided by treaty.'	July 1991: Passed by the Supreme Soviet
'On the Principles of the Taxation System in the Russian Federation'	Outlines the general principles of the Russian Federation taxation system including taxes, levies, duties, the rights and obligations of the tax payers and tax offices.	27 December 27 1991: Passed by the Supreme Soviet
The Subsurface Resource Law or the Law on Underground Mineral Resources.	Establishes a legal framework for mining operations (petroleum production) in the Russian Federation. The 'State' is established as the 'exclusive' owner of all mineral resources. Private and state-owned entities are permitted lease exploration and production rights from the state. Licences are to be issued through public commercial tenders. The leasing procedure involved collaboration with regional authorities of the territory under consideration. The licence is issued jointly 'by the agency of representative power of the Republic within the Russian Federation, territory, region or autonomous formation and the State agency for the administration of the subsoil found or its territorial subdivision'.	21 February 1992: Passed by Russian Parliament (the Supreme Soviet). May 1992: Effective as a law.

Law on Revising and Amending the Russian Federation Law on the Underground (Amendments to the Subsurface Resource Law)	A new version of the Law of the Subsoil, containing approximately 20 amendments to the original version. 'The major change [is] the establishment of ownership rights exclusively at the Federal Level, although still embodying the concept of joint management with the regions.' 'Foreign (and domestic) investors are permitted to obtain exploration licences through direct negotiations, rather than the cumbersome tender process required to date.' 'Subsoil users [are permitted] to establish a separate subsidiary and transfer the licence for a specified area so that the new enterprise may acquire a foreign shareholder and legally continue the development of reserves which originally were part of the parent company's assets. Prior to the new Act there was no provision regulating assignments of development rights to a third party.' 'The new Act contains provisions regarding Production Sharing Agreements. Article No. 19-6 states that "the subsoil user may be exempted from all taxes and charges, except the tax on profit from its share of production due to the user in accordance with production terms and conditions".'	November 1994: Approved by the State Duma January 1995: The first version is vetoed by President Yeltsin. Mid-1995: A second version is endorsed by President Yeltsin.
Regulations on the Procedures for Licensing the Use of the Subsoil	The State Committee for Geology and Use of the Subsoil (Roskomnedra) is established as the Federal body in charge of, and responsible for issuing licences in coordination with the appropriate regional authorities, and legislative bodies.	July 1992: Passed by the Supreme Soviet.
Order No. 65. On Production Licences	Russian companies are permitted to transfer exploration and production licences to the Joint Ventures in which they participate.	May 1995: Passed by Roskomnedra

283

Legislation	Purpose	Status
Oil and Gas Law	Designed to complement the 1992 Law on the Use of the Subsoil by 'establishing new – and in many cases more restrictive – licensing and operating rules for the oil and gas industry'. The legislation is expected to clarify the legislative framework, and licensing procedures for oil extraction, pipeline transportation, oil and gas operations, geological surveying, and underground storage. The latest version, the draft tax code, was submitted to the Duma in the spring of 1997. The draft contains a number of proposals for changes in existing petroleum taxation, including: (i) proposals related to the implementation of effective PSA legislation, (ii) a blueprint for the key elements of a comprehensive and stable tax regime including bonuses and royalties, and (iii) a proposal to replace the (wellhead) excise tax with an additional profits tax (the R-Factor tax). The provision would apply to new projects, while the excise tax would continue to apply to existing projects.	1991: Drafting procedures initiated. 1993: A draft is approved by the Russian parliament. The draft is not passed into law as the Parliament is dismissed by Yeltsin. Mid-1994: A second version is passed by the State Duma. The draft is not successful. January 1995: A third version is rejected by the State Duma. Spring 1997: Draft tax code submitted to the Duma. July 31 1998: Part I of the Tax Code (RF Law N. 146) was passed by the Duma, which increases the chances for its 'special parts' to be adopted. October 1998: Various versions have been passed by the parliament, and subsequently rejected by President Yeltsin.

| Edict (Presidential Decree) No. 2285. 'On Matters of Production Sharing Contracts in the Use of Underground Resources'. | 'Provides for the distribution between parties of the agreement of produced mineral resources, such distribution replacing taxes, tariffs and duties, including customs duties, excises and other mandatory payments provided for in the current legislation (hereinafter referred to as taxes), except the tax on profit and payments for the use of subsoil resources, which can be paid both in the value form or in-kind, as agreed by the parties.' | December 1993: Presidential Decree |
| Russian Federal Law on Production Sharing Agreements | Reflects the principal provisions of Edict No. 2285.

Contains a provision for 'exemption from all Federal taxes, including customs and excise duties apart from income tax. With respect to local taxes, if exemptions are not provided for by local Governments, an adjustment is to be made to the state share of profit oil to compensate for such local taxes in the event that they are required to be paid. Provision is made for payment of signature and/or production bonus, rent, royalty and payments for the use of the land'.

While the law was passed in December 1995, the document contained a number of contradictions to existing legislation, i.e. The Production Sharing Law and draft Law on Oil and Gas provide for the transfer of subsoil rights to third parties. The Law on the Subsoil permits such a transfer if and only if existing Federal laws permit the transactions.

The law left a number of important issues unresolved. Specifically, a number of other pieces of legislation were required before the law could be implemented. These include: (i) a new federal law on Production Sharing Agreement Taxes; and (ii) the revision of seven other tax laws: VAT (to provide for a refund mechanism), Profits tax (to provide for special rates for Production Sharing Agreements), Land Payments, Road Fund, Property Tax, Excise tax (exemption provision), and customs tariff (exemption provision). | June 14, 1995: Approved by the Duma
December 1995: Passed by the Russian Parliament |

285

Legislation	Purpose	Status
Law on Amendments to the Production Sharing Law	Contains critical amendments to the Law on Production Sharing. Specifically, the legislation attempts to 'make the law more investor friendly and remove [the contradictions, and ambiguities that have acted as] roadblocks to the implementation of large scale investment projects'.	July 1996: Rejected by the Russian Parliament
Law of Lists	Lists the fields that are eligible for development by Production Sharing Agreements.	June 1997: Approved by the Russian Duma
	The law lists only seven fields, only one of which involved the participation of a foreign entity. The fields include:	
	(i) the Prirazlomonye field – Barents Sea (Rosneft, Gazprom and BHP), (ii) Samotlor, (iii) Krasnoleninsk, (iv) Romashkinskoye, and (v) A number of fields off Sakhalin Island.	

Sources: World Bank (1997); IEA (1995); Grey (1998); Kryukov and Tokarev (1998).

286

implementation of the tax code has been delayed by a number of political and economic disturbances. To cite only two examples: (1) In March 1997, the Minister of Fuel and Energy Pyotr Rodionov resigned, leaving the position vacant, (2) on 24 April 1997, President Yeltsin appointed First Deputy Prime Minister Boris Nemtsov to the 'additional' position of Minister of Fuel and Energy. Nemtsov, who had called for the radical liberalization of the energy sector, was appointed in an effort to accelerate the reform of the Russian energy industry.

While Part I of the draft tax code was adopted by the Duma on 31 July 1998, further progress was delayed by the emergence of an acute financial and economic crisis. On 17 August 1998 Russia abandoned efforts to maintain a stable rouble in the face of mounting pressure, a spill over from the Asian economic crisis, and a dramatic reduction in hard currency revenues from oil sales. Days later, on 23 August 1998 President Yeltsin dismissed the entire cabinet when three attempts to reappoint former Prime Minister Viktor Chernomyrdin were rejected by the Duma. Sergei Generalov was appointed to the position of Acting Minister of Fuel and Energy. Debate and controversy surrounding the current financial and economic crisis disrupted Parliamentary discussions and proceedings throughout 1998. In the mean time the passage of critical petroleum legislation, including the draft tax code and legislation facilitating the approval of Production Sharing Agreements, remained in limbo (USEIA, 1998).

As mentioned above, years of aggressive monetary policies, price liberalization, and the natural ageing of the oil industry have resulted in a significant increase in production and transportation costs throughout the Russian Federation. According to Khartukov (1998), nominal crude oil production costs reached 200 000 rbls/tonne in 1997, nearly seven times higher than the 29 600 rbls/tonne reported in 1994.

The pipeline transportation costs for crude oil sales have also risen significantly (see Table 7.5). To cite only two examples: (1) the transportation costs for domestic oil sales reached US$5.60/ tonne in 1997, a 246 per cent increase over the $1.62 reported in 1994 (transportation costs are calculated as an average for the Russian Federation), and (2) pipeline transportation costs from West Siberia to the Black Sea reached US$22.25/tonne in 1997, a 43 per cent increase over the US$15.60/tonne reported in 1994.

There have also been considerable market distortions created by the 'non-payments problem'. Economic distortions, the failure of institutions and economic reform accompanying the dissolution of the unique Stalin planning model have resulted in massive non-payments problems throughout the Russian Federation.

The non-payments problem may be defined as the shortfall in cash flow resulting from: (1) the inability to operate a profitable business, and (2) customer payment problems.

In the case of the Russian oil industry, the non-payments problem appears to have resulted from inconsistent price liberalization policies, and the plethora of taxes and levies that were imposed on the industry in 1992. As mentioned above, the gradual elimination of price controls in 1991–92 led to rapid increases in 'supplier' prices (crude oil production costs) throughout the Russian Federation. At the same time, the retail prices of energy were held virtually constant at 1990 levels. The imbalance reached crisis proportions in 1991 when average production costs reached an unsustainable 98.2 per cent of the wholesale enterprise price. While oil prices were raised slightly in September 1992, customers were either unwilling or unable to pay their bills, so that the shortfall in cash flow continued to increase. According to the World Bank (1993, p. 27): 'Arrears to oilfield suppliers rose dramatically in response to these cash shortfalls, exacerbating the problem of oilfield equipment shortages'.

The non-payments problem has been aggravated by the following considerations: (1) the lack of uniform accounting procedures and bankruptcy laws (USEIA, 1995); (2) friction in the banking sector which has resulted in delays of up to six months for payment settlements; (3) reluctance to pay under conditions of high inflation; (4) financial difficulties of customers (World Bank (1994); (5) the lack of reliable short-term financing instruments such as letters of credit and exchange (USEIA, 1995); (6) non-payment by public authorities (USEIA, 1995) and (7) heavy reliance on barter as a means of managing cash flow (World Bank, 1993). The banking system in the Russian Federation remains a major stumbling block in transition as it provides neither credit to bridge transactions nor an expeditious means of cheque clearing. As a result, transactions have to be financed out of cash flow and often made in cash. This considerably raises the costs of undertaking transactions (Hobbs et al., 1997).

As might have been anticipated, the non-payments crisis has had significant implications for the Russian oil industry. In 1997, a mere 10–20 per cent of domestic oil deliveries were paid for in cash. The balance, approximately 80–90 per cent of domestic oil sales was paid for by barter or short term letters of credit, promissory notes, treasury and municipal bills, state and corporate bonds, offsetting tax obligations and so on. A significant portion of the domestic delivery bill was simply not paid. According to

Khartukov (1998, p. 2): 'Instant cash payments are typically rewarded by generous price discounts – in some cases as much as 50% off listed crude and product prices'.

In the final analysis, the non-payments problem appears to have placed an effective cap on the level of domestic crude oil and product prices throughout the Russian Federation. In addition, the crisis has exacerbated a number of existing economic distortions, frustrating the economic reform process, and delaying the evolution of competitive markets in the Russian oil industry.

In particular non-payment provided a barrier to price liberalization. The non-payments problem acts as an effective barrier to price liberalization policies, and the attainment of parity between domestic and world oil prices. It also provides a barrier to the industrial profitability of domestic oil sales. As a result of the non-payments problem Russian oil companies and JV producers continue to be dependent on hard currency exports for cash flow, and financial liquidity.

More importantly, non-payment imposes a barrier that discourages domestic investment. According to the USEIA (1998, p. 4):

> Non-payment of bills by domestic customers has forced companies to rely on external funding for development projects, but banks have been reluctant to make even export-backed loans following the 90-debt moratorium and treasury bill defaults, causing companies to scale back expenditures.

Clearly, the non-payments problem in the oil industry is a result of the failure of economic reform, policies and institutions throughout the economy of the Russian Federation. The solution must be found in the coordination of the reform process including macroeconomic policy, institutional and political reform, and energy policy. Among the basic requirements for reform are: (1) the adoption of stable, non-inflationary, monetary policies; (2) banking reform, and the development of reliable institutions for financial intermediation; (3) the adoption and enforcement of a comprehensive bankruptcy law; and (4) the adoption of clear, and well defined accounting procedures. A further requirement is the clear definition and vigorous enforcement of property rights (Hobbs and Kerr, 1999). As can be seen, property rights need to be defined not only to protect property from criminal elements and from corrupt bureaucrats and politicians but the state itself. Sub-federal levels of government in the Russian Federation have been particularly notorious for moving to confiscate any profits that arise from investments in their jurisdictions. Many foreign companies were burnt once by changing tax and regulatory regimes and subsequently withdrew from Russia. They will be hard to entice back (Kerr, 1993).

One final major problem for the Russian economy is the high degree of monopolization that arose as a result of Soviet economists' predisposition to very large-scale enterprises and the high transaction costs associated with widening markets in the post command period (Gaisford, Kerr and Hobbs, 1994). This problem is particularly acute in the petroleum sector.

In the early reforms the old monolithic state monopoly was dismembered and replaced by three vertically integrated companies: LukOil, Yukos and SURGUT-neftgaf. All oil production, refining and production supply industries were restructured as joint stock companies. Ownership of shares, however, rather than being widely held, tends to be concentrated in the hands of the former Soviet managerial class and ex-Communist Party officials. The Soviet pipeline network co-ordination mechanism, Glavstransneft, was transformed from the monopoly purchaser, transporter and disseminator of all crude oil produced in the former USSR into Transneft, which provides services, at least in theory, on an open access basis. As suggested above, crowding on pipelines has led to charges of favouritism and graft in the allocation of space. Transneft remains in state hands. Downstream transportation and distribution remain the responsibility of Rosnefteproduct.

In the domestic market of the Russian Federation, the transition to a price allocation system necessitates the existence of markets. After a decade of transition from the command system, only the most rudimentary markets exist (Gaisford et al., 1995). As a result, progress to a smoothly functioning domestic energy system has been frustrated and is central to understanding the foot-dragging on price liberalization and political retreats toward non-market allocation detailed above. Until such time as the domestic energy system is able to pay its own way, neither domestic nor foreign investment will be sufficiently secure to induce the large amount of funds required to modernize the system. As discussed above, export sales have had to be restricted, largely to prevent the domestic market from being starved as a result of oil and gas producers seeking the higher returns available from hard currency transactions. If those higher returns were invested in the domestic system, the short term pain might have been politically acceptable but a large portion of the profits tend to end up in 'nest eggs' located in foreign banks. Enforcement is difficult and costly as rising amounts of petroleum products have been channelled into illegal export opportunities and black markets.

The fundamental reason why downstream markets have not evolved can be found in the very nature of the old hierarchical structure. In a Soviet-type command economy each state enterprise was tied, without choice, to both supplier(s) and customer(s) (Hobbs et al., 1993). No alternatives to the officially designated suppliers or customers were allowed. In essence, the command system might be viewed as a single 'tube' running from the reservoir to the consumer. In this situation, the consumer is tied inexorably to one source of supply.

The monolithic nature of this system led to an emphasis on the development of a few, large-scale transportation and refining networks. In fact the entire CIS refining system – with a capacity of some 9–10 mmb/day serving 290 million people – is concentrated in just 15 large refining complexes which are, in effect, geographic monopolies. This system of geographic monopolies was endemic in the command economy (McNeil and Kerr, 1997) but is more difficult to dismantle in the energy sector due to the dedicated nature of the transportation system. Many republics and Russian regions are totally dependent on just one pipeline and/or rail network with no excess capacity available.

Given the unique design of energy infrastructure in the Russian Federation, if the downstream distribution system was successfully privatized and prices freed, industrial users and consumers would simply be faced with purchasing from a monopoly. In the absence of a reliable and comprehensive communications system (at even the most basic level of telephones and telephone directory services) and without transportation alternatives, even large industrial end users would be hard pressed to seek out alternative sources of supply. Without competitive pressure or regulatory oversight similar to that existing in modern market economies, such monopolies have little incentive to invest in the massive capital projects necessary for the revitalization of the Russian oil and gas industry – clearly a recipe for stagnation, not success.

Under these circumstances, it is unlikely that foreign corporations will be willing to invest in those sections of the transportation system that serve the domestic market. Of course, most proposals relating to transportation investment in the post-command era have tended to focus on establishing direct pipeline links to foreign customers: for example, Turknevia to Turkey, Baku to the Black Sea, and possibly Turknevia to Pakistan and Western Siberia to the Barents Sea (at Murmansk). If foreign investment was truly earmarked for the domestic market, there would undoubtedly be more interest in improving the domestic transportation system. New pipelines dedicated to the export market can only lead to further diversion of volumes previously earmarked for domestic consumption. While producer organizations continue to be enticed by the clear benefits of increased exports, they are unlikely to be willing to invest in upgrading the domestic system.

The challenge to the Russian government is to deregulate the downstream supply system in such a way so as to ensure that industrial users and consumers have alternative suppliers with which they can deal. It is important to note that this can only be dealt with in the long run by investing in alternative supply systems for individual markets. If this bottleneck is not overcome, the funds necessary to revitalize the Russian petroleum industry will not be forthcoming.

7.3 YELTSIN'S LEGACY

At the Twenty-Eighth Party Congress (June 1990), Boris Yeltsin – the newly elected President of the Russian Republic – delivered a short but concise speech announcing his resignation from the Communist Party of the USSR. As he strode out of the hall, leaving 4,700 stunned delegates to contemplate the 'implications' for traditional Soviet politics – the act was hailed as a dramatic – albeit effective – act of political theatrics. Indeed, from that moment on, the momentum in Russian politics would clearly favour the President. Today, more than nine years later, Yeltsin and his flair for the dramatic, have once again left a lasting 'epitaph' for surprised Russian – and Western – politicians.

On 31 December 1999, Boris Yeltsin shocked the world by announcing his resignation – to be effective immediately – as the first (and only) President of the Russian Federation. His last act, the appointment of Prime Minister Vladimir Putin as Acting President, was hailed as Yeltsin's final – and perhaps his most brilliant – political ploy. In short, the surprising 'New Year's resignation left Yeltsin with a number of milestones for the millennium: (1) the solidification of his own personal status as a patriot, and champion of Russian Democracy; (2) full immunity against prosecution for charges of corruption; and (3) enhanced political prospects for his chosen successor – Vladimir Putin.

As expected, Putin won the presidential election on 26 March 2000. He was purposely vague regarding his policies during the election. Instead of engaging in political debate or media campaigns he chose to outline his vision for Russia in an open letter to the electorate. The manifesto was published in a variety of Russian newspapers on 24 February 2000.

The letter, which revealed little in the way of well-defined policy proposals, included three major promises to the Russian population: a 'worthy life', a 'strong state', and a 'dictatorship of the law'. On the economic front, Putin vowed to defend the Russian economy from illegality: that is, illegal intervention by bureaucrats, gangsters and criminals, and to create the foundations for an honest and efficient private sector. To this end, the letter advocated 'a large-scale inventory of the country'. As Russia has virtually no idea of the value of resources that have been (and are available to be) plundered, the inventory is seen as a necessary 'starting point' in the battle against illegality. In the words of President Putin: 'We have a very bad idea of what resources we possess today'.

The exact meaning of Putin's words have been subject to a variety of colourful interpretations. The dichotomy of public opinion is summarized effectively by Putin's main campaign slogan: 'A dictatorship of the law'. As reported by Harding (2000): 'His enemies are convinced he will be a dictator. His fans insist he will bring law and order to a turbulent country'.

At the same time, the need for a comprehensive, and independent, legal system in the Russian Federation, is beyond dispute. While the Russian government has made considerable progress towards the creation of a free market economy, physical and administrative distortions arising from decades of central planning continue to influence the development of the Russian economy. These include the artificial separation of domestic and export markets for strategic commodities, including energy supplies, and difficulties associated with the absence of the legal and institutional framework that forms the foundation for a successful and efficient free market economy.

In short, during Yeltsin's tenure as President the Russian Duma failed to make significant progress on the issue of domestic legal reform, and the creation of a 'law-based' state remains a distant, and elusive, goal. Its adoption is an absolute prerequisite to the attainment and preservation of a market-based economy. In the meantime, reform will continue to be frustrated, and economic policy rendered impotent by the perpetuation of policies and institutions that are reminiscent, in some respects, of the Soviet era.

The failure to provide adequate administrative institutions, and a 'law-based' state has particular relevance to the Russian upstream petroleum industry, which will continue to flounder in a system where the property rights are not well defined, and contracts remain unenforceable. An excellent example is provided by the Duma's ill-fated attempts to liberalize domestic crude oil prices described in detail above. Remarkably, and despite the best efforts of the Duma since 1995, full price liberalization has yet to be achieved. Domestic crude oil prices in the Russian Federation averaged US$14.97 per barrel in the week ended 17 March 2000, a mere 54 per cent of the US$27.86 reported for Siberian light exports. The failure of what would otherwise appear to be straightforward economic policy has been attributed to a number of factors, including a shortage of physical export capacity (and/or discriminatory tariff policies by Transneft), the complex (revenue-based) tax system, the 'temporary' reintroduction of the export duty, the absence of a comprehensive and stable legal framework for petroleum taxation and the licensing and operations of Russian and International oil companies, and market distortions created by the non-payments problem.

The non-payments crisis reflects severe economic dislocations, and the failure of macroeconomic policies and institutions throughout the Russian Federation. In the case of the upstream petroleum industry, the solution can only be found in the coordination of the reform process – including macroeconomic policy, institutional and political reform, and energy policy. The policy prescriptions for the energy sector (that is, price liberalization) will fail without the adoption and enforcement of banking reform, a comprehensive bankruptcy law, and well-defined accounting procedures. All these remain Yeltsin's legacy and Putin's challenge.

The inefficiencies inherent in the petroleum–fiscal regime currently governing the development of the Russian upstream petroleum industry may be summarized briefly as follows: (1) the failure of price liberalization policies, and/or artificially low domestic crude oil prices; (2) the existence of an onerous (revenue based) tax system; (3) difficulties associated with the non-payments problem, including non-compliance and the failure to collect 'notional' tax revenue from Russian oil enterprises; (4) the failure to meet the criteria specified by multilateral lending agencies, and the resulting 'failure' of the funds provided by these agencies to generate positive, long-term implications for the Russian oil industry; (5) the failure to develop and complete an acceptable and comprehensive legal and contractual framework for oil operations – including Russian enterprises, PSAs, and Joint Ventures.

In the years ahead, the challenge facing the Putin government will be to provide a stable environment within which these problems and inefficiencies might be resolved. Effective reform will require the coordination of macroeconomic policies, and the provision of independent and stable legal and judiciary institutions throughout the Russian Federation.

Putin insisted in his campaign that he be judged by his actions. The actions that Putin and the Duma have taken towards the resolution of the difficulties affecting the Russian upstream petroleum industry can be listed as follows:

1. *Increased state control over export/import activities, and currency transfers.* The new legislative initiatives were introduced to address the problem of capital flight, and include the following proposals: (a) the mandatory state registration of all import (and export) transactions involving goods and services by Russian residents, and (b) an extension of the term of a bank's liability to freeze foreign currency transactions that are under investigation from 5 days to thirty days. The draft regulations were passed by the Duma on 15 March 2000 (Zabotkine, 2000)

2. *An increase in the number of initiatives designed to stimulate foreign investment in the Russian Federation.* The draft legislation, and proposals include a decree increasing the allowable deductions for business expenses of foreign firms, and the creation of a new federal committee to address the issue of tax disputes with foreign entites.

3. *An increase in the export duty.* In December 1999, Putin signed a resolution raising the 'temporary' oil export duty to 15 euros per tonne (approximately US$2.00 per barrel).

4. *The introduction of harsh 'confiscatory' penalties for tax arrears.* In January 1999, Putin signed a resolution addressing the non-payments (non-compliance) problem in the oil industry. Under the new rules of engagement, Russian oil companies were required to pay all federal tax debts by 1 February 2000. The penalty for non-compliance is a mandatory 20 per cent reduction in oil exports.

With the exception of the policy measures designed to stimulate foreign investment in the oil industry, the majority of these policies can be assumed to be detrimental to the future development of the Russian petroleum industry.

In the final analysis, more than rhetoric and imagination will be required to revitalize the ailing Russian petroleum industry. The interests of the Russian Federation will be best served by the development of an 'efficient' tax regime and legislative system that will enable the oil industry to operate efficiently, thereby facilitating the exploitation of all reserves that confer social benefits. Yeltsin's legacy is Putin's challenge.

8. Russian Oil in the 21st Century

The Russian oil economy is a microcosm of the Russian economy. Just as the petroleum sectors of modern market economies must be analysed using conventional economic tools based on market incentives, and responses to them constrained by the regulatory environment within which they operate, the technology available and the resource base, the Russian industry must be analysed within the context of incentives and constraints. When economists analyse the future prospects for oil in modern market economies they tend to dwell on topics such as reserves, regulations and incentives. This is because the markets within which they operate function efficiently; that is, transactions costs are low, and as a result economic actors respond in predictable ways. While the predictions of economists analysing market-based systems are not always correct, the reasons for their failings most often lie either in unanticipated events or the rudimentary nature of economics as a developing social science.

When forecasts are made regarding the oil economies of developing countries, in contrast, less emphasis is placed on reserves, regulations and incentives, and, rather, it is politics and the functioning of supply chains that come to the forefront of analysts' minds. This is because knowing that reserves exist does not mean they can be exploited successfully; understanding regulations does not mean that they are applied, and while individuals will respond to incentives, they may not be sufficiently transparent to allow predictions to be made. Often, those hired to make predictions are specialists in political risk rather than economics.

Understanding the functioning of supply chains becomes important because the transactions costs associated with doing business in developing countries are so high (Hobbs and Kerr, 1999). Transaction costs are often classified into: (1) information costs; (2) negotiation costs; and (3) monitoring and enforcement costs (Hobbs et al., 1997). Information costs are incurred prior to a transaction and relate to the resources that must be expended to identify potential transaction partners, verifying the reputation of potential partners, determining the product characteristics produced by/desired by potential transaction partners and appropriate prices. Negotiation costs relate to the resources that must be expended to facilitate the transaction actually taking place. These consist of the time and effort it takes to prepare for and undertake negotiations, legal costs associated with drawing up contracts, fees paid to intermediaries and, if necessary, bribes. Monitoring and enforcement costs are those that must be assumed to ensure that once an agreement on a transaction is made, it is actually carried out. Monitoring costs include those associated with testing products to ensure they comply with agreed quality standards, financial instruments such as documentary letters of credit that ensure one gets paid, the costs of settling disputes such as quality arbitrations and equipment or systems that can be used to identify the party at fault if there is a breakdown in product flow or deterioration in quality. Enforcement costs are those associated with either ensuring compliance or acquiring compensation in cases where the terms of the transaction are breached. Typically, there are many transactions along a supply chain from the extraction of natural resources to the final consumer. In well functioning market economies these transaction costs are often sufficiently small to be ignored when undertaking economic analysis. In developing countries with limited communications infrastructure, poorly defined and enforced property rights, rudimentary commercial legal systems and high degrees of corruption, these costs become sufficiently large that they cannot be ignored and often come to dominate formal economic analysis. In modern market economies, if the transaction costs along the supply chain become sufficiently high, a more efficient alternative may be for firms to integrate vertically along their supply chains internalizing transactions within the firm. The cost savings associated with vertical integration must be weighed against the additional managerial costs associated with coordinating additional activities within the firm. One suspects that this is the reason multinational oil companies that operate in developing countries are so often vertically integrated and would be even more so if allowed to be by host governments.

Command economies also faced very high transaction costs. These were at the heart of the command economies' difficulties. The information requirements of the central planners were enormous and, in fact, far exceeded the resources available for the collection and processing of information. While command economies, by definition, save on negotiation costs because transactions are accomplished through bureaucratic orders emanating from the central planners, these savings were never sufficient to offset the inability of the planners to acquire sufficient information to match supply with demand along thousands, or hundreds of thousands, of supply chains. Further, as we have seen, there was virtually no monitoring of quality or resources put into enforcement (except periodic use of threats to an individual's livelihood and draconian retribution for failure, neither of which had any sustained influence on improving the system). The transaction costs associated with the centrally planned economy eventually constrained the Soviet Union's development to a sufficient degree that it had become stagnant by the time Gorbachev assumed the reigns of power. In a similar way, high transaction costs can be seen as limiting the ability of economies to develop (Hobbs and Kerr, 1999).

No amount of tinkering with the command system could have shaken it out of its malaise no matter what the 'good' intentions of the Gorbachev reforms. There was no way out of the allocation problems identified by von Mises (1981) at the dawn of the experiment with command economies. The Russian oil economy did find, develop and deliver large quantities of oil and oil-based products during the 70-year experiment in command economics. It represents a major accomplishment and, as laid out in Chapter 1, it has left a legacy of impressive reserves that are available for both exploitation and future development. The transition decade that started with Yeltsin's assumption of power had only limited impact on the sector's ongoing stagnation.

In one view, the Yeltsin years represent a wasted decade of foregone opportunities that frittered away the chance to modernize and revitalize a sector with huge economic potential. When viewed from the perspective of modern market economies, this is a reasonable conclusion. Russia and its oil economy, however, has not made the transition to a modern market economy, and hence the judgement is too harsh. Just as it is not useful to apply the standards of modern market economies to developing countries, it is not useful to apply them to economies in transition.

A careful reader will have noticed that, unlike the previous chapters, Chapter 7 which describes the post-command years has little to say about reserves, technologies or exploration activities. The focus was on politics, the functioning of supply chains and the evolving structure of incentives. It is clear that the Russian Federation is not a market economy although it has abandoned central planning and many aspects of allocation by command. As a result, the factors that will affect how the Russian oil economy will evolve in the 21st century relate to the constraints imposed by politics, the functioning of the supply chain and the ability to respond to incentives. The latter depends as much on the stability of the incentives system as it does on the structure of the incentives themselves. It should be clear from the discussions in Chapter 7 that there was no stability in the incentive systems during the Yeltsin era. The end result was little response to the incentives provided and short-run perspectives on the management and development of what are long term resource exploitation questions in the petroleum sector.

Transaction costs are high in post command supply chains (Bolger et al., 2001; Hutchins et al., 1995). This is because the underlying market institutions required for the operations of markets do not yet exist in Russia and many other transition economies (Middleton et al., 1993). Information costs are high because communications systems are poor, particularly in more remote parts of the country where much of the oil industry's future potential is located. This raises the cost of identifying potential transaction partners. Further, institutions that are common in modern market economies such as 'Better Business Bureaux' and that help in the verification of potential transaction partner's reputation do not, for the most part, exist. The net effect is that considerable time and effort must be expended in building trust through personal relationships. This reduces the set of potential transaction partners for any business deal and, as a result, reduces competition. Often, transaction partners may be restricted to family members to reduce the cost of building trust – with, of course, the associated efficiency loss.

High information costs help to re-enforce the local monopoly nature of the infrastructure inherited from the Soviet system. Even if alternative supply channel options for petroleum products such as trucking are available, it is difficult to identify market opportunities in distant places (Kerr, 1996). As a result, competition is stifled and local monopolies remain. This means that markets have reduced contestability resulting in an absence of incentives for the existing monopolies to increase their efficiency or to re-invest in improvements to their existing infrastructure. Instead, profits can be skimmed for personal consumption or sequestered in foreign banks or domestic safe havens.

Negotiation costs are also high in post-command supply chains. In modern market economies, payment can largely be taken on trust. It is not necessary to have a trust-based relationship with transaction partners. Most transactions take place over the phone organized by clerks with 60 to 90 days to pay. Banks facilitate this system with bridging credit. If transaction partners are tempted to renege on their obligation (to pay or supply), the commercial legal system is, for the most part, a sufficient threat to prevent such 'opportunistic behaviour'. In post-command economies business people are forced to expend a great deal of their time simply organizing transactions to ensure that payment is made or received. Often, payment must be made in cash, which increases the costs of security and requires that businesses be entirely dependent on cash flow financing. Again, the need for trust reduces the set of potential partners for any transaction and inhibits competition.

Monitoring and enforcement costs are also high in command economy supply chains. The long tradition of 'shoddy quality' in the command era has been difficult to change. As a result, rejection rates are high and far more resources must be put into testing and checking inputs and outputs than is typical in modern market economies. The absence of an effective commercial legal system means that other methods must be found to enforce contracts or to obtain compensation in the case of a breach. Again, this often requires prevention in the form of relationships based on trust that are costly to build and act to limit competition.

Crude oil is an internationally traded commodity. From the outside this makes the extensive Russian reserves appear to be ripe for development and subsequent sales to the international market. This view of the Russian oil economy sometimes manifests itself in the corporate offices of multinational oil companies in plans to build dedicated pipelines to export points. Beyond the political difficulties associated with finding a secure route through now independent former Soviet Republics that are often politically unstable or subject to large scale banditry, it is not possible to isolate the oil sector from the rest of the Russian economy.

The domestic oil industry in the Russian Federation also cannot be separated from the development of the Russian economy. The major reason why it is not possible to isolate the development of the oil economy from the broader influences of the evolution of the Russian economy is poor protection of property rights and the absence of a well functioning commercial legal system. Given the widespread malaise that exists in the Russian economy, any sector that begins to prosper will attract the attention of politicians, bureaucrats, industrial oligarchs and criminal elements, all of whom will attempt to appropriate any gains for themselves. Defending one's interests against such an onslaught will be very costly and will often exhaust the available rents. This scenario has already been played out in oil field development where a number of international companies and consortia initially attempted to become involved in exploration, oil field development and pipeline construction only subsequently to withdraw.

After over half a century of command era mismanagement and a decade of stagnation, the Russian oil economy can be characterized as having over-exploited reserves and ageing infrastructure. In the case of the former, there is little difference between the command era's emphasis on getting as much production from the reserves as fast as possible in order to provide fuel for the rapid rates of growth planned for other sectors of the economy, and the post-command era's incentive to exploit reserves as fast as possible given poorly defined property rights – get as much as possible out of the resource before someone confiscates your property.

In the case of ageing infrastructure, particularly pipelines and refineries, large scale investments are now required. To be competitive, new investment will require access to and transfers of Western technology. Technological transfer to Russia is, however, fraught with difficulties (Davies et al., 1996). While at the time transition began, Russia had a highly skilled and educated technical cohort such as scientists and engineers, the education system has been starved of funds so that the number of new graduates has fallen and the universities and training institutions, for the most part, do not have sufficient resources to keep their curricula current. Upgrading the skills of those trained in the command era has been particularly difficult (McNeil and Kerr, 1995). Over time, it has become more and more difficult to find qualified Russians to facilitate the transfer of technology. This is particularly the case in the far-flung oil fields which still suffer from poor quality housing and social amenities – capable engineers and scientists gravitate to Moscow and other major centres.

Poor or non-existent protection for intellectual property rights make Western firms cautious about sharing their technology. Further, given the high degrees of risk that are associated with doing business in Russia, firms are unwilling to invest in assets that may become stranded, preferring more mobile forms of technology, often embodied in human capital which can be easily withdrawn in times of financial turmoil (Middleton, et al., 1993). Given the dedicated nature of many of the assets in the petroleum industry, this bias in technology transfer may considerably inhibit modernization of the sector. Again, overcoming this difficulty will require fundamental changes to the operation of the Russian economy.

What are the prospects for the Russian economy over the medium term? It has been common to assume that the end point of the transition process will be the transformation of the Russian economy into a replica of modern market economies. In other words, the end point of the transition process is a market economy. Given this assumption then, the important question becomes: how long will the process of transition take and what is the most efficient path to get there? One thing is clear, the Russian economy is not yet a market economy. It is also clear that the process of transition to a market economy has proved far more difficult that was imagined either by those in the West providing advice on how to proceed or those in Russia charged with shepherding the process in the optimistic days after the fall of the Gorbachev government. In the beginning, experts talked in terms of years, now multiple decades seems a more common term. The process has also not been linear with lots of fits and starts – some examples in the case of the petroleum sector were provided in Chapter 7. As a result, particularly over the last few years, it is hard to discern if progress is still being made. Concrete examples are few.

This brings us to an alternative hypothesis. Is the Russian economy actually evolving into a modern market economy or is it becoming something completely different? The assumption that the end process of transition would be a modern market economy has its roots in the Cold War. While the Cold War was, to a considerable degree, a military competition, it also had an ideological component. This ideological element was couched in terms of capitalism versus socialism. It was touted as a struggle for supremacy between economic systems. With the collapse of the Soviet Union and the admission by China, at least by its actions if not formally, that socialism has been a failure, victory was claimed for capitalism. As the 'superior' system, it was natural to assume that it would be adopted by countries that had been the protagonists of socialism. In fact, this is the result they desired and in Russia it became the official policy. In this bi-polar world, alternatives were not even considered. Given the failure of the Russian economy to become a market economy, are there alternative models that might better characterize its evolutionary pattern?

While economies are never static, they can take on stable characteristics (or reach a stable equilibrium in economists' terms). This means that they return to a recognizable set of characteristics over time after receiving a shock. For example, modern market economies retained their essential characteristics after the OPEC induced oil price shocks of the 1970s so that their mode of operation was not much different in the 1990s than it was in the 1960s. While there is a certain comfort in stability, it also suggests that reforms can be very difficult.

The economy of the Russian Federation and its petroleum sector may be gravitating to what was characterized as a 'licensing' economy in Chapter 1 (Kerr and MacKay, 1997). In the absence of secure property rights and a functioning and enforceable commercial legal system, government bureaucrats retain considerable economic power through their ability to license economic activities. While central planning has been abandoned, government interference in the economy of the Russian Federation remains high. The tight central control exercised by the Communist Party has been broken but many of the old personal relationships upon which Party control was based remain intact and those who wielded power in the Soviet era continue to do so. The loosening of control at the centre, however, has meant that those with the right to license economic activity have been able to use their power in entrepreneurial ways. The ability to 'license' economic activity is the basis of corruption. Given the poor state of government finances, government employees are poorly and often only intermittently paid. Thus, income from corruption has become the major source of bureaucrats' income. They are dependent upon it. The ability to license economic activity comes in a variety of forms including not only the requirement of various forms of permission but also access to credit, foreign exchange or communications systems, permitting the avoidance of taxation, and protection from intimidation. The ability to license activity also provided the mechanism whereby many of the public assets created during the Soviet era were transferred to former Party officials or the former cohort of the professional managerial elite.

Without well defined property rights, however, those in control of productive assets have no security of their tenure for the assets they control. This makes them vulnerable to rent extraction by those who license activity directly. It also means that productive assets are not seen as building blocks for future wealth creation, but rather as opportunities for 'mining' in the short run to acquire cash that can be transferred into assets that are relatively safe from confiscation if the individual has control of the firm taken away from him. Large foreign bank accounts and hoards of convertible foreign currencies are only the two most obvious manifestations of being able to exploit the right to license economic activity. This means that instead of being reinvested in activities leading to business growth, profits are siphoned off. In the absence of a banking and financial system, this is a major inhibitor of economic growth. It is also the reason why so much foreign capital fails to find its way into productive assets.

The high corruption costs imposed by the 'licensing' economy greatly increase the cost of doing business for firms while the siphoning of any profits and the mining of existing assets reduces investment. Together, they are a recipe for stagnation over the long run even if the removal of the controls imposed by central planning allowed for considerable opportunities for profit, particularly due to the heavy monopolization of supply chains (Hobbs et al., 1997). For example, fortunes were certainly made by the 'oligarchs' that controlled much of the domestic oil and gas distribution systems.

Given high costs and low levels of investment in productive assets, why isn't there a strong movement for reform of the system? The problem is that those who have the rights to license economic activity have a direct stake in the continuation of the system. Their livelihood depends upon it.

It is no coincidence that the most difficult reforms in the Russian Federation have related to the definition, much less enforcement, of property rights. Freeing prices was largely accomplished, although, as illustrated in Chapter 7, it was often a tortuous process. Privatization has been more difficult but considerable progress has been made and there are few barriers to new private enterprises (as long as you obtain a license). While it took some hard lessons to bring a measure of macroeconomic stability, that too has been accomplished to a considerable degree. On the other hand, little progress has been made on establishing an effective property rights system. In particular, land and resource markets have failed to develop. While commercial law is now on the books, it is not enforced. Extending property rights would put at risk the power of those who have the ability to license.

Given that the ability to personally exploit the right to license is endemic throughout the bureaucracy of the Russian Federation, there is a built-in resistance to changing the status quo. This is why the periodic anti-corruption campaigns are bound to fail. While these campaigns can be used to punish a few individuals who have fallen out of favour or whose corrupt practices have become so lucrative or overt so as to cause political embarrassment, they cannot be used to end widespread corruption. Only the formal extension of property rights and the introduction of competition into the provision of licenses can accomplish this change (Kerr and MacKay, 1997).

It is interesting to note that in May 2001 the Putin government removed the boss of Gazprom, who was once touted as the most powerful businessman in Russia, from its board and was questioning the head of the oil company Sibneft regarding theft and fraud. According to *The Economist* (9 June 2001, p. 52)

> The stamp of Kremlin involvement is plain. Where the rich and powerful are concerned, Russia's legal system is a political weapon, not a dispenser of justice. Likewise, state shareholdings in companies like Gazprom are mainly for politics not profits.

While some would laud the move as ridding the economy of two corrupt and powerful business oligarchs, it goes to prove the point that there is no security of tenure and protection of property rights. It would be like the US government deciding to remove the head of Standard Oil or the British government harassing the chairman of BP through the courts on the order of the Prime Minister.

Part of the problem may be that those in power in Russia simply do not understand what is required to move the economy from its current licensing mode to one based on markets. Freeing prices and privatization are easy concepts to understand, and ones that were politically popular. It was also what constituted the naïve advice given by Western economists at the start of the liberalization process (Hobbs et al, 1997). The more complex institutions required to underpin the operation of modern market economy are much more difficult to explain, much less to create. Property rights, the rule of. law and mechanisms to reduce information, negotiation and monitoring/enforcement costs are difficult concepts to grasp for those with no experience of market economies. They are also not the types of concepts that catch the voters' interest in the same way as anti-corruption campaigns or privatization schemes. Given considerable and widespread resistance to the further extension of property rights by those dependent on corruption for their livelihood and a failure to understand the workings of a market economy on the part of Russia's political leaders, progress on further reform may prove illusive.

Russia is also geographically relatively isolated from modern market economies. The best success stories in the transition process, the Czech Republic, Poland and Hungary, share long borders with Western European countries. Competition from these countries reduced the monopoly power of those with the ability to license by simply providing the alternative of crossing the border to obtain goods and services. As one moves east, this form of competition is reduced and the transition process has not progressed as far and, instead, economies may be settling into being licensing economies; this is the case not only in the Russian Federation but also across the constituent Republics of the former Soviet Union.

It may seem strange that a chapter concluding a book on the Russian oil economy has had so little to say about oil. One might expect a host of statistics on reserves, infrastructure and trade (in fact much of this information is in Chapter 1). One might have expected forecasts based on standard economic analysis of market trends. Of course, the point of the book is that this type of forecasting is not appropriate and not reliable. This is because since its inception, the Russian oil economy has not been and is not a market economy. While this should not be a surprise for the long years when central planning held sway, it is probably less clear for the post-command period. We have been at pains to make the case that the Russian Federation is not a market economy, and may not become one. This means that oil in Russia has been produced, and continues to be produced, without markets. The result is that nothing is at it appears.

While Russian oil reserves indicate the existence of oil, they have been exploited on the basis of non-market decision processes so that their management has been entirely different from that which takes place in market driven oil fields. The technology available for exploring, developing and extracting oil cannot be compared to that in the West in terms of efficiency and reliability. Similar caveats apply to the transportation and refining industries. The legal/regulatory/financial environment of the Russian Federation means that decisions relating to depreciation, reinvestment and capital availability cannot be modelled based on international norms. Despite a decade of reform, many decisions affecting the oil industry are political rather than economic in nature. Competitive forces don't exist and the domestic market remains isolated from international prices. Corruption adds to costs and leads to inefficiencies associated with maintaining secrecy. Economic forecasting is based on a number of simplifying assumptions about the behaviour of economic actors. Almost none of those apply to the Russian oil industry.

Thus, we eschew making predictions about the Russian oil industry in the 21st century – it is only good economic practice to know your limitations. In the short run, it can be safely said that the Russian oil economy is unlikely to live up to its potential. In the long run, its performance will depend upon whether the Russian economy can move beyond its current 'licensing' economy stage to become a true market economy. At the moment, the end point of the second great economic experiment that began at the end of the 20th century, the dismantling of the work of the century's first great experiment of constructing a command economy, remains opaque.

Appendix A: The Early Regional Development of the Russian Oil Industry, 1860–1975

Region	Location	History
The Caucasus – Caspian		
Baku/Azerbaijan	The oil fields of Azerbaijan are located on the Ashperon Peninsula near Baku, the Aspheron archipelago, and the Kura river fields.	Baku is the oldest producing region in the CIS. In the years before the October Revolution it was the undisputed center of Russian oil production. Under the direction of the Soviets, Baku crude oil flows rose steadily from the 7.8 million tonnes reported in 1916 to approximately 22 million tonnes in 1940.
		In the decade of the 1940s production decline in the older fields – forestalled to some extent, by offshore drilling and a number of new discoveries – began slowly to diminish regional crude oil flows. Production fell gradually, to a mere 17 million tonnes in 1950.
		While crude oil flows would not regain their 1940 peak, a series of new discoveries, and the use of water injection, would propel the level of Azerbaijan production to a post-World War II high of 21.7 million tonnes in 1966. After this point, the crude oil flows from ageing fields slowed gradually to only 19 million tonnes in 1972.
		By the early 1970s there were approximately 60 oil fields in Azerbaijan. These included: (i) Lobatinskoe – located approximately 10 km. to the southwest of Baku; (ii) Karadag – located approximately 43 km. to the southwest of Baku; (iii) Neftyanye Kamni – located in the Caspian Sea, approximately 120–140 km. southeast of the Aspheron peninsula; and (iv) Sangachaly-Duvannyi – located in the Caspian, approximately 41 km. southwest of Baku.
		In 1971, some 35 per cent of total Azerbaijan production originated from the prolific offshore Neftyanye Kamni oil fields. It is interesting to notice that, in the same year (1971), the average costs of production for the Neftyanye Kamni oil fields were only 33 per cent of those estimated for the mainland, Azerbaijan.

Region	Location	History
Maykop/ Krasnodar Kray	The Maykop oil region is located on the northern edge of the Western Caucasus. It extends from Maykop (East) to Kerch on the Crimean Peninsula (West).	Maykop oil deposits were first discovered in 1908. Crude oil production from this region rose to 100 000 tonnes in 1911, before declining rapidly in the years immediately preceding the October Revolution. Once considered a major Russian producing region, Maykop was quickly displaced by the prolific 'Second Baku'.

The Maykop producing area was revived in 1951 when a number of significant discoveries were made in the Kuban. In the 1960s, a number of new oil fields were discovered when 90 exploratory wildcats were drilled at an average depth of over 4000 m. Crude oil flows rose steadily over the fifth, sixth, seventh and eighth five-year planning periods to approximately 5.5 million tonnes in 1972.

In the early 1970s, the largest oil field was Anastasievsko-Troitskoe – located approximately 130 km. to the northwest of Krasnodar. |
| Dagestan ASSR | The Dagestan oil district is located on the coast of the Caspian Sea, in close proximity to northern Baku. | Production from the Dagestan oil producing region was first recorded in the nineteenth century, when crude oil was extracted from shallow pits with winches and buckets. The first oil well was completed in the year 1898. Despite these promising developments, however, the Dagestan producing area was not seriously explored until the second five-year plan. The first commercial Dagestan oil well was completed in 1934. Production rose steadily from the modest flows recorded in 1936 to 250 000 million tonnes in 1939.

In 1943 the Makhachkalinskoe oil field was discovered on the Caspian coast of Dagestan. By 1965, a total of eleven oil fields had been discovered, yielding crude oil flows of an impressive 2 million tonnes in 1970. |
| Georgia | The Georgian oil region is located in the Soviet Socialist Republic of Georgia, to the east of the capital city of Tbilisi. | The first oil production from the Georgian oil region was recorded in the early 1930s. By 1939, the crude oil flows from two major Georgian fields – Shiraki and Mirzaani – had reached an impressive 330 000 tonnes. After this point the region experienced a gradual period of decline, which sent crude oil flows falling to only 120 000 tonnes in 1950, and a disappointing 24 000 tonnes by 1970. In the early 1970s, the sparse crude oil flows from Georgia were considered too small to be anything more than a local concern. |

310

The Russian Platform

Ukraine

The Ukraine contains three major producing areas: (i) the Western Ukraine oil region – located on the eastern slope of the Carpathians between Przemysl and Chernovtsy; (ii) the Eastern Ukraine – located between Kharkov, Kiev, and Dnepropetrovsk; and (iii) the Southern Ukraine – located by Kerch in the Crimea.

The Western Ukraine is one of the world's oldest petroleum producing regions. While modest levels of production are reported as early as 1860, systematic, or commercial production did not begin until 1874. Production from the commercial centre of Borislav – which rose steadily to a peak of 2 million tonnes in 1909 – was disrupted by years of war, revolution, and German occupation. Flows declined gradually, and inexorably, to a mere 140 000 tonnes in 1944.

The development of the Eastern Ukrainian oil region was impaired by the 'early' success of the Western development programme. By the early 1950s, the entire Eastern Ukrainian oil industry was centred around two small fields at Romny and Poltava. Production from these two fields reached 30 000 tonnes in 1951. The region – possessing limited value as a commercial oil area – was primarily a local concern.

After 1965, a number of new oil fields were discovered in the Eastern Ukrainian producing region. The largest of these – Gnedintsevskoe and Lelyakovskoe – were responsible for 7.3 million tonnes of production in 1970 – 54 per cent of the total for the Ukraine. As 44 per cent of the Ukraine was estimated to contain oil bearing territory, exploration efforts continued throughout the eighth and ninth five-year plan at drilling depths between 4000 m. and 7000 m.

In the post-World War II period, Ukrainian crude oil flows rose steadily to 2.2 million tonnes in 1960, and an impressive 14.5 million tonnes in 1972. Well over 80 per cent of the total, some 11.6 million tonnes, originated from Eastern Ukrainian oil fields.

Belorussia

The Soviet exploration effort was initiated in Belorussia as early as 1961. Four years later, in 1965, some 39 000 tonnes were extracted from a number of fields, including the prolific Rechitskoe oil field located in close proximity to the town of Gomel. Crude oil flows rose rapidly to 5.8 million tonnes in 1972.

Volga-Urals
The 'Second Baku'

The Volga-Urals producing area is located in the eastern portion of European Russia. It is bounded by the Ural Mountains (on the East) and the Volga River (on the West). The major Volga-Urals oil province includes Tartar ASSR, Bashkir ASSR, Udmurt ASSR, and the Kuybyshev, Volgograd, Saratov, Orenburg, and Perm Oblasts.

The first Volga-Urals oil deposit was discovered (near Molotov) by Soviet petroleum geologists in 1929. Large-scale commercial development was delayed until the third five-year plan. In short, the strategic importance of the Second Baku region did not become evident to Soviet planners until oil flows in the Ukraine, Maykop, and parts of Grozny were disrupted (placed under German control) during the Second World War. Post World War II production rose steadily to an impressive 10.6 million tonnes in 1951. In the early 1950s the major Volga-Urals oil fields were located in close proximity to the cities of Molotov (Perm), Ufa, Bugulma and Kuybyshev.

After this point the prolific Volga-Urals producing region entered a period of rapid growth, and expansion that would send crude oil flows soaring to 224.9 million tonnes in 1975. By the early 1970s, however, production declines in the major producing areas – Tartar ASSR, Bashkir ASSR, and the Kuibyshev Oblast – signified the end of the region's pre-eminence as the major oil producing region of the USSR.

Region	Location	History
Volga-Urals The 'Second Baku'	Bashkir ASSR	The first commercial exploitation of the Bashkir producing region was achieved in 1932, with the rapid development of the newly discovered oil field by the city of Ishimbaevo. By the early 1970s over 100 oil fields had been discovered in the Bashkir producing region. These included: (i) Tuimazinskoe – located approximately 180 km. to the west of Ufa; (ii) Shkapovskoe – located approximately 35 km. to the south of Belebei; and (iii) Arlanskoe – located approximately 160 km. to the northwest of Ufa. An aggressive Soviet development programme sent crude oil flows soaring to an impressive 44.4 million tonnes in 1968. After this point, the effects of rapid development – ageing and in some cases literally exhausted oil fields – and technical 'breakdowns', began to take a toll on physical output. During the eighth five-year plan crude oil flows slowed to only 40.7 million tonnes in 1970. The reduction would, subsequently, be attributed to failure on the part of the Bashkir producing union. Labour discipline in Bashkiria had been reported, and there were considerable losses due to accidents and breakdowns. Trade unions were attacked for not improving working conditions and social amenities sufficiently.
	Kuibyshev Oblast	The first commercial exploitation of the Kuibyshev producing region was achieved in 1936. By the early 1970s, over 100 new oil fields had been discovered, including: (i) Mikhailovsko-Kokhanovskoe – located approximately 22 km. southeast of the city of Tolkai; and (ii) Kuleshovskoe – located 140 km. to the southeast of Kuibyshev. Crude oil production rose steadily from 200 000 tonnes in 1940 to 35.4 million tonnes in 1971. In June 1972 their 500 millionth tonne was produced.
	Volgograd Oblast	While a Soviet exploration programme was initiated in 1940, the commercial exploitation of the Volgograd Oblast was not achieved until 1951. The first significant volumes of commercial crude oil supplies were produced from the Korobkovskoe oil field (located to the northwest of the town of Kamyshin). Crude oil flows from the Volgograd Oblast rose steadily to a peak of 6 million tonnes in 1965, before entering a period of decline.
	Saratov Oblast	The Soviet exploration effort was initiated in the Saratov Oblast in 1935. By the early 1970s, approximately 40 oil fields – including the prolific Sokolovogorskoe oil field – had reached the development stage. Approximately 30 million tonnes of crude oil were extracted from the Saratov Oblast in the years 1935 to 1970, inclusive.
	Orenburg Oblast	The Soviet exploration effort in the Orenburg Oblast was rewarded with the discovery of the Buguruslanskoe oil field in 1937. By the early 1970s, a total of thirty oil fields had been discovered in the promising producing area. These included Ponomarevskoe on the Dema river, Pokrovskoe, Bobrovskoe, and Sorochinsko-Nikolskoe. Crude oil flows in the Orenburg Oblast rose steadily to 2.6 million tonnes in 1965. After this point the Soviet exploration and development programme was accelerated in all promising Volga-Urals producing areas, and Orenburg crude oil flows rose to 7.4 million tonnes in 1970 and 9.4 million tonnes in 1972.

312

Udmurt ASSR	The Soviet exploration effort in the Udmurt producing region was initiated in 1945. By the early 1970s, a total of only 20 oil fields had been discovered. These included the Arkhangelskoe, Kiengopskoe, and Gremikhinskoe oil fields – all located in the southern territories of Udmurt. Crude oil flows rose slowly, but steadily to over 1 million tonnes in 1972.
Emba/Kazakhstan The 'Pre-Caspian Basin'	The first commercial production from the Emba oil region was achieved in 1912. Despite periodic disruptions due to war and revolution, crude oil flows rose steadily to 1.3 million tonnes in 1950. The relatively slow development of this highly promising producing area has been attributed to a critical shortage of water in the desert steppe on the north coast of the Caspian Sea. By 1965, the crude oil flows from over 20 fields in the Emba area had reached only 2 million tonnes. By the early 1970s the major oil fields in Kazakhstan included: (i) Koschagyl – located approximately 100 km. inland from the northern shore of the Caspain; (ii) Prorva – located on the coast of the Caspian; (iii) Kenkiyak – located approximately 125 km. to the southeast of Aktyubinsk; and (iv) Uzen – located in the Mangyshlak peninsula. Crude oil flows from Kazakhstan registered a large increase during the eighth five-year plan, rising steadily from the 3.1 million tonnes reported in 1966 to 13.2 million tonnes in 1970. By the late 1960s, the water problem had been solved successfully by the construction of a 250 km. pipeline from Shevchenko to Novyi Uzen – a small oil town located in the arid desert near the prolific Uzen oil field.
The vast Emba producing region is located to the South of the Second Baku, on the north coast of the Caspian Sea (Kazakstan).	
Ukhta-Pechora/ Komi	Oil deposits were first sighted in the Ukhta-Pechora district at the time of Peter the Great, however, commercial development was delayed until the beginning of the twentieth century. While Soviet geologists indicated an interest in the area in the early 1920s, the drilling programme was not undertaken seriously until the mid-1930s. Production rose steadily from 100 000 tonnes in 1937, to a 'wasteful' peak of 1.2 million tonnes during the oil 'crisis' of World War II. As the political and economic situation stabilized, production was gradually reduced to a more efficient level – 800 000 tonnes in 1950. In the 1930s and 1940s, a number of new oil fields were discovered between the Ukhta and Pechora rivers. However, it was not until 1959, that the major Zapadnyi-Tebuk oil field would be discovered approximately 60 km. to the east of the city of Ukhta. By the early 1970s, a total of 20 oil fields had been discovered in the Komi producing region. In the eighth five-year plan, significant progress was made in the development of the Komi producing area, and crude oil flows rose significantly from 2.2 million tonnes in 1965 to 5.6 million tonnes in 1970.
The Ukhta-Pechora oil district is located in the northeastern corner of European Russia. The name is derived from its proximity to the city of Ukhta and the Pechora River.	

313

Region	Location	History
The West Siberian Plain		
West Siberia	The West Siberian Plain is located between the Urals and the Yenisey River.	While Soviet exploration efforts were initiated in 1930, the first West Siberian gas field would not be discovered until the 1950s. In 1953 a number of barges transporting rigs on the river Ob' were delayed by the village of Berezov. To pass the time, the passengers decided to drill a test well in a promising area by the river bank. Their efforts were rewarded by a tremendous explosion of gas and water, and the discovery of the prolific Berezov gas field. Despite the unprecedented, and fortuitous occasion, the systematic exploration of the West Siberian producing region would be delayed until the late 1950s. Years later, in 1960, the first West Siberian oil field was discovered by the town of Shiam on the Konda River. The Soviet West Siberian exploration programme was intensified in the eighth and ninth five-year planning periods. By April 1973, 113 oil fields had been discovered in the vast, and often inhospitable permafrost, of the West Siberian Plain. A major research institute 'Giprotyumenneftegaz', was established in Tyumen to solve the problems presented by severe northern winter conditions, and the vast expanses of roadless marshland, and taiga. In 1966 experimental hovercrafts were used to transport 160 ton drill rigs over the bogs surrounding Shiam. As a result, the construction capacity of an average West Siberian rigging crew was increased by five times, from the 10 rigs completed in 1965 to 50 rigs by the year 1970. By 1972 twelve West Siberian oil fields had started production. These included: (i) Trekhozerone and Teterevo-Mortyminskoe in Shiam; (ii) Ust-Balykskoe, Zapadno-Surguitskoe, Pravdinskoe, Mamontovskoe, Fedorovskoe, and Solkinskoe in Surgut; (iii) Megionskoe, Vatinskoe, Sovetskoe, and Samotlorskoe in Nizhenevartovsk. The use of the latest technology – including the maintenance of formation pressure via local underground reservoirs of thermal mineral water – sent crude oil flows soaring from 209 thousand tonnes in 1964 to 31.416 million tonnes in 1970. In March 1973, the Soviet government announced that over 200 million tonnes of crude oil had been produced from West Siberian oil fields.

The Far East

| Sakhalin | The Sakhalin producing region is located on Sakhalin Island – the subject of an ongoing territorial dispute between Japan and Russia. The Southern portion of the 'Russian' Island was lost to Japanese forces in the Russo-Japanese War of 1905. Japan, in turn, lost South Sakhalin to Soviet forces at the end of the Second World War. | In the years following World War I, the Japanese leased portions of North Sakhalin in an attempt to supplement domestic oil supplies. Production from the North Sakhalin region was initiated, on a small scale, in 1921. A steady increase in crude oil flows inspired active Soviet exploration, forming a haphazard partition of the fields controlled by Japan and the Soviet Union. The Japanese concessions were soon completely surrounded by Russian oil fields. By the end of World War II, Japan had lost all its North Sakhalin concessions, and the South portion of the Sakhalin Island. Crude oil flows reached 1.2 million tonnes in 1950.

In the 1950s, the strategic importance of Sakhalin oil supplies was undisputable. In the words of Hassmann (1953, p.77), 'In the first place, it serves as the oil supply of the army in the Far East, of the Far Eastern industrial district on the Amur River, and of coastal and ocean navigation in the Pacific. Without the oil from Sakhalin the army, industry, and shipping would have to depend upon imports of crude oil…[These] shipments would have to travel a distance of 4,400 to 5,000 miles over the Trans-Siberian Railroad.'

Sakhalin production rose steadily in the fifth, sixth, seventh and eighth five-year planning periods, and the crude oil flows from over thirty oil fields reached 2.472 million barrels in 1970. |

Sources: Elliot (1974); Hassmann (1953).

315

Appendix B: Reserve Classifications of the Soviet Union

The precise determination of petroleum reserves is a complex and inexact science posing a variety of problems for scientists attempting to evaluate even mature producing regions such as the United States, and Great Britain. In the case of the USSR, official estimates were simply unavailable. Still listed under the rigid provisions of the State Secrecy Act of 1947, the volume of petroleum reserves in Russia remains a state secret to this day. Any published evaluation of oil reserves must be based on an unofficial, or casual appraisal, of probable resources. As a general rule of thumb, these estimates have been calculated as 'an analysis of resource exploitation within the Soviet Union and a projection of the likely oil potential of the sedimentary basins of the Soviet Union measured against the experience of comparable sedimentary basins elsewhere in the world' (Goldman, 1980, p. 115).

An excellent example of such an appraisal is provided in Elliot (1974). He states:

> In fact, the USSR claims 37 per cent of the earth's potential oil-bearing areas. Working from estimates that the average density for the potential areas of the world is 16,400 tons of oil per square kilometer would give the USSR, with 11,900,000 sq. km. of oil-bearing territory, about 195,160 million tonnes of ultimate oil reserves. This compares well with figures for the ultimate reserves of the United States of around 100,000 million tons, although these are more densely contained, within an area of less than 5 million sq. km (p. 81).

To further complicate matters, the estimation of USSR reserves was confused by the existence of a unique Soviet classification system for mineral resources, a system for which there was no exact equivalent in Western terminology. To be specific, Soviet mineral resources were divided into the following categories:

Category A: Including gas [oil] reserves 'which have been fully explored in areas outlined by productive wells and reliably established by experimental exploration' (Elliot, 1974, p. 19).

Category B: Including gas [oil] reserves 'in areas where drilling has given favorable indications of commercial gas [oil] possibilities, borne out by a commercial flow from at least two wells' (Elliot, 1974, p. 19).

Category C_1: Including gas [oil] reserves 'in locations that geological and geophysical data show to be favorable to the accumulation of gas [oil]. Porosity and permeability have been established by drilling or by analogy with neighboring deposits that have already been explored. A commercial flow of gas [oil] must be obtained from at least one well in the estimated area' (Elliot, 1974, p. 19).

Category C_2: Including gas [oil] reserves in 'new structures of gas[oil]-bearing provinces with strata of a type that are known to be productive in other deposits. It also [includes]... reserves of known fields situated in unexplored tectonic blocks and strata that favorable geological and geophysical data suggest are likely to prove productive' (Elliot, 1974, p. 19).

Two last categories include "predicted" (*prognoznye*) oil and gas reserves.

Category D_1: 'Covers reserves in horizons that have already given indications of being productive and that may be assumed from geological prospecting to contain certain quantities of gas [oil]' (Elliot, 1974, p. 19).

Category D_2: 'Refers to gas [oil] in possibly productive horizons that have shown no definite indications as yet because of insufficient prospecting' (Elliot, 1974, p. 19).

According to Elliot (1974), the simple mathematical sum of $A+B+C_1+C_2$ is roughly equivalent to the Western reserve classification of 'Proved' plus 'Possible' oil reserves. However, few calculations involving Soviet statistics are as simple as they might appear.

According to Goldman (1980, p. 116), there was some argument

among Western specialists as to just which Soviet category corresponds to which American category. As interpreted by the CIA, Soviet *A* reserves are those established through drilling or by fields completely surrounded by wells. *B* reserves are those surrounded, but not completely, by three producing wells. *C_1* reserves are marked by at least two wells in the producing zone or else the area must be directly adjacent to reserves of a higher category.... Some authorities, such as the CIA, include part of *B* along with all of *A* in the category 'Proved' [Geologist A. A.] Meyerhoff's definitions are slightly different. He points out that the CIA and others use the outdated pre-1971 classification system, which tends to understate what we, in the West, regard as "Proved" reserves. According to Meyerhoff, the CIA includes only A category and one-half of B Category as Proved reserves. It should have included all A and all of B reserves. Proved, and Probable reserves, says Meyerhoff, should refer to A plus B plus part of C_1, which the CIA does not do. Given Meyerhoff's definition, the CIA seems to have understated Soviet reserves. In sum, even the straightforward is complex.

Hewett (1984) defines the sum of Soviet categories $A+B+C_1$ as 'Proven' plus 'Probable' plus some 'Possible'. Table B.1 illustrates the subtle differences between Western and Soviet reserve classifications.

Table B.1: Reserve classifications in the Soviet Union

	(Post 1971)	
Western Classification	Soviet Classification	CIA
Proved	$A+B$	A+part of B
Probable	$A+B$+part of C_1	$A+B$
Possible	$A+B+C_1+C_2$	
Predicted	$A+B$+part of C_1+C_2+D	

It is interesting to note that, despite the tremendous influence of *perestroika* and reform in the early 1990s, the State Secret Act of 1947 is still in force – has not yet been repealed – at the time of writing.

Appendix C: Long-Distance Oil Pipelines in Russia, 1908–1988

Original Destination	Commenced Operations In	Diameter (mm)	Length (km)
Early 1900s			
Khadyzhensk-Krasnodar	1908	125	89
Third Five-Year Plan			
Zolnoe-Syrzran	1940	300–530	141
Fourth Five-Year Plan			
Severokamsk-Perm	1947	250	61
Tuimasy-Ufa I	1947	350	158
Baitugan-Klyavlino	1949	200	40
Krymskaya-Krasnodar	1949	300	101
Sokolovaya Gora-Saratov	1949	250–350	55
Bavly-Kuybyshev I	1950	300–350	308
Tuimasy-Ufa II	1950	350	158
Total			881

Original Destination		Commenced Operations In	Diameter (mm)	Length (km)
Fifth Five-Year Plan				
Karabash-Bavly I		1951	350	69
Bavly-Kuybyshev II		1953	530	308
Karabash-Bavly II		1954	530	69
Romashkino-Kuybyshev		1954	530	250
Tuimasy-Ufa III		1954	530	155
Zolnoe-Syrzran	Loops 38-100	1954	250	104
Krotovka-Kuybyshev		1955	530	97
Kuybyshev-Saratov		1955	530	357
Omsk-Tuimasy I		1955	530	1,332
Pokrovka-Syrzan		1955	350	100
Total				2,841
Sixth Five-Year Plan (Abandoned in 1958)				
Shkapovo-Subkhankulovo		1956	530	94
Smolenskaya-Krasnodar		1956	200–300	34
Subkhankulovo-Aznakaevo-Almetevsk I		1956	530	111
Kaltasy-Ufa I		1957	300	209
Sernye Vody-Krotovka		1957	300–350	79
Zhirnovsk-Volgograd		1957	350–530	314
Total				841
Seventh Five-Year Plan				
Mukhanovo-Kuybyshev		1958	720	108
Subkhankulovo-Aznakaevo-Almetevsk II		1958	720	112
Severokamsk-Perm	Loops	1959	250	28
Ishimbai-Orsk		1960	530	333
Chekmagush-Yazykovo		1961	720	51

Original Destination		Commenced Operations In	Diameter (mm)	Length (km)
Seventh Five-Year Plan (continued)				
Gorkiy-Ryazan I		1961	720	395
Kaltasy-Chekmagush		1961	720	109
Ozek-Suat-Groznyy		1961	530	193
Ufa-Subkhankulovo III		1961	820	161
Yazykovo-Salavat		1961	720	164
Almetevsk-Kuybyshev I		1962	820	274
Almetevsk-Nizhny Novgorod II		1962	820	580
Kamennyi Log-Perm	Loops 0-27	1962	400	27
Kamennyi Log-Perm		1962	400	68
Ryazan-Moskva		1962	530	197
Gorkiy-Yaroslavl		1963	820	358
Kuleshovka-Kuybyshev		1963	530	91
Tikhoretsk-Tuapse		1963	530	239
Kuybyshev-Saratov	Loops 241-323	1964	530–720	82
Kuybyshev-Unecha I (Druzhba)		1964	1,020	1,274
Omsk-Irkutsk	Loops	1964	720	234
Omsk-Irkutsk		1964	720	2,108
Tikhoretskaya-Novorossiysk I		1964	530	239
Unecha-Mozyr I		1964	820	106
Unecha-Polotsk I		1964	820	136
Shaim-Tyumen		1965	530	410
Total				8,077
Eighth Five-Year Plan				
Aznakaevo-Almetevsk II		1966	530	51
Naberezhnye Chelney-Almetevsk II		1966	530	95
Shkapovo-Ishimbai		1966	530	137
Chekmagush-Aznakaevo-Urussu		1967	530	125

Original Destination		Commenced Operations In	Diameter (mm)	Length (km)
Eighth Five-Year Plan (continued)				
Chernushka-Kaltasy		1967	720	119
Kaltasy-Ufa II		1967	720–530	204
Osa- Perm II		1967	530	107
Ozek-Suat-Malgobek		1967	530	169
Ust-Balyk-Omsk		1967	1,020	964
Almetevsk-Kuybyshev II		1968	1,020	275
Chernushka-Osa		1968	350	108
Gorkiy-Ryazan I	Loops 0–30	1968	530	30
Kushkul-Ufa		1968	300	80
Aleksandrovskoe-Nizhnevartovst		1969	720	47
Almetevsk-Nizhny Novgorod III		1969	1,020	575
Nizhnevartovsk-Ust-Balyk		1969	720	196
Yaroslavl-Kirishi		1969	720	524
Buguruslan-Syran		1970	400–530–720	332
Gorkiy-Ryazan II		1970	530	396
Unecha-Mozyr II		1970	1,020	106
Total				5,136
Ninth Five-Year Plan				
Kuybyshev-Unecha II (Druzhba)		1971	1,220	1,310
Lopatino-Samara II		1971	1,020	23
Alexandrovskoe-Anzhero-Sudzhensk		1972	1,220	818
Alexandrovskoe-Anzhero-Sudzhensk	Loops 349–409, 418–455	1972	1,020	52
Nozhovka-Mishkino-Kiengop		1972	300–530	104
Omsk-Tuimasy I	Loops 345–366	1972	530	52
Tuimasy-Omsk II	Loops 0–72, 313–365, 610–645	1972	820–720	172
Tuimasy-Omsk II		1972	720	1,334

Original Destination		Commenced Operations In	Diameter (mm)	Length (km)
Ninth Five-Year Plan				
(continued)				
Naberezhnye Chelny-Almetevsk Refinery		1973	720	113
Ust-Balyk-Kurgan-Ufa-Almetevsk		1973	1,220	1,813
Ust-Balyk-Nizhnevartov		1973	1,020	278
Anzhero-Sudzhensk-Krasnoyarsk		1974	1,020	691
Kuybyshev-Tikhoretskaya		1974	820	1,279
Michurinsk-Kremenchug		1974	720	453
Pokrovka-Krotovka II		1974	530	173
Tikhoretskaya-Novorossiysk II		1974	820	237
Ukhta-Yaroslavl		1974	820	1,133
Unecha-Polotsk II		1974	820	136
Usa-Ukhta		1974	720	407
Lisichansk-Tikhoretskaya I		1975	720	313
Samotlor-Aleksandrovskoe		1975	1,020	66
Subkhankulovo-UKPN II		1975	530-400	14
UKPN III-Subkhankulovo		1975	400-530	14
Total				10,985

Tenth Five-Year Plan

Original Destination		Commenced Operations In	Diameter (mm)	Length (km)
Nizhnevartovsk-Kurgan-Kuybyshev		1976	1,220	2,245
Nizhny Novgorod-Surgut		1976	820	266
Samotlor-Nizhnevartovsk		1976	1,220	65
Kuybyshev-Lisichansk		1977	1,200	926
Omsk-Pavlodar		1977	1,020	95
Kiengop-Naberezhnye Chelny		1978	530–720	260
Surgut-Perm		1978	1,220	1,268
Tikhoretskaya-Novorossiysk II	Loops	1978	820	32

Original Destination		Commenced Operations In	Diameter (mm)	Length (km)
Tenth Five-Year Plan (continued)				
Yaroslavl-Moskva		1978	720	314
Kaltasy-Kuybyshev		1979	820–720	347
Perm-Gorkiy		1979	1,020	820
Gorkiy-Yaroslavl II		1980	1,020	370
Kiengop-Naberezhnye Chelny	Loops 25–207	1980	530	186
Urevskie-Yuzhnnii Balyk		1980	1,220	196
Total				7,390

Eleventh Five-Year Plan

Original Destination		Commenced Operations In	Diameter (mm)	Length (km)
Bavly-Kuybyshev II	Loops 45–46, 69–70	1981	530	2
Drain pipe to Nizhnekamskii Refinery		1981	720	29
Guriev-Kuybyshev		1981	720	149
Tyumen-Yurgamysh		1981	530	245
Voznesenskaya-Groznyy		1981	720	82
Yaroslavl-Polotsk		1981	1,020	833
Drain pipe to Saratov Refinery		1982	530	82
Kaltasy-Kuybyshev	Loops 109–172	1982	820	61
Kuzmici-Volgograd Refinery		1982	530	74
Perm-Almetevsk		1982	1,020	446
Groznyy-Baku		1983	720	385
Krasnoyarsk-Irkutsk II		1983	1,020	849
Drain pipe to Almetevsk Refinery		1984	820	15
Kuchiminskaya-Perm		1984	1,220	1,203
Shaim-Konda		1984	530	108
Khokhryakovskoe-Tyumen		1985	530	94
Lisichansk-Tikhoretskaya II		1985	720	308
Perm-Nizhny Novgorod		1985	1,220	926

Original Destination		Commenced Operations In	Diameter (mm)	Length (km)
Eleventh Five-Year Plan (continued)				
Povkh-Pokachi-Urevskie		1985	720	155
Saratov-Kuzmichi		1985	530	338
Total				6,384
Twelfth Five-Year Plan				
Buguruslan-Syran	Loops	1986	530	22
Kholmogory-Kuchiminskaya		1986	1,220	296
Lyantorskoe-Fedorovskoe		1986	720	73
Priraslomne-Karkateevy		1986	400	36
Tarasovske-Purpe-Muravlenkovskoe		1986	720	169
Telepanovo-Kalmash		1986	530–400	55
Bakhilovskoe-Khokhryakovskoe		1987	720	124
Gerasimovskoe-Luginetskoe-Parabel		1987	530	228
Krasnoleninsk-Shiam-Konda		1987	820	345
Vyngapur-Belosernyi		1987	820	86
Vyngayakhinskoe-Khanymei		1987	350	43
Yuzhno-Talinske-Krasnoleninsk		1987	530	28
Urengoi-Kholmogory		1988	1,020	236
Vat-Yegan-Aprelskaya		1988	530	37
Yaroslavl-Kirishi	Loops	1988	720	349
Total				2,127
Total Russia				44,892

Appendix D: Internal Oil Pricing Policies of the Soviet Union

1. *The Period Of War Communism, 1918–21.* The Bolshevik victory was solidified by the occupation of the Winter Palace on the night of 7 November 1917. One day later, on 8 November 1917, the Congress of Soviets adopted the infamous decree of Land and Farming. All land was to be nationalized immediately, and the right to use it transferred to the peasants. No person could hold more land than they alone could cultivate as the hiring of labour was to be forbidden (Nove, 1972). The nationalization of all heavy industry, banks, railroads, merchant marine and shipping operations followed shortly thereafter. The official statement on an internal pricing policy was brief and to the point. All state industrial enterprises were instructed to 'deliver their products to other state enterprises and institutions on the instructions of the appropriate organs of VSNKh without payment, and in the same way...obtain all the supplies they require'. Railways and the State Merchant Fleet were required to 'transport gratis the goods of all state enterprises. In making this proposal, the congress expressed the desire to see the final elimination of any influence of money upon the relations of economic units' (Venediktov, 1957, p. 519).

    The establishment of a 'moneyless' system remained a common theme in Party negotiations throughout the period of War Communism. In a best case scenario, production from all industries was to be assigned value – a common denominator for State accounting purpose – in terms of labour units. This continued until the economy virtually collapsed, and the policies of War Communism were replaced.

2. *The New Economic Policy (NEP), 1921–27.* This period represented a temporary retreat into the realm of a mixed market economy. The policies of the NEP required a stable unit of account or currency, and were accompanied by the creation of a State Bank in October 1921. A

number of smaller banks were created in 1922 to provide financial credits to industry, for example, Electrobank was established to finance rural electrification. Limited financial stability was established under the prudent supervision of Narkomfim, the People's Commissariat of Finance.

While the decree nationalizing all small-scale industry was revoked on 17 May 1921 all heavy industry, including the oil sector, banks and railroads, remained Socialized. On 5 August 1921 the Prices Committee was established to set the wholesale and retail prices for all goods manufactured by state enterprises.

3. *The Five-Year Plans (1928–65).* The idea of a State-planned or command economy found its most distinct expression in the conception and implementation of the five-year plans. The main elements of the unique Stalin planning model may be summarized as follows: the system of physical controls which had already existed during the 1920s was greatly extended. Prices were fixed, and there was no market for producer goods. Instead, materials and equipment were distributed to existing factories and new building sites through a system of priorities, which enabled new key factories to be built and bottlenecks in existing industries to be widened. The plan set targets for the output of major intermediate and final products, and the physical allocation system was designed to see that these were reached. In short, the prices for inputs (costs) and outputs (production) of an industry were determined directly by the central planning authorities and strictly controlled. In a noticeable departure from the Western style free market economy, all questions of resource allocation were decided by Gosplan – on the basis of central or state priorities – and not chosen 'rationally'; that is, via commodity arbitrage and the profit motive, according to an objective economic efficiency criterion. As a result, the Soviet 'fiat' or command prices played no role in the allocation of scarce economic resources, and were maintained and calculated primarily for accounting purposes (Hobbs et al., 1997).

The Soviet oil pricing system, which evolved gradually throughout the Stalin era, was based on the 'cost-plus' method of centralized price control. In the initial stages of the planning process, the Committee for Prices (Goskomtsen) would consult with the appropriate oil-related Commissariats (Ministries) in an effort to establish individual 'target' prices for crude oil and petroleum products. All target prices were based on average production costs plus a small mark up for profits. In the secondary, and final stages, the target prices were modified to accommodate variations in production costs, transportation tariffs, and the quality of goods supplied by enterprises in different regions of the USSR. In the case of the oil industry, the Soviet Union was divided into a number of cost zones reflecting natural differences in the quality of crude oil and petroleum products, and average production costs in different regions of the country (IEA, 1993). An average wholesale enterprise price was established for each cost zone.

The entire Soviet pricing structure contained only three types of state-controlled prices:

A) **Wholesale Enterprise Prices** – roughly comparable to Western producer prices. These were the prices paid to energy producers. For example: The wholesale enterprise price for crude oil included basic production costs, a small allowance for the depreciation of capital, and a small mark-up for profits;

B) **Wholesale Industry Prices** – roughly comparable to Western wholesale prices. These were the prices that industrial and agricultural users paid for energy. For example: The wholesale industry price for crude oil was equal to the wholesale enterprise price plus transportation charges and a small profit to the intermediary selling organization. The wholesale industry prices for petroleum products were equal to the wholesale enterprise price plus transportation tariffs, applied sales taxes (the turnover tax if levied), and a surcharge for wholesale distribution; and

C) **Retail Prices** – roughly comparable to Western retail prices. These were the prices charged to consumers. For example: The retail prices for petroleum products were equal to the wholesale industry price plus a surcharge reflecting average operating expenses, and a small profit margin for retail distributors (gas stations, hardware stores and so on).

According to Khartukov (1995, p. 1) 'Overall, these state-controlled prices were characterized by striking stability and naturally so, for in an 'economy' that was tantamount to a nation-wide command distribution system, prices were never intended to reflect changing conditions of the virtually non-existent, overcontrolled marketplace. The prices were revised in the infrequent price reforms but remained basically unchanged during decade-long interim periods.' As a result of this inflexible centralized pricing system, it was possible for an enterprise to incur profits (losses) simply by having lower (higher) than average production costs for its own particular price zone. Profits from state enterprises boasting lower than average production costs were automatically taxed (confiscated) by the central planning authorities, while enterprises suffering losses from higher than average production costs were heavily subsidized (IEA, 1993).

In the years 1930–65 all prices in the producer and consumer goods industries included the following basic components:

- **A: Costs of Production** – including wages, the costs of raw materials, overheads, a small allowance for depreciation, but no interest charge on capital.
- **B: Profit** – A small markup for profits.
- **C: Trade and Transportation Costs**
- **D: A Tax on Profit** – whereby 'profits were partly taxed, partly retained by the firm or industry. A high proportion of all profits in excess of the plan was placed at the disposal of the firm in the Director's Fund, later known as the Fund of the Firm' (Davies, 1979, p. 215).
- **E: Applied Sales Taxes** – A turnover tax which was a simple markup on the wholesale price of consumer goods and levied in an effort to equate the supply and demand for various products. The turnover tax was adopted on 2 September 1930.

4.   *The Kosygin Reforms and the Brezhnev Era (1965–90)*. The Kosygin reforms (1966–67) were introduced in an effort to improve the efficiency of the basic Stalin Command economy. The primary objectives were: (1) to increase the powers of managers; and (2) to replace the traditional method of output targets with a more sophisticated methodology – utilizing elaborate incentive schemes and detailed economic calculations – thereby reducing the number of directives to be issued by central command. The salient pricing policies may be summarized as follows:

> **A:** Profits gained new significance as an indicator of success. The small mark-up for profits was increased significantly, and varied in a manner that would discourage firms from understating the potential of the firm, that is, the portion of profits to be retained by the factories was higher for planned profit (if achieved) than for actual profit.
>
> **B:** The Director's Fund was replaced by three large Economic Incentive Funds: (1) the Production Development Fund or

Investment Fund, (2) the Social–Cultural Measures and Housing Fund, and (3) the Material Rewards Fund or Bonus Fund. All three funds were financed, partially or in their entirety, by retained profits.

C: A capital charge, approximately 6 per cent was levied on the fixed and working capital stock of an enterprise (Davies, 1979, p. 229).

D: Prices were 'reformed' (adjusted) to reflect the costs of production. It is important to notice that prices were still calculated according to the Marxist theory of the 'price of production', cost plus a fixed percentage of the value of a firm's capital assets (Nove, 1972).

The effect of the Kosygin economic reforms on the domestic or internal oil pricing system may be summarized as follows:

A: Wholesale Enterprise Prices – In the 1966–67 price reform, the wholesale enterprise prices (in every cost zone) were adjusted to include the following elements: (1) a higher mark up for profits; (2) a 6 per cent charge on capital assets; and (3) a finding fee for the fuel.

The finding fee, or geological exploration cost, was added to wholesale energy prices on 1 July 1967. That is to say, for the first time in the history of the Soviet Empire, the crude oil producing enterprises were to be charged an exploration fee – approximately 25 per cent of the geological exploration cost – as a basic or essential cost of production. In practice, only a fraction of these costs were actually recovered. According to Goldman (1980, p. 44) as late as 'mid-1976 miners were charged only 60 per cent of all their geological exploration costs'.

The inclusion of both the finder's fee and capital charge was intended to improve the information content or planning value of oil prices. In theory, the new wholesale enterprise price should have been equal to the true marginal cost price in each cost zone. In reality, and despite the best efforts of Gosplan, the new prices were 'not the true marginal cost price(s) because the capital charge was in fact too low, and the rent charges were not true rents' (Hewett, 1984, p. 134).

In short, while an exploration fee may have been included in the costs of production, rent payments (for the use of the land), royalties (for the raw mineral deposits), and charges for the outright purchase of land, were not. Mineral deposits and land were essentially free goods to the Soviet oil producers. The discrepancy led to a pricing system that, according to Goldman (1980, p. 45) failed to:

> reflect the full economic costs at an early stage of production...[As a result], raw materials [tended] to be underpriced in the Soviet Union. . . . Accordingly the Soviet Union [expended] more fuel per kilowatt of electric power and per ton of open hearth steel smelted, and more metal per unit of engine power than the United States. . . . Given the planning system with its emphasis on output and the tendency to understate or ignore the true costs involved, it was inevitable that there would be waste and inefficiency in Soviet mining practices.

The Kosygin price reforms led to an increase in the average wholesale enterprise, or producer price to 10.53 roubles per tonne (US$1.59/Bbl) by the year 1970, a 134 percent increase over the 4.50 roubles per tonne (US$0.68/Bbl) reported in 1965. Discounting questions raised by the absence of a workable rouble/$ exchange rate, the new prices compare favourably with world market prices prevailing at that time. To cite only one example: the price of Saudi Light 34° averaged $1.80 a barrel in 1967. In the same year, the Soviet enterprise wholesale price of crude ranged from 8 roubles to 22 roubles a tonne among eight producing zones. The higher price was 3.00 roubles per barrel, or (at official exchange rates of that time) US$3.33, the lower price 1.10 was roubles ($1.22).

The coincidence of Soviet and global oil prices would be short lived. As mentioned above, the Soviet Wholesale Enterprise prices were never intended, or expected, to reflect the realities of the dynamic international oil market. In the interest of general price stability, Brezhnev chose to ignore the two major world oil shocks of the 1970s – the Yom-Kippur war and Arab oil embargo of 1973 and the Iranian revolution of December 1978. By the year 1981, Soviet producer prices (US$2.22/Bbl) looked ridiculous in comparison to a world oil price of over US$34.00/Bbl. Despite the growing disparity between domestic and international oil prices, producer prices were revised upwards only once in the 20-year time period 1970–90. In 1982, Soviet wholesale enterprise prices were set at 24.79 roubles per tonne (approximately US$5.00/Bbl), a 135 per cent increase over the 10.56 roubles per tonne (US$2.22/Bbl) reported in 1981.

This rigid system of centralized price control was maintained throughout the early years of Gorbachev and *perestroika*. As a result, the wholesale enterprise price crept up only hesitantly from 24.79 roubles per tonne reported in 1982 to 25.70 roubles per tonne (US$5.94/Bbl) in 1990, an aggregate increase of only 3.67 per cent in 8 years. At the same time the costs of production had been rising steadily. The gradual migration of the centre of Soviet oil production to the harsh region of West Siberia, and an accompanying deterioration of operating conditions, led to a twofold increase in the level of production costs from 9.4 roubles per tonne (US$1.97/Bbl) in 1982 to 21.13 roubles per tonne (US$4.87/Bbl) by 1990.

According to Khartukov (1995, p.2):

> As a result, in the absence of any dampening profit-related taxes, the formal after-tax profitability of Soviet oil production dropped to 0.1% (as opposed to the normative rate of 36.5% established in 1982). Consequently, on the threshold of the 1990s, the national oil-producing industry was being subsidized by the state for one-third of its total capital requirements. The necessity and urgency of another price reform became apparent.

B: Wholesale Industry Prices – In the 1967 price reform, the wholesale industry prices for petroleum products were raised to levels that were, by most accounts, roughly equivalent to world market prices (Gustafson, 1989). To cite only one example:

> the price of automobile gasoline of a medium octane averaged about 95 rubles a ton [in 1967], 3.8 times the higher-priced Soviet crude. That price works out to 31 kopecks a gallon, or $0.34 at official exchange rates, which was probably in line with world prices (excluding the fact that European governments collected large taxes from consumers through gasoline prices) (Hewett, 1994, p. 135).

In the years that followed, however, wholesale industry prices, like producer prices, were maintained at levels which failed to reflect either the full cost of production or (equivalently) the opportunity cost of energy supplies. In short, the aggregate Soviet price indices suggest only a minor 28 per cent increase in the industrial wholesale prices for petroleum products throughout the decade of the 1970s (Hewett, 1994, p. 136).

Gosplan's continued (stubborn) refusal to respect the opportunity cost of domestic energy supplies has been well documented. According to Gustafson (1989, p. 239):

> The two world oil shocks . . .hardly caused a ripple in the perceived costs of energy to Soviet consuming enterprises and their ministries. Only in 1982 were wholesale prices raised again, coal and gas by an average of 42 and 45 per cent, respectively, and oil by 230 per cent.

By 1990, the price of a medium octane gasoline had reached only 195 roubles per metric tonne – a 105.3 per cent increase over the 95 roubles per metric tonne reported in 1967. Utilizing the closest Soviet approximation to a free market exchange rate – the official Auction market (VEB) rate of 19.34 roubles per US dollar in 1990 – the Soviet wholesale gasoline price works out to approximately US$0.0327 per gallon or less than one cent (US) per litre (see Table D.1).

Table D.1: Moscow inter-bank foreign currency exchange

(Moscow foreign exchange auction market (VEB) rates)		
	Period Average	End of Period
1989	8.92	8.92
1990	19.34	22.88
1991	61.95	169.20
1992	227.90	414.50
1993	1,018.00	1,247.00
1994	2,212.00	3,550.00
1995[a]	4,560.00	4,537.00

Notes: Currency Equivalents: Unit of Currency: Rouble
Roubles Per US Dollar
[a] As of 16 November, 1995.

Source: World Bank (1997).

This is a preposterous valuation, when compared to the Western prices prevailing at that time. To cite only a few examples: the refiner sales for resale price for unleaded mid-grade motor gasoline in the United States averaged 81.3 cents (US) per gallon in 1990. At the same time, the spot market product prices for unleaded motor gasoline in New York Harbor and Rotterdam averaged 73.38 cents per gallon and 71.91 cents per gallon, respectively.

Once again it is necessary to draw attention to the confusion generated by the lack of a free Soviet exchange system, or in other words, a workable rouble/$ exchange rate. Assuming the official Soviet exchange rate for 1990 – $1.69 US$/Rbl – the Soviet wholesale industry price for medium octane gasoline averaged approximately US$1.07 per gallon, 35 cents (US) above the 71.91 cents (US) reported in Rotterdam.

The source of general confusion – a glaring discrepancy between 'official' and 'auction market' (VEB) rates – may be found in years when Soviet monetary policies were based more on fantasy than reality. For example, the official foreign exchange value of the rouble was held constant at 0.9 Rbls/US$ for a full decade following 1961, a year which was distinguished by Nikita Khrushchev's heroic attempt to reorganize the entire Soviet monetary system. Afterwards, discounting a few brief episodes of 'controlled' depreciation, the rouble was appreciated, slowly but steadily from 0.90 Rbls/US$ ($1.11 US$/Rbl.) in 1971 to 0.59 Rbls/US$ ($1.69 US$/Rbl) in 1990.

Needless to say, and as a direct result of what can adequately be described as a sustained episode of 'planned' or forced appreciation, official markets were plagued by frequent and often acute shortages of foreign exchange. By the mid-1980s there was no question, even in Gosplan's judgement, that the official exchange rate was grossly overvalued.

The first steps towards the liberalization of the Soviet exchange system were taken in the late 1980s. For the first time in the history of the Soviet Union, enterprises were permitted to conduct a limited amount of foreign trade independently, that is, without the direct intermediation of a Foreign Trade Organization. Vneshekonombank (VEB), the State monopoly of foreign currency transactions, was abolished in 1989. Months later, in November 1989, Gosplan established a system of currency auctions permitting a limited number of 'pre-authorized' buyers and sellers to purchase foreign currency at State auctions and, later, interbank exchange markets. According to Goldberg (1993, p. 7): 'This marked the start of a period of exchange rate movements that, in principle, would be driven by market forces'.

These 'official' auction exchange rates, however, were rigorously controlled and, as such, provide only limited information on the 'free market' or competitive exchange value of the rouble. In short, the state auctions were not open to all agents engaged in foreign exchange transactions. Indeed, the number of Soviet enterprises permitted to 'trade' in foreign currency was strictly limited. As stated by Goldberg (1993, p. 7): 'Authorized buyers and sellers of foreign exchange were chosen using less than transparent criteria, and licenses for currency purchases at auctions (and later in interbank markets) were not distributed openly'.

Still, Gosplan was determined to transform the rigid Soviet exchange system. On 22 July 22 1990, internal trade in foreign currencies was legalized. State auctions, and the new flexible exchange rate system were 'used to gradually and steadily lower the real value of the rouble against the dollar' (Goldberg, 1993, p. 9). The effects of even this partial liberalization of the exchange system were startling. The Moscow foreign exchange auction (VEB) rates reached levels as low as 0.052 US$/Rbl (19.34 Rbls/US$), a mere 3 per cent of the value implied by the official exchange rate of $1.69 US$/Rbl.

The glaring discrepancy between 'official' exchange rates – and their grim implications for the value of the rouble, and for capital flight – was widely recognized. In November 1990 Gosplan introduced a new commercial exchange rate, which would replace the official Soviet exchange rate for most transactions. The new commercial exchange rate was, according to Granville (1995, p. 71) 'fixed in terms of a basket of currencies and set at a level three times as depreciated as the official exchange rate'. The system of state currency auctions was abandoned on 9 April 1991. It was replaced by the Moscow Currency Exchange (MICEX), and a simplified system of interbank currency transactions (Goldberg, 1993). Despite these revisions, however, both state currency auctions and interbank currency markets continued to suffere from a severe shortage of hard currency supplies. According to Goldberg (1993, pp. 8–9):

Foreign currency auctions in the Soviet Union were in practice characterized by a distinct shortage of exporters willing to voluntarily sell foreign exchange. In part, this shortage can be ascribed to the system of foreign exchange surrender and taxation of export earnings that led to massive under-invoicing of exports and widespread avoidance of foreign exchange repatriation. Indeed, for 1991, some estimates suggest that capital flight from Russia may have been as large as 10 to 15 billion dollars. Among those export earnings that actually were repatriated, only a small share were sold in interbank markets. Of the officially-recorded 30 billion dollars in export earnings from convertible-currency countries in 1991, less than 5 per cent passed through the interbank foreign currency markets.

The non-existence of a forward exchange market and the negative domestic real interest rates are other important reasons for avoidance of foreign exchange repatriation.

In light of this severe shortage of hard currency supplies, equilibrium, if feasible, on state auction and interbank exchange markets required detailed prescriptions of state intervention (control). Indeed, from January to September 1991, there is clear evidence that both the system of state auctions, and interbank exchange markets were manipulated in a vain attempt to stabilize the dollar value of the rouble. As described by Goldberg (1993, p. 9): 'During this period of 1991 there is evidence that the controlled price of dollars in these currency markets was achieved by rationing importers' access to foreign currency. The procedures implemented for rationing and allocating foreign currency to importers were not transparent'.

Gosplan's unique rouble stabilization programme included the following market manipulations: (1) direct intervention through the sale of dollars on foreign exchange markets by the Central Bank of Russia; and (2) surrender requirements on raw materials. A presidential decree issued on 30 December 1991 outlined severe hard-currency stablization requirements:

- The mandatory sale of 40 per cent of all hard currency export earnings to the 'state' hard currency reserve at the special commercial exchange rate of 55 Rbls/US\$. The list of goods and services subjected to the new 40 per cent surrender requirement included raw materials and products (that is, crude oil and petroleum products), transportation, freight, tourism and financial services. All surrendered revenue was to be used to service foreign debt, stabilize the dollar value of the rouble via direct market intervention, and purchase those imports deemed necessary for the general health of the economy; and

- The mandatory sale of an additional 10 per cent of export earnings (from all exports) to the hard-currency stabilization fund at the CBR, market exchange rate – also known as the quasi market exchange rate (approximately 90–110 Rbls/US$ from January to March 1992). The hard currency stalilization fund was reserved for direct intervention in official currency markets (Granville, 1995).

The coincidence of an abundance of direct and indirect stabilization policies distorted domestic price signals, disturbing even the simplest relationship between official and 'black' market exchange rates. Indeed, in the months from January to April 1992, according to Goldberg (1993, p. 2) the foreign exchange surrender requirements alone:

> effectively taxed exporters at a rate of at least 30 percent of export earnings. The foreign exchange rate regime prevailing at this time also taxed importers heavily. For importers, the cost of obtaining foreign exchange in official (interbank) markets was approximately 25 per cent higher than it was if the foreign exchange was purchased in black markets. . . . That system encouraged non-repatriation of export earnings, international barter activity, and avoidance of legal foreign exchange markets by both exporters and importers.

As a result, posted nominal exchange rates failed to reflect information critical to buyers and sellers of foreign exchange, and to some extent, Gosplan policy makers.

To cite only one example: in January–April 1992, official interbank exchange rates were held relatively constant at approximately 150 Rbls/US$. At the same time, the effective 'post-taxation' exchange rate had reached levels as low as 106 Rbls/US$ for exporters, and a mere 55 Rbls/US$ for foreign investors (the special commercial exchange rate). Harsh trade restrictions and excessive taxation had completely severed the traditional relationship between official and black market (cash) exchange rates, resulting in a 'negative' black market premium, or discount, on foreign exchange. As a result, US dollars could be purchased on the black market for only 125 Rbls/US$ in June 1992, a significant 17 per cent rebate from the 150 Rbls/US$ required on official markets (the interbank exchange market).

The policy encouraged Russian importers to bypass official exchange markets altogether – inspiring a conspicuous increase in the level of international barter activity. Exporters, on the other hand, received only 106 Rbls/US$ for repatriated capital, a significant deduction from the 125 Rbls available on the black market. The economic disincentives, to limit the volume of official transactions to the bare minimum, were unmistakable as were the dire implications for Russia, a country suffering from a large balance of payments deficit, and capital flight.

Gosplan's final solution, sweeping currency reforms and a new unified exchange rate system, was initiated on 3 July 3 1992. It is important to note that there are no official statistics representing a true (free market) Rbl/US$ exchange rate throughout this chaotic period of reform. All calculations, including those employing the black market exchange rate, must take into account the following distortions: (1) direct intervention by the CBR; and (2) numerous and binding trade restrictions, including those preventing effective arbitrage between cash and non-cash/legal and black markets (Goldberg, 1993).

C: Retail Prices – Over the period 1967–90, Gosplan retained a rigid retail pricing policy, literally guaranteeing the stability of prices for all consumer necessities (Goldman, 1980). Energy and petroleum products were classified as indispensable, and their prices were held constant for prolonged periods of time. This 'zero' inflation policy would oon prove unsustainable. By the early 1980s, there was clear evidence that consumer energy prices had fallen to levels below the industrial wholesale price, necessitating significant payments, or subsidies, to retail energy enterprises (Hewett, 1984).

The Soviet intolerance of domestic inflation, and a failure to recognize, or respect, the 'true' costs of production, led to gross economic distortions, and inefficiency and waste in the consumption of energy products. These tendencies, and a resultant predilection to squander valuable energy resources, were aggravated by deficiencies in the pricing of energy products.

Hewett (1984, p. 136) noted the following examples of sluggish or inefficient retail energy pricing: 'In many instances energy going to consumers is not even metered. For example, most residential consumers of natural gas simply pay a flat fee. Electricity charges to collective farms are based on the rated power of the equipment, not actual use'.

The results of these unusual pricing policies, and a host of administrative procedures encouraging the inefficiency utilization of Russia's energy reserves, were widley publicized. According to Goldman (1980, p. 143):

> Building insulation is poor and drafts are endemic. S. Orduzhev, the Minister of the Gas Industry, complained that it takes 60 per cent more fuel to heat buildings erected in the 1970s than those built twenty-five years earlier. If better insulation were to be used, savings of 15 billion cubic meters of gas a year could be made. In other cases, centrally heated buildings are often too hot. Operators of heating plants are evaluated on the basis of how much heat they generate, not how much energy they save. Since Soviet radiators do not always function as advertised, and most do not have individual control valves, many residents are forced to open the windows if they want to cool their apartments, even in the middle of winter. . . . Similarly, because drivers in most of the Soviet Union's motor pools are awarded premiums, not for the task performed, but on the kilometers driven, drivers both in the cities and on the farms are constantly criticized for wasting gasoline. This may explain the incredible sight of herds of elephant-like water trucks roaming Moscow streets with spray spewing from their trunks even in the rain.

The Soviet transportation market provides an excellent example of excessive stability in the pricing of petroleum products. To cite only two examples:

1. The official retail price for regular 76 RON gasoline was held constant at 0.3 roubles per litre throughout the years 1986–91 (IEA, 1993). That is to say – assuming the official interbank exchange rate – the price of a medium octane, unleaded gasoline averaged only 1.55 cents (US) per litre in 1990. Discounting considerations raised by the lack of an appropriate Rouble/US$ exchange rate, this figure may be compared to 1.182 Deutsche Marks (DM) per litre of unleaded gasoline ($0.729 US per litre of unleaded gasoline) in

Germany, and $0.307 per litre in the US (Reinsch et al., 1994); and
2. The official retail price of diesel fuel averaged only 61 roubles per metric tonne in 1990, seven roubles below the wholesale industry price of 68 roubles per metric tonne.

D) Planning Prices – As early as 1964 one finds reference to energy related planning prices or, to be precise, the 'set of indicators of current and capital costs' of different kinds of fuel delivered to the various regions, produced by The Council for the Study of Productive Forces (SOPS) for the Gosplan (Campbell, 1983). The new 'planning prices' were intended to provide useful accounting 'guidelines' for major investment decisions, including the location, level of technology, and capital intensity of new production facilities.

Unfortunately, early estimates were crude at best – simple mathematical calculations involving only current production costs and capital costs (with no interest charge on capital) – and, as a result, failed to reflect the full or true marginal costs of production. Despite these obvious shortcomings, there were some slight advantages. The planning prices were revised frequently, and according to Gustafson (1989, p. 239) 'never allowed to slip as far behind the world level as transactions prices did'.

The sophistication of planning prices grew hesitantly with the size and complexity of the Soviet economy. In 1974, after a full decade of debate, Gosplan switched to a system of *zamykaiushchie zatraty* (ZZ), or shadow prices. The event was hailed as a major achievement for Gosplan. The revised planning prices would surely reflect the national economic costs of energy in the main economic region of the USSR. Despite a number of important improvements – including the introduction of finding fees and capital charges (1976), and the use of complex, and fairly comprehensive linear programming models – they did not.

Table D.2 illustrates the ZZ and representative transactions prices (wholesale industry prices) for Mazut in selected regions of the USSR for the years 1975–80. It is important to notice that, for most regions, the wholesale industry price had been set, and fixed, at levels well below the ZZ or planning price. While a number of explanations have been offered for this apparent pricing paradox, the most likely is the Soviet aversion to 'profit' and 'inflation', and

The Russian Oil Economy

a general reluctance to raise wholesale energy prices so soon after the 1967 reform. According to Campbell (1983, p. 256):

> I suspect that one of the main reasons is that the price setters still operate with a cost-of-production bias and would have been uncomfortable with the profits that would be generated for the oil and gas industry by transactions prices closer to the ZZ. It may also be that the transactions prices, set in 1967, were based on earlier, and more optimistic views of costs than emerged by the time the ZZ were calculated (in the early 1970s).

Table D.2: Wholesale industry prices vs. ZZ planning prices for selected regions of the USSR

Regions	Mazut ZZ 1975–80 Rbl./t	Mazut Wholesale Industry 1975–80 Rbl./t	Mazut ZZ Post-1980 Original	Mazut ZZ Post-1980 Revised
Northwest	22–25	16.70	21–23	46.2–47.6
Murmansk Oblast'	26–28	16.70	24–26	na
Komi ASSR	16–19	16.70	14–17	na
North Caucasus	20–22	16.70	20–22	46.8–48.9
Middle Volga	20–22	16.70	18–21	44.8–46.9
South Ural	17–20	16.70	17–19	44.8–46.9
Novosibirsk & Tomsk Oblast'	14–17	18.90	13–16	44.8–46.9
Krasnoiarsk Krai	15–17	18.90	13–16	44.8–46.9
West Ukraine and Moldavia	22–25	16.70	23–25	46.8–48.3
Latvia & Estonia	24–26	16.70	23–26	30–34
Armenia & Azerbaijan	19–23	16.70	20–22	46.8-48.3
West Kazakhstan	16–18	16.70	16–18	44.8–46.2
Amurst Oblast'	17–19	21.70	17–19	48.3–49.6

(roubles per tonne)

Note: All figures in roubles per standard tonne. The wholesale industry (transactions) prices are FOB the delivery terminal. In the years 1975–80 the price of Mazut was differentiated among three cost zones – the entire European territory of the USSR plus Central Asia, Siberia, and the Far East. In addition, mazut prices were differentiated by sulphur content and viscosity. The prices reported above are for grade 40, sulphurous (0.5–2 per cent sulphur) mazut.

Source: Campbell (1983).

The message to end users – to ignore the real domestic opportunity cost of mazut and other pretroleum products, utilizing as much energy as Gosplan would allow – was clear. To Gosplan policy makers, the ZZ could only complicate an already obscure internal pricing policy. Not only were wholesale industry prices set well below planning prices, but there is clear evidence that the ZZ themselves failed to reflect the full marginal costs of production.

To cite only a few of the more prominent deficiencies:

1. As mentioned above, the finding charges introduced in 1967 simply failed to reflect the true geological exploration costs (Campbell, 1983);
2. Land acquisition costs, rent charges, and the costs of aquiring raw materials were never adequately represented within the Soviet pricing system. While the concept of marginal cost pricing, and a centralized system of rent collection, was adopted in 1976, it was rapidly abandoned. Apparently, 'average costs rose enough that from 1979 centrally collected rent charges were dropped altogether, and rent was redistributed within the branch to cover high cost producers' (Campbell, 1983, p. 258);
3. The use of an artificial, or contrived interest charge on capital. In 1979, the fixed interest charge on capital was reduced from 6 per cent to 3.9 per cent. The specific reduction of the interest charges on 'oil' capital to levels well below the national average, reflects Gosplan's clumsy attempt to mask, or disguise, the rising cost of petroleum production;
4. The calculation of ZZ by means of a linear programming model which failed to consider the opportunity cost inherent in the high level of international oil prices.; and
5. Frequent and sizeable upward revisions in the ZZ.

Table D.2 illustrates two distinct estimates of ZZ for the post-1980 planning horizon. Remarkably, the revised estimates are over two times as high as the original forecast (see Table D.2). The major discrepancy – a radical departure from Gosplan's established tradition of rigid price and planning stability – suggests that the original forecast suffered from 'overoptimistic perspectives on energy costs' (Campbell, 1983, p. 263).

As illustrated in Table D.2, the price of Mazut fuel oil was originally expected to either fall, or remain constant in most regions of the USSR throughout the tenth and eleventh five year planning periods. By 1979 these hopes had been abandoned completely. Instead, according to Campbell (1983, p. 264) a: 'growing relative scarcity of mazut meant that it would increasingly have to be reserved for the highest priority applications (basically to cover peak load needs in power generation) where its advantage over coal (the closing fuel) was greater than in the closing uses originally envisaged and underlying the 1975 to 1980 ZZ values (base-load condensing stations)'.

Soviet planners, however, were well aware of the fact that the prices for energy were set well below both domestic and international opportunity costs, undermining a widely publicized and essential campaign for the conservation of energy supplies. At the same time, the idea that the international oil price should play any role, however minor, in the formation of domestic energy policy drew sharp resistance from the established planning community.

The opposition, to both an implied loss of control and the sheer volume of work required by a significant revision to the 5-year plan, was clear. To cite only one example: a 1978 proposal to adopt global export prices as the relevant opportunity cost for the ZZ was debated at some length, only to be dismissed as heretic. According to Campbell (1983, p. 263):

> One of the Gosplan representatives, for example, [could not] quite accept the logic that the world market price should determine the extensive and intensive margin in oil development, and that this would require new ZZs and prices for coal and all other energy reosources, which would then force still more cost and price changes throughout the whole economy.

An excellent opportunity to amend the 'obscure' Soviet pricing dilemma was presented by the 1982 price reform, and indeed, some progress was made in a positive direction. To be precise, a form of rent charges was, once again, included in the revised ZZs for gas, oil and petroleum products (Gustafson, 1989) (see Table

D.2). Many more obvious 'pricing' deficiencies were overlooked, and Gosplan chose simply to ignore the opportunity costs inherent in the global oil price.

To the detriment of Soviet energy planners, the 'price changes introduced in 1982 were confined primarily to enterprise and industrial wholesale prices' (Hewett, 1984, p. 138). In the case of petroleum products, transactions prices were raised only hesitantly, and maintained at levels well below the ZZ, or planning price. This feat was accomplished 'by squeezing out rent in the pricing of crude oil' (Campbell, 1983, p. 271), thereby erasing a significant portion of the benefits to energy conservation and efficiency.

To cite only one example: the ZZ for the northwestern region of the USSR ranged from 46.2–47.6 roubles per tonne in the post 1980s, 10 roubles per tonne higher than the 36 roubles per tonne proposed by the 1982 price reform. Years later, and indeed for the period 1986–90 , the official wholesale industry price of heavy fuel oil averaged only 42.38 roubles per tonne, significantly below even what the ZZ envisioned for the eleventh five year plan 1980–85.

In short, while some progress was made on the issue of Soviet energy pricing, the sheer enormity of the problem – reorganizing and recalculating the entire Soviet planning matrix – would prove insurmountable. Instead, prices maintained their status as simple accounting tools. Problems such as the allocation of scarce economic resources towards competing ends, and energy conservation – which are simplified by the role of prices in a competitive or free market economy – became multi-headed hydras, with each head representing an additional source of confusion to consumers and policy makers alike. The dilemma was captured in a timely and prophetic warning by Goldman (1980, p. 49):

> Obviously, if wholesale prices of energy in the Soviet Union were allowed to reflect world prices, the rate of profit of the various production ministries would be much higher, but then Soviet planners would be tempted to divert all of their output to foreign markets. This would cause chaos in the domestic planning system. Yet, if prices were higher, Gosplan, the Soviet planning agency, might be more willing to increase the amount of investment allocated for the development of new but seemingly unprofitable fields. As matters now stand, the planners are getting mixed signals. Domestic petroleum prices and profit ratios

tell them they should divert investment resources elsewhere. Foreign prices and profit ratios tell them the opposite. Any shadow prices they adopt to provide them with a more realistic but unofficial view of costs and opportunities will have to be very carefully designed to head off what otherwise could be a disaster in the way Soviet investment resources are allocated. As we shall see, until 1977 confusion negatively affected efforts to increase investment efficiency.

As we have seen, the shadow prices or ZZ were not calculated correctly, and failed to reflect the full opportunity cost of even domestic prodution. A growing inventory of difficulties – which included waste and inefficiency in the consumption of energy products – was dealt with in the traditional Soviet fashion using the imposition of ever more targets, bonuses, and production quotas. In short, the first faltering steps towards conservation were taken in 1976, at which time energy-savings targets were included in short-term (annual) plans. However, according to Gustafson (1989, p. 240): 'But since in most cases no one knew how much energy each enterprise should consume, the targets were arrived at through bargaining, surrounded by all the usual tactics familiar to students of command economies. The predictable result was bogus energy savings.'

The solution was clear. At the earliest possible opportunity (1982), Gosplan organized an interagency programme to determine exactly how much energy (fuel) a typical enterprise should consume. In phase I, energy-efficiency norms were established for each of the major production processes, mechanisms, engines, appliances, machines and all gadgets using energy. In the second phase of the programme, the individual normatives were consolidated into central or State standards and employed as a basis for the 'plan. It was according to Gustafson (1989, p. 240): 'of course a herculean task, and the standards writers . . . sensibly focused on a handful of high-priority targets, emphasizing processes over products and producer goods over consumer goods'. Still, Gosplan had yet to exhaust the assortment of physical targets, minor adjustments and bandages through which it hoped to restore order, and some semblence of efficiency, to the wasteful Soviet energy industry.

In the spring of 1983, Andropov put his official stamp of approval on a comprehensive new energy programme. The programme, authenticated by a multitude of definitive 'energy' studies, provided a 17-year blueprint for conservation – including a 'dangerous transition to a consumption-oriented approach' (a crude form of demand side management) – 'while simultaneously protecting the economy's energy supply' (Gustafson, 1989, pp. 242–3).

On the demand side alone, the physical recommendations were staggering. These included, but were by no means limited to; 'improvements in measurement of consumption, oversight, accountability, and incentives (including price increases); retirement of obsolete equipment, capture of waste heat, and improvement of coal quality. Finally, as a short-term tactical aim, the commission urged rapid fuel switching, especially displacement of oil by gas in power plants (Gustafson, 1989, p. 244).

The new energy policies would prove more burdensome than the dreaded reorganization, and tedious re-estimation, of the entire Soviet input–output matrix. By the late 1980s there were more than one hundred energy-saving norms in effect in the Ministry of Construction Materials alone. To further aggravate matters, the complex incentive programme was not having the desired effect on industry. Managers, having once satisfied their annual energy-saving target, had no more incentive to save energy (Gustafson, 1989).

At the same time, Soviet enterprises were faced with a legion of conflicting, and in some cases mutually exclusive, targets, production quotas, and energy-savings norms. Encountering ceaseless pressure from Gosplan, Soviet managers had little alternative but to sacrifice everything for the production target. As Hewett (1984, p. 141) observed:

> No matter how many targets are added to the enterprise plan, it is the output target to which the enterprise director will pay the closest attention. It is still (1984) as it has been for a long time, acceptable for an enterprise director to tell his ministry, or the ministry to tell Gosplan, that the output plan was hard to meet, but that it was nevertheless done, even at the cost of violating the input plan. It is far more difficult for an enterprise director to tell his ministry that production has fallen, but that costs fell even farther, and profits were up.

By 1985, Gosplan had exhausted its repertoire of ad hoc targets, repairs, economic bandages, and inadequate incentive policies. The time was ripe for Gorbachev, and a complete overhaul of the Soviet energy industry. Price liberalization, a critical ingredient to the reform process one would have thought, would be discreetly delayed till the early 1990s, and beyond. The predictions of von Mises (1981) continued to dog the command economy until its end.

References

Anon (1991), *Soviet Energy: An Insider's Account*, London: The Centre For Global Energy Studies.

Anon (1995), *Yesterday, Today and Tomorrow of the Russian Oil and Gas Industry*, Moscow.

Aslund, A. (1991), *Gorbachev's Struggle for Economic Reform*, Ithaca, New York: Cornell University Press.

Barton, R. (ed.) (1998), *The Almanac of Russian Petroleum: 1998*, London: Petroleum Intelligence Group.

Bolger, B., J.E. Hobbs and W.A. Kerr (2001), 'Supply Chain Relationships in the Polish Pork Sector', *Supply Chain Management*, **6** (2), 74–82.

Braudel, F. (1995), *A History of Civilizations*, Harmondsworth: Penguin Books.

Campbell, R.W. (1983), 'Energy Prices and Decisions on Energy Use in the USSR', in P. Desai (ed.), *Marxist Central Planning and the Soviet Economy: Economic Essays in Honor of Alexander Erlich*, Cambridge, MA: MIT Press.

Cohon, G. (with D. Macfarlane) (1997), *To Russia With Fries*, Toronto: McClelland and Stewart.

Coleman, F. (1997), *The Decline and Fall of the Soviet Empire: Forty Years that Shook the World, From Stalin to Yeltsin*, New York: St. Martin's Griffin.

Considine, J.I. and W.A. Kerr (1993), 'Russian Re-Centralization of Energy', *Geopolitics of Energy*, **15** (1), 7–10.

Davies, A.S., L.C. Cronberg and W.A. Kerr (1996), 'Economic Constraints on Technology Transfer to the Transitional Economies: The Example of the Agriculture and Food Sectors', in J. Kirkland (ed.), *Barriers to International Technology Transfer*, London: Kluwer Academic Publishers, pp. 109–133.

351

Davies, R.W. (1979), 'Economic Planning in the USSR' in M. Bernstein (ed.) *Comparative Economic Systems: Models and Cases*, Homewood: Richard D. Irwin.

Ebel, R.E. (1961), *The Petroleum Industry of the Soviet Union*, Washington, DC: The American Petroleum Institute.

Ebel, R.E. (1970), *Communist Trade in Oil and Gas: An Evaluation of the Future Export Capability of the Soviet Bloc*, New York: Praeger Publishers.

Ebel, R.E. (1994), *Energy Choices in Russia*, Washington D.C: Center for Strategic and International Studies.

Elliot, I.E. (1974), *The Soviet Energy Balance: Natural Gas, Other Fossil Fuels, and Alternative Power Sources*, New York: Praeger Publishers.

Ericson, P.G. and R. S. Millar (1979), 'Soviet Foreign Economic Behavior: A Balance of Payments Perspective' in *Soviet Economy a Time of Change*, Vol. 2, Joint Economic Committee, US Congress, Washington, 10 October.

Freeland, C. (1995), 'Democracy on the Critical List', *Financial Times*, 31 October, p. 17.

Friedman, M. and R. Friedman (1980), *Free to Choose*, New York: Harcourt Brace Jananovich.

Gaisford, J.D., J.E. Hobbs and W.A. Kerr (1995), 'If the Food Doesn't Come – Vertical Coordination Problems in the CIS Food System: Some Perils of Privatization' *Agribusiness: An International Journal*, **11** (2), 179–186.

Gaisford, J.D., W.A. Kerr and J.E. Hobbs (1994), 'Non-Cooperative Bilateral Monopoly Problems in Liberalizing Command Economies' *Economic Systems*, **18** (3), 265–279.

Goldberg, L.S. (1993), *Foreign Exchange Markets in Russia: Understanding the Reforms*, Washington: International Monetary Fund.

Goldman, M.L. (1980), *The Enigma of Soviet Petroleum: Half-Empty or Half-Full*, London: George Allen and Unwin.

Granville, B. (1995), *The Success of Russian Economic Reforms*, London: The Royal Institute of International Affairs.

Grey, D.F. (1998), '*Evaluation of Taxes and Revenues from the Energy Sector in the Baltics, Russia and Other Former Soviet Union Countries*', IMF Working Paper (WP/98/34), Washington: International Monetary Fund.

Gustafson, T. (1989), *Crisis Amid Plenty: The Politics of Soviet Energy Under Brezhnev and Gorbachev*, Princeton: Princeton University Press.

Harding, A. (2000), 'Why is Putin Popular?' *Media Reports*, BBC News, London, 8 March.

Hassmann, H. (1953), *Oil in the Soviet Union*, Princeton: Princeton University Press.

Henderson, R.D. and W.A. Kerr (1984/85), 'The Theory and Practice of Economic Relations Between CMEA Member States and African Countries', *Journal of Contemporary African Studies*, **4** (1/2), 3–35.

Hewett, E.A. (1984), *Energy Economics and Foreign Policy in the Soviet Union*, Washington, DC: The Brookings Institution.

Hewett, E.A., B. Roberts and J. Vanous (1987), 'On the Feasibility of Key Targets in the Soviet Five-Year Plan' in US Congress, Joint Economic Committee, *Gorbachev's Economic Plans*, Washington, DC: USGPO.

Hobbs, J.E and W.A. Kerr (1999), 'Transaction Costs' in S. Bhagwan Dahiya (ed.), *The Current State of Economic Science* Vol. 4, Rohtak, India: Spellbound Publications PVT Ltd, , pp. 2111–2133.

Hobbs, J.E., J.D. Gaisford and W.A. Kerr (1993), 'Transforming Command Economy Distribution Systems', *Scottish Agricultural Economics Review*, **7**, 135–140.

Hobbs, J.E., W.A. Kerr and J.D. Gaisford (1997) *The Transformation of the Agrifood System in Central and Eastern Europe and the New Independent States*, Wallingford: CAB International.

Hourani, A. (1991), *A History of the Arab Peoples*, Cambridge Mass: Harvard University Press.

Hutchins, R.K., W.A. Kerr and J.E. Hobbs (1995), 'Marketing Education in the Absence of Marketing Institutions: Insights from Teaching Polish Agribusiness Managers', *Journal of European Business Education*, **4** (2), 1–18.

The International Petroleum Encyclopedia (1994) Tulsa: Pennwell Publishing Company.

IEA (1993), *Energy, Prices, Taxes, and Costs in Russia as of Mid-1993*, Paris: International Energy Agency.

IEA *Weekly Petroleum Status Report*, Paris: International Energy Agency, (various issues).

IMF (1992) *Economic Review: The Economy of the Former USSR in 1991*, Washington DC: International Monetary Fund.

Kemp, A.G. (1987), *Petroleum Rent Collection Around the World*, Ottawa: The Institute for Research on Public Policy.

Kemp, A.G. and P.D.A. Jones (1996), *Investment, Taxation and Production Sharing in the Russian and Azerbaijan Petroleum Industries*, North Sea Study Occasional Paper, No. 55, Department of Economics, Aberdeen: The University of Aberdeen.

Kerr, W.A. (1993), 'Domestic Firms and Transnational Corporations in Liberalizing Command Economies – A Dynamic Approach' *Economic Systems*, **17** (3), 195–211.

Kerr, W.A. (1996), 'Marketing Education for Russian Marketing Educators', *Journal of Marketing Education*, **18** (2), 39–49.

Kerr, W.A. and E. MacKay (1997), 'Is Mainland China Evolving Into a Market Economy?' *Issues and Studies*, **33** (9), 31–45.

Kerr, W.A., J.E. Hobbs and J.D. Gaisford (1994), 'Privatization of the Russian Agri-Food Chain: Management Constraints, Underinvestment and Declining Food Security', in G. Hagelaar (ed.), *Management Studies and The Agri-Business: Management of Agri-Chains*, Department of Management Studies, Wageningen: Wageningen Agricultural University, pp. 118–128.

Khartukov, E.M. (1995), *Russia's Oil Prices: Passage to the Market*, East West Center Working Papers, Energy and Minerals Series, No. 20, Honolulu: East West Center.

Khartukov, E.M. (1998), 'Low Oil Prices: Economic Woes Threaten Russian Oil Exports', *Oil and Gas Journal*, **96** (23), 25–30.

Koen, V. and S. Phillips, S. (1993), *Price Liberalization in Russia: Behavior of Prices, Household Incomes and Consumption During the First Year*, IMF Occasional Paper No. 104, Washington, DC: International Monetary Fund.

Kort, M. (1993), *The Soviet Colossus: The Rise and Fall of the USSR*, Armonk, New York: M.E. Sharpe Inc.

Kryukov, V. and Tokarev, A. (1998), *Comparative Analysis of Taxation Schemes Applicable for Development of Oil Reserves in Western Siberia*, Moscow: Russian Academy of Sciences – Siberian Branch.

McNeil, A.O. and W.A. Kerr (1995), 'Extension for Russian Agricultural Complexes: Lessons from a Dairy Project', *European Journal of Agricultural Education and Extension*, **2** (1), 49–58.

McNeil, A.O. and W.A. Kerr (1997), 'Vertical Coordination in a Post-Command Agricultural System - Can Russian Dairy Farms Be Transformed?', *Agricultural Systems*, **53** (2–3), 253–68.

Middleton, J., J.E. Hobbs and W.A. Kerr (1993), 'Poland's Evolving Food Distribution System - Joint Venture Opportunities for British Agribusiness, *Journal of European Business Education*, **3** (1), 36–45.

Moynahan, B. (1994), *The Russian Century: A History of the Last Hundred Years*, New York: Random House.

Nove, A. (1972), *An Economic History of the USSR*, Harmondsworth: Penguin Books.

OECD/IEA (1995), *Energy Policies of the Russian Federation: 1995 Survey*, Paris: Organization for Economic Cooperation and Development/International Energy Agency.

Rakowska-Harmstone, T, and A Gyorgy (ed.) (1979), *Communism in Eastern Europe*, Indiana: Indiana University Press.

Reinsch, E.A., K.J. Brown, and J.O. Stanford (1988), *Stability Within Uncertainty: Evolution of the World Oil Market*, Study No. 28, Calgary: Canadian Energy Research Institute.

Reinsch, A.E. and J.I. Considine (1991), *After the Crisis: World Oil Market Projections 1991–2006*, Study No. 39, Calgary: Canadian Energy Research Institute.

Reinsch, A.E., J.I. Considine and E. J. MacKay (1994), *Taxing the Difference: World Oil Market Projections 1994–2009*, Study No. 59, Calgary: Canadian Energy Research Institute.

Reinsch, A.E., I. Lavrosky and J.I. Considine (1992), *Oil in the Former Soviet Union*, Study No. 48, Calgary: Canadian Energy Research Institute.

Riva. J.P. (1994), *Petroleum Exploration Opportunities in the Former Soviet Union*, Tulsa: Pennwell Publishing Company.

Sipovsky, Y. (1996), 'Transneft Shows Little Desire to Control Costs as it Plans Another Transportation Tariff Hike: Higher and Higher', *Russian Petroleum Investor*, August, 31–32.

Smith, A. (1993), *Russia and the World Economy: Problems of Integration*, London: Routledge.

Smith, J. (1997), 'Taxation and Investment in Russian Oil', *Journal of Energy Finance and Development* 2 (1), 1–17.

Stern, J.P. (1987), *Soviet Oil and Gas Exports to the West: Commercial Transaction or Security Threat?*, Energy Papers No. 21, London: Royal Institute for International Affairs.

Stern, J.P. (1992), *Oil and Gas in the Former Soviet Union: The Changing Foreign Investment Agenda*, London: Royal Institute of International Affairs.

Stone, R.W. (1996), *Satellites and Commissars: Strategy and Conflict in the Politics of Soviet-Bloc Trade*, Princeton: Princeton University Press.

Tchurilov, L. (1996), *Lifeblood of Empire: A Personal History of the Rise and Fall of the Soviet Oil Industry*, New York: PIW Publications.

The Oil and Gas Industry of the USSR: Statistical Handbook, Moscow: The Petroleum and Gas Ministries of the USSR, (various issues).

The Economist (2001), 'Russia and Its Oligarchs: Who's Next' 9 June.

USEIA (1995), *Energy Policies of the Russian Federation: 1995 Survey*, Washington, DC: United States Energy Information Administration.

USEIA (1997), *Russia: Recent Developments*, Washington, DC: United States Energy Information Administration, August (www.eia.doc.gov/emeu/cabs/russia\.html).

USEIA (1998), *Russia: Energy Situation Update*, Washington, DC: United States Energy Information Administration, September (www.eia.doe.gov/emeu/cabs/russia\html).

Venediktov, A. (1957), *Organizatsiya Gosudarstvennoi Promyshlennosti v SSSR*, Vol. 26, Leningrad.

VICES (1990), *Comecon Data 1989*, Vienna Institute for Comparative Economic Studies, New Haven: Greenwood Publishing Group Inc.

von Mises, L. (1981), *Socialism*, Indianapolis: Liberty Classics.

Watkins, C.G. (1993), *Unraveling a Riddle: The Outlook for Russian Oil and Gas*, paper delivered to the 7th Symposium on Pacific Energy Cooperation, Tokyo, Japan, February.

White, S. (ed.) (1991), *Political and Economic Encyclopedia of the Soviet Union and Eastern Europe*, London: Longman.

World Bank, (1993), *Staff Appraisal Report: Russian Federation Oil Rehabilitation Project*, Report No. 11556-RU, Washington, DC: World Bank, 26 May.

World Bank (1994*)*, *Staff Appraisal Report: Russian Federation, Second Oil Rehabilitation Project*, Report No. 12943-RU, Washington, DC: World Bank, 13 June.

World Bank (1997), *Taxation of Oil Production in the Russian Federation*, Review Paper, Washington, DC: World Bank, 24 February.

Yergin, D. (1991), *The Prize, The Epic Quest for Oil, Money and Power*, New York: Simon and Schuster.

Yurko, J.R. and V. M. Reitman (1990), 'Observations and Results of the Recovery of Bitumen and Heavy Oil by Enhanced Oil Recovery (EOR) Methods in the USSR', *USSR Exchange Tour Report*, Calgary: Alberta Oil Sands Technology and Research Authority.

Zabotkine, A. (2000), 'The Duma Passes New Import/Export and Currency Regulations in the First Reading', *Russia Morning Comment*, Moscow: United Financial Group, 16, March.

Index